Dr Sigee presents a contemporary view of research on bacterial plant pathology, bringing together:

Bacterial structure and function
Taxonomy
Environmental microbiology
Infection and development of plant disease
Resistance mechanisms
Molecular genetics
Disease control

The author has a unified approach to the field, emphasising the contributions made by recent and exciting developments in cell and molecular biology. The book is written in a clear and concise manner, illustrated with numerous tables, diagrams and photographs.

T0215260

BACTERIAL PLANT PATHOLOGY:
CELL AND MOLECULAR ASPECTS

BACTERIAL PLANT PATHOLOGY:
CELL AND MOLECULAR ASPECTS

DAVID C. SIGEE

Department of Cell and Structural Biology, University of Manchester

CAMBRIDGE
UNIVERSITY PRESS

PUBLISHED BY THE PRESS SYNDICATE OF THE UNIVERSITY OF CAMBRIDGE
The Pitt Building, Trumpington Street, Cambridge, United Kingdom

CAMBRIDGE UNIVERSITY PRESS
The Edinburgh Building, Cambridge CB2 2RU, UK
40 West 20th Street, New York NY 10011–4211, USA
477 Williamstown Road, Port Melbourne, VIC 3207, Australia
Ruiz de Alarcón 13, 28014 Madrid, Spain
Dock House, The Waterfront, Cape Town 8001, South Africa

http://www.cambridge.org

First published 1993
First paperback edition 2005

A catalogue record for this book is available from the British Library

Library of Congress cataloguing in publication data

Sigee, D. C. Bacterial plant pathology: cell and molecular aspects/David C. Sigee.
p. cm.
Includes index.
ISBN 0 521 35064 6 (hardback)
1. Bacterial diseases of plants.
2. Phytopathogenic bacteria. 3. Phytopathogenic bacteria–Control. I. Title.
SB734.s54 1992
632′.32–dc20 92–18077 CIP

ISBN 0 521 35064 6 hardback
ISBN 0 521 61967 X paperback

Contents

Acknowledgements

I am grateful to my colleague Dr Harry Epton and to research workers Nidhal Suliaman, Abdul Al-Issa, Razak Al-Rabaee, Hisham El-Masry, Nicola Walker, Mark Wilson, Nigel Hodson, Sarah Nicholson and Dominic Howitt who have kindly allowed me to include illustrative material arising from our joint research. I am also grateful to Drs T. A. Chen, C. A. Clark, Y. P. Jiang and S. W. Matthews, and to the Editors of *Microbios*, *Physiological and Molecular Plant Pathology* and *Phytopathology* for permission to publish original micrographs. I would like to thank Dr Alan Vivian for critically reading Chapter 8 and for his helpful comments.

1

Bacteria as plant pathogens

Bacteria–plant associations

The origin and evolutionary development of higher plants has occurred in environments that were already colonised by bacteria, resulting in the co-evolution of a range of bacteria–plant associations. The associated microbes may be broadly considered in two main categories: epiphytic bacteria (present on the outside of the plant) and internal bacteria (infecting the plant tissue).

Epiphytic bacteria

These are associated with the plant surface, which is generally divided into root (rhizosphere) and aerial (phyllosphere) regions. A wide range of bacteria are adapted to various microenvironments at the soil and air interface, and are important in such aspects as nutrient uptake, frost damage, and biological control of plant pathogens. Many of these epiphytic bacteria are saprophytes, obtaining complex nutrients from the plant. Some epiphytic bacteria are also parasites, spending part of their life cycle on the plant surface, and part within the plant tissue.

Infective bacteria: parasites and symbionts

Parasitic bacteria are able to invade plant tissue, where they grow and multiply, and cause localised or general deterioration in the health of the plant. The great majority of these parasites are extracellular, multiplying within intercellular spaces but not penetrating plant cell walls or entering protoplasts. The relatively few parasitic bacteria that are able to penetrate the higher plant cell include members of the genus *Agrobacterium* (with the ability to transfer part of the genome into the plant cell) and *Rhizobium* (where the

1

whole organism enters the plant cell). These associations ultimately lead to symbiotic interaction, where both partners benefit from the combination, and where the distinction between parasitism and symbiosis is not always clear. Thus, although agrobacteria typically have an adverse effect on their infected plants, some strains of *Agrobacterium rhizogenes* have no overt effect on their host. Similarly, some strains of the normally symbiotic *Rhizobium japonicum* produce toxins that adversely affect plant growth.

This book is concerned primarily with bacteria that are parasites and act as plant pathogens. Epiphytic bacteria are also considered specifically where they are potential plant pathogens (facultative parasites) or where they are important in the biological control of plant pathogens.

Evolution of bacterial plant pathogens and the origins of disease

One of the surprising aspects of bacterial plant pathology is that very few species of bacteria have evolved the ability to invade and grow in healthy plant tissues. These bacteria are present in only five major taxonomic groups, comprising the genera *Agrobacterium*, *Erwinia*, *Pseudomonas*, *Xanthomonas* and the coryneform bacteria (*Arthrobacter*, *Clavibacter*, *Curtobacterium* and *Rhodococcus*). Gram-positive bacteria in particular, which constitute a substantial part of soil microflora, have no pathogens of significance – with the exception of the coryneform bacteria. The ability of bacteria to cause disease is thus a relatively unusual event, and the origin of this type of bacteria–plant interaction depends on quite separate evolutionary pressures on bacteria and plants.

Evolution of bacterial pathogenicity

The internal environment of healthy plants (e.g. leaf mesophyll tissue) is not particularly conducive to bacterial growth, since, although it is humid and protected from environmental extremes, it is low in nutrient. Bacterial growth can only occur when the nutrient status is increased, and this has been achieved during evolution in various ways:

1. Plant cells may be perturbed by bacteria in such a way as to release water and nutrients without killing the cells. This is effected by altering the permeability of the plasmalemma, and is mediated by special pathogenicity (*hrp*) genes (see Chapter 8). These *hrp* genes are highly conserved and appear to have arisen on a discrete number of occasions, which may explain why the number of pathways leading to bacterial phytopathogenicity seems to be limited.

2. Bacteria in the genus *Agrobacterium* are able to alter the nutrient level of the internal plant environment in a very specialised way by redirecting the plant synthetic machinery to produce and secrete specific microbial nutrients (opines). This is achieved by injecting bacterial DNA into the plant cells via a highly evolved and complex sequence of events (see Chapter 8).

3. Some prokaryotes (Mollicutes and some bacteria) are adapted to survive permanently within plant conducting tissues (wet and high nutrient), and have evolved specific invertebrate vector associations for direct transmission between vascular sites.

Other plant pathogenic bacteria (particularly those causing soft rot diseases) have evolved a role as secondary invaders, being able to survive in plant tissue only when natural degeneration or other pathogens have caused cell damage and the release of nutrients.

The ability of bacteria to grow in plant tissue has further evolved with the acquisition and development of many features which promote a more rapid colonisation of the plant tissue: such as cell wall degrading enzymes and toxins. These features lead to more pronounced disease symptoms and are referred to as virulence factors (see Chapter 7).

Evolution of plant resistance: compatibility and incompatibility

As with other types of host–parasite interaction, success of the parasite depends on causing minimal damage during its growth in the host, so there is intense selection pressure for compatibility on the part of the parasite. Survival of the plant, on the other hand, depends on its ability to isolate and restrict the growth of the parasite by localised reaction against it, so there is pronounced selection pressure for incompatibility on the part of the host.

In the case of the pathogen, evolution of compatibility has occurred in relation to a particular host or group of host plants, and has involved the development of a set of genes which closely relates to these plants. Host range varies considerably between different bacteria, being highly specific with some of the foliar phytopathogens (e.g. *Pseudomonas syringae* pv. *tabaci*/*Nicotiana tabacum*), but quite broad in other cases (e.g. *Erwinia amylovora*/over 130 species in the family Rosaceae).

Whereas evolution of plant pathogenic bacteria has involved adaptation to specific hosts, evolution of plants has involved the development of a general incompatibility mechanism against a very wide range of phytopathogenic bacteria. This incompatibility mechanism (resistant response) shows some similarity to the immune system of higher animals, with recognition of the foreign organism followed by the production of anti-bacterial compounds (see Chapter 6). In the case of plants, however, the resistant response is localised, directed specifically against phytopathogens (i.e. not all foreign bodies) and plant anti-pathogen compounds (phytoalexins) are very different from the antibodies of the vertebrate immune system.

The association between bacteria and plants is a continuously evolving interaction, with the periodic origin of new types of phytopathogen, leading to new types of plant disease or more virulent forms of existing disease. In 1984, for example, a new foliar disease of citrus (citrus bacterial spot) caused by *Xanthomonas campestris* was identified in a Florida nursery, and has since been widely isolated and implicated in extensive outbreaks of the citrus disease (Graham & Gottwald, 1990).

Bacteria and plant disease

A plant pathogen (phytopathogen) may be defined in very broad terms as a biological agent which causes a deterioration in the health of the plant. Where this deterioration is apparent as clear and reproducible symptoms it is typically referred to as a named plant disease.

A wide range of biological agents, plus various environmental chemical and physical factors, are capable of causing disease in plants. Some indication of this diversity is provided by recent lists of named diseases (with their causal agents) of agricultural and ornamental crops published by the American Phytopathological Society (see Hansen, 1985; Smiley, 1988). Table 1.1 summarises information from these lists for four different types of crop. The percentage occurrence of different categories of disease agent followed a similar pattern in different crops, with approximately 50–65% of all diseases caused by fungi, 10–20% caused by viruses and 5–10% caused by bacteria. Other biological pathogens include Mollicutes (mycoplasma-like organisms and *Spiroplasma*), nematodes, algae and the higher plant *Cuscuta*. Other factors causing disease include genetic defects and various environmental aspects such as pollution (e.g. photochemical oxidants, sulphur oxides), physical factors (e.g. extremes of temperature, pH) and mineral deficiency (e.g. lack of calcium, boron or manganese). A small but variable proportion of diseases were caused by completely unknown factors.

Bacterial and fungal pathogens

Although fungi are generally more important than bacteria as plant pathogens – both in terms of number of diseases caused and overall economic losses – bacterial phytopathogens are highly successful as disease agents,

Table 1.1. *Percentage occurrence of different agents in lists of named diseases of crop plants*

	Soft fruit	Cereals	Root crop	Ornamentals
Biological agents				
Fungi	49	58	55	64
Bacteria	6	6	9	8
Mollicutes	2	1	2	1
Viruses	14	15	12	8
Graft-transmissible	4	—	—	3
Nematodes	13	20	17	16
Algae	*1			
Higher plant (*Cuscuta*)	*1	—	—	1
Genetic defects	2	*1		
Physical agents				
Pollution	—	*1	*1	—
Physical	3	—	2	—
Mineral deficiency	1	*1	*1	—
Cause unknown	4	*1	2	—

*1, less than 0.5% of total list.
Data are collated from Hansen (1985) for soft fruit (apple, citrus, pear, tomato), cereals (barley, corn, oats, wheat), root crops (beet, potato, onion, sweet potato) and ornamentals (rhododendron, rose, carnation, chrysanthemum). Graft-transmissible diseases in this listing are probably caused by undetermined virus or Mollicute pathogens.

causing disease throughout the whole range of families of higher plants. In comparison with fungal pathogens they have a number of apparent biological disadvantages:

1. The bacterial cell is thin-walled and fragile and is easily damaged by desiccation, irradiation and high temperature. Plant pathogenic bacteria are therefore relatively susceptible to adverse conditions outside the host plant, although many can be found in the general environment throughout most of the year (see Chapter 4).
2. Plant pathogenic bacteria do not generally form spores and thus differ from fungi in not being able to form resistant structures. Under unfavourable conditions, bacteria depend instead on protection provided by the host plant.
3. Plant pathogenic bacteria generally lack an efficient means of long-distance dispersal (independent of vectors). Unlike fungal spores, which are typically light and windblown, bacteria are sticky and are not easily airborne. Physical dispersal of bacteria is normally via aerosols or rain-splash, and there is greater dependence on biological dispersion via vectors or transport of infected plant tissue.
4. Unlike fungi, bacteria are not able to directly penetrate the plant cuticle. Entry

into the host plant can therefore only occur via natural openings (such as stomata) and wounds.

These limitations are counterbalanced by a number of advantages that bacteria have over fungal pathogens. These include a rapid rate of reproduction, rapid entry into infection courts and independent motility. In general, once bacterial pathogens are established in a particular area or host plant, the progression of disease and the localised spread of pathogen are very rapid.

Criteria of pathogenicity

One of the first steps in determining whether a particular disease is bacteria-induced might be to attempt the isolation and culture of bacterial cells from diseased tissue. However, plant pathogenic bacteria are not restricted in their occurrence to diseased plants, but are widely present in the environment as epiphytes, so that association of a particular organism with a diseased plant does not imply causality of disease. Further experimental evidence is required before a particular bacterial isolate can be designated as the responsible agent.

The need for some experimental confirmation of pathogenicity was recognised by Burrill (1880), the first investigator to identify a bacterium as the causal agent of a plant disease. Working on fireblight disease of pear, Burrill observed motile bacteria within mucilaginous fluid from diseased tissue, and published a description of the bacterial species, which he named *Micrococcus amylovorus* and which he thought was the cause of the disease. Burrill went on to demonstrate that fireblight could be transferred by inoculation of healthy plants with diseased material, showing that it was caused by a transmissible agent.

Criteria for the designation of a disease agent were initially defined by Koch (1880), and are subsequently referred to as **Koch's postulates**. These are:

1. The organism must be consistently associated with the diseased tissue.
2. The organism must be isolated and grown in pure culture, with no other organisms present.
3. The cultured organism must be inoculated into healthy plants of the same species from which isolation originally occurred, and must produce the same disease as originally observed.
4. The organism must be reisolated and reinoculated into healthy plants to produce the same disease.

A recent example of the use of these criteria to confirm bacterial phytopathogenicity is provided by the work of Brown and Michelmore (1988) and Van

Bruggen *et al.* (1988) on corky root of lettuce. This is a serious disease of lettuce in California, causing deterioration in the root system of infected plants, and has been variously attributed to a variety of fungal and bacterial pathogens as well as different abiotic factors. Studies by the above workers (Fig. 1.1) led to the isolation of rod-shaped Gram-negative bacteria with a single lateral flagellum, which were grown in pure culture and reinoculated to reproduce the original disease in accordance with Koch's postulates. Procedures for the isolation and *in vitro* culture of bacteria are clearly a major aspect of disease investigation, and are discussed in Chapter 3.

Although Koch's postulates are generally applicable to bacterial diseases, and have been widely used in confirming the pathogenic status of many bacteria, a number of problems may arise. They are not strictly applicable, for example, in the case of obligate pathogens, where culture *in vitro* is not possible and where inoculation of healthy plants by a suitable direct isolate must be substituted. Problems with implementing Koch's postulates may also arise where symptoms are not well defined and clear diagnosis of disease is difficult. An example of this is given by ratoon stunting disease of sugar-cane caused by *Clavibacter xyli* subsp. *xyli* (Davis *et al.*, 1988), where stunting is the only overt symptom of the disease, but is also typical of a number of other diseases.

Other criteria may also be useful in determining pathogenic status in addition to those proposed by Koch. These include:

1. Microscopic examination of diseased tissue. This will reveal multiplication and invasion of particular organisms. It may also demonstrate a localisation of the pathogen in the tissue which relates directly to the symptoms, e.g. the localisation of bacteria to xylem vessels in bacterial wilt diseases.
2. Remission of symptoms with antibiotic therapy, where specific elimination of an organism relates to loss of particular disease symptoms. This approach is especially useful where a bacterial pathogen is difficult to isolate and culture, and was employed by Bennett *et al.* (1987) in confirmation that Sumatra disease of cloves was caused by a xylem-limited bacterium.

Plant pathogenic bacteria and crop monoculture

In natural environments there is typically a great diversity of species within the plant community and a corresponding diversity of plant pathogens, all at relatively low level. The destruction of such communities, and the development by man of large-scale culture of single crops (monoculture) has led to the selection of particular crop-related pathogens and major outbreaks of

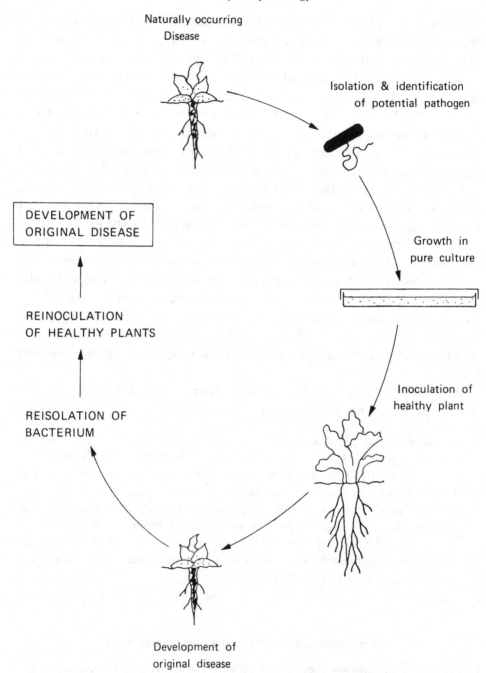

Fig. 1.1 Identification of the causal agent of corky root disease of lettuce using Koch's postulates. Control inoculation of healthy lettuce plants with sterile medium (water or nutrient broth) did not result in the development of corky root disease, and the disease bacterium could not be isolated from these plants.

disease (epidemics). Monoculture not only leads to larger quantities of host plant being available for infection, but also promotes spread of the pathogen over both long distances (wind or vector transport from major sites of infection to major sites of potential infection) and short distances (spread from plant to plant within a crop). The latter is particularly relevant to bacterial pathogens, where localised dispersal is of prime importance (see above).

With many bacterial diseases, the region and date of origin (or initial observation) can be historically defined, and the geographical spread of the pathogen within a particular crop documented. A good example of this is provided by bacterial leaf blight of rice, caused by *Xanthomonas campestris* pv. *oryzae* and one of the most destructive diseases of rice in Asia (Ou, 1984). This disease was initially observed in Japan in 1884. By 1908 it was common in the south-west part of the country, and in 1926 was recorded in the north-west. After 1950, the disease increased markedly and by 1960 was known to occur throughout Japan, with the exception of the island of Hokkaido. In India, bacterial leaf blight was initially recorded in the Bombay area in 1951, and has since become widespread throughout the country with the introduction and cultivation of new high yielding but susceptible rice cultivars. In recent times, bacterial leaf blight has also been reported as a major disease of rice in China, Korea, the Philippines, Taiwan, Thailand and Vietnam, and is also present in Australia and the Caribbean.

Other examples where the origin and spread of a particular bacterial pathogen has been recorded include *Erwinia amylovora* (see below) and *Pseudomonas avenae*. The latter pathogen, which is an important agent of foliar disease in oats and maize in warmer climates, was first observed as a series of epidemics in the USA (1890–1908) and has now spread world-wide (Shakya *et al.*, 1985).

Economic importance

A number of bacterial diseases are of major economic importance, with direct financial loss due to decreased agricultural production and indirect loss due to the implementation of expensive control measures. The financial loss in agricultural production typically arises both due to a direct effect on the quality and quantity of the agricultural product as well as an overall deleterious effect on the plant itself. Post-harvest damage to stored plant products by bacterial pathogens (particularly soft rot bacteria) is also an area of major economic importance.

Table 1.2. *Economically important bacterial plant pathogens*

Bacterium	Host	Disease	Geographic location
Agrobacterium tumefaciens	Rosaceae, chrysanthemum grapevine	Crown gall	Temperate & mediterranean regions
Erwinia amylovora	Apples, pears & some ornamentals	Fireblight	World-wide in temperate zones
Erwinia carotovora	Potato	Soft rot	Temperate zones
Pseudomonas solanacearum	Over 200 species, e.g. banana, potato, tobacco, tomato	Bacterial wilt	World-wide in tropics & subtropics
Xanthomonas campestris pv. *citri*	Citrus fruit	Citrus canker	Asia, S. America, N. America (Florida) New Zealand
Xanthomonas campestris pv. *oryzae*	Rice	Leaf blight	Asia, S. & N. America, Australia, China

Some of the more widespread and economically important bacterial plant pathogens are shown in Table 1.2. As can be seen from this table, economically important phytopathogenic bacteria comprise a wide taxonomic range of organisms and result in a variety of diseases. More detailed descriptions of some of these pathogens are given below:

Agrobacterium tumefaciens   This is a very common and economically important pathogen of dicots, with a very wide host range including many members of the Rosaceae (stone fruits, pome fruits and Rubus species). Symptoms are both localised (gall formation) and more general, with frequent loss of overall vigour and yield. Assessment of disease is often complicated by the fact that infected plants may be more susceptible to invasion by other pathogens (fungal, bacterial and insect), which may cause more damage than the crown gall disease itself.

Xanthomonas campestris pv. *oryzae*   This bacterium causes bacterial leaf blight of rice (see above), and has the effect of reducing the overall growth and maturation of the host plant; resulting in poor development and lowered grain quality, with an increased number of undeveloped grains. According to Ou (1984), over 300 000 to 400 000 hectares have been affected annually in Japan in recent years, with yield losses of 20 to 30% in infected fields. Losses are higher in the Philippines and Indonesia, and in India millions of hectares are severely infected.

Erwinia amylovora This pathogen is the causal agent of fireblight disease of rosaceous plants (Van Der Zwet & Keil, 1979). It has proved extremely destructive to the apple and pear industries of a number of countries, and also to the cultivation of various ornamentals. With apples and pears, the disease not only destroys the current season's crops but may also lead to loss of branches and whole trees, leading to long-term devastation of orchards. Originating in the New York area of the USA, fireblight caused destructive epidemics in its westward spread across the USA, and major economic losses after its appearance in New Zealand and Europe. In the UK, for example, between 1958 and 1969, fireblight resulted in estimated losses of 20 000 pear trees, 20 000 hawthorn, 15 000 *Cotoneaster* and 2000 *Pyracantha* shrubs (Great Britain Ministry of Agriculture, 1969).

Erwinia carotovora This is a major soft rot pathogen of temperate climates, where it is particularly important in causing potato losses – either in the growing crop or in post-harvest storage.

Crop losses are generally considered to be substantial, though there is little clear information available. In the USA, Kennedy and Alcorn (1980) estimated that soft rot caused a loss in income of $14 million in 1976 and $11 million in 1977, with the number of States reporting the disease rising from 11 in 1971 to 16 in 1978. On a world-wide basis, Perombelon and Kelman (1980) have estimated total losses on all susceptible crops to be between $50 million and $100 million annually.

References

Bennett C. P. A., Jones P. & Hunt P. (1987). Isolation, culture and ultrastructure of a xylem-limited bacterium associated with Sumatra disease of cloves. *Plant Pathol.* **36**, 45–52.

Brown P. R. & Michelmore R. W. (1988). The genetics of corky root resistance in lettuce. *Phytopathology* **78**, 1145–50.

Burrill T. J. (1880). Anthrax of fruit trees: or the so-called fire blight of pear, and twig blight of apple trees. *Am. Assoc. Adv. Sci. Proc.* **29**, 583–97.

Davis M. J., Dean J. L. & Harrison N. A. (1988). Quantitative variability of *Clavibacter xyli* subsp. *xyli* populations in sugar-cane cultivars differing in resistance to ratoon stunting disease. *Phytopathology* **78**, 461–8.

Graham J. H. & Gottwald T. R. (1990). Variation in aggressiveness of *Xanthomonas campestris* pv. *citrumelo* associated with citrus bacterial spot in Florida citrus nurseries. *Phytopathology* **80**, 190–6.

Great Britain Ministry of Agriculture, Fisheries and Food (1969). *Fire Blight of Apple and Pear*. London: H.M.S.O. (Ministry of Agriculture, Fisheries and Food Adv. Ref. Publ. 571).

Hansen J. D. (1985). Common names for plant diseases: The American
     Phytopathological Society. *Plant Dis.* **69**, 649–76.
Kennedy B. W. & Alcorn S. M. (1980). Estimates of U.S. crop losses to procaryote
     plant pathogens. *Plant Dis.* **64**, 674–6.
Koch (1880). *Investigations into the Etiology of Traumatic Infective Diseases.*
     Translation by W. Watson Cheyne. London: The New Sydenham Society.
Ou S. H. (1984). *Rice Diseases*, 2nd edition. Slough: Commonwealth Agriculture
     Bureaux.
Perombelon M. C. M. & Kelman A. (1980). Ecology of soft rot erwinias. *Ann.*
     Rev. Phytopathol. **18**, 361–87.
Shakya D. D., Vinther F. & Marthur S. B. (1985). World-wide distribution of a
     bacterial stripe pathogen of rice identified as *Pseudomonas avenae.*
     Phytopathol. Z. **114**, 256–9.
Smiley R. W. (1988). Common names for plant diseases: The American
     Phytopathological Society. *Plant Dis.* **72**, 567–74.
Van Bruggen A. H. C., Grogan, R. G, Bogdanoff C. P. & Waters C. M. (1988).
     Corky root of lettuce in California caused by a Gram-negative bacterium.
     Phytopathology **78**, 1139–45.
Van der Zwet T. & Keil H. K. (1979). *Fireblight: A bacterial disease of rosaceous*
     plants. Agriculture Handbook 510. Washington D.C.: U.S.D.A. Science and
     Education Administration.

2

Bacterial structure and function

Plant pathogenic bacteria are typically motile, single-celled organisms, for which a range of light and electron microscope techniques are available to investigate aspects of structure and morphology (Sigee, 1989). Determination of the chemical and structural organisation of plant pathogenic bacteria is important to an understanding of cell function and host–pathogen interactions, and is also an important factor in bacterial taxonomy and pathogen identification. Phytopathogenic bacteria can be examined either during growth in sterile medium (*in vitro* culture, Fig. 2.1) or in association with higher plants, where they may be present on the plant surface or within infected tissue (growth *in planta*, see Fig. 2.3c).

Characteristic morphology and fine structure

The small size (0.5–2.0 μm diameter) of plant pathogenic bacteria places them close to the limits of resolution of the light microscope, and the examination of bacterial preparations with this instrument normally involves the use of an oil immersion objective to obtain maximum detail. These organisms are also quite difficult to see in terms of their optical contrast, and light microscope examination normally involves the use of stained preparations or phase-contrast microscopy (see Fig. 6.17a). Under optimal conditions of light microscopy, general features of morphology such as size, shape, and the presence of flagella and a capsule may be resolved, but little further detail can be determined.

At the level of the electron microscope, examination of whole cell preparations can provide useful information on general morphology (Fig. 2.3b,d), but normally reveals little detail of internal fine structure. Fig. 2.1 shows the appearance of fully hydrated cells in the frozen state, which is as close to the living condition as possible for conventional transmission electron micro-

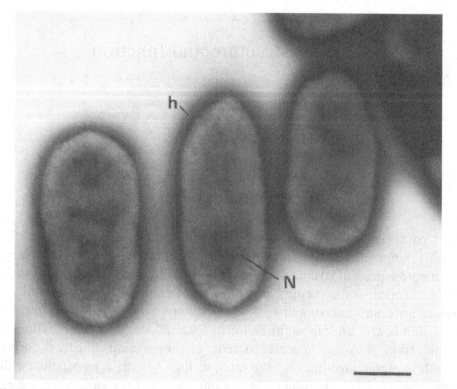

Fig. 2.1 Transmission electron micrograph of whole cells of *Erwinia amylovora*. The cells (taken from nutrient broth culture) are maintained on a cold stage ($-170°C$) in a fully-hydrated state and show internal regions of electron-dense nucleoid (N) and electron-transparent peripheral protoplasm. The bacteria are surrounded by a dense halo of frozen medium (h). Bar scale: 0.5 μm. (Photograph taken in collaboration with N. Hodson.)

scope observation. Some detail of the electron-dense central nucleoid can be seen in these cells, but surface structures such as flagella cannot be observed due to the lack of contrast enhancement.

The fine structure of a typical plant pathogenic bacterial cell is shown diagrammatically in Fig. 2.2 which incorporates information from a range of cell preparations, including sectioned material (Fig. 2.3*a,c*) and negative-stain preparations (Fig. 2.3*b,d*). In common with other prokaryotes, the cells of phytopathogenic bacteria may be divided into three main regions: a central nucleoid zone, a peripheral ribosomal area and a cell surface complex (Murray, 1978).

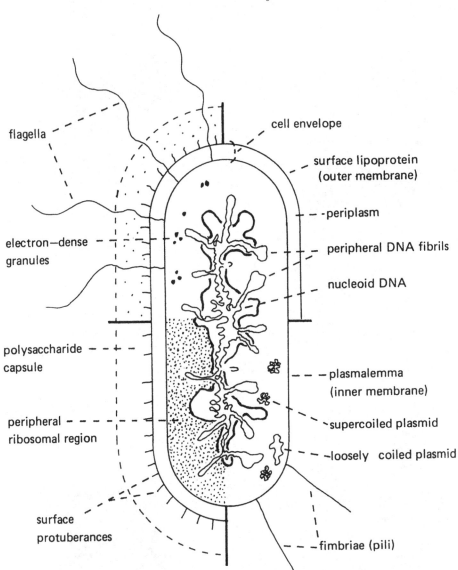

Fig. 2.2 General fine-structural features of a Gram-negative plant pathogenic bacterium. The different structural aspects shown in the quadrants of this composite diagram are revealed by a range of different electron microscope techniques. The central nucleoid area is delineated by a continuous line. (Adapted from Sigee, 1989).

Central nucleoid

The central nucleoid can be seen clearly in conventionally prepared ultrathin sections of plant pathogenic bacteria as a central electron-transparent area (Fig. 2.3*a*). This distinct central region is not an artefact of fixation, since it can also be seen under the electron microscope in chemically untreated cells

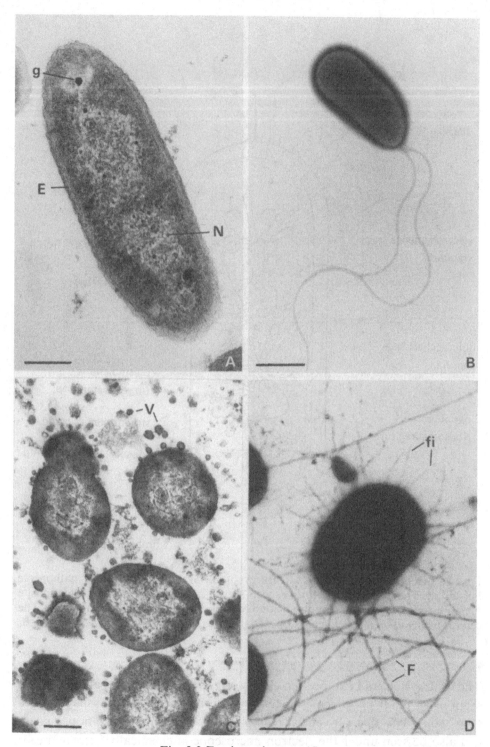

Fig. 2.3 For legend see p. 17.

(ultrathin cryosections, freeze-fractured preparations, frozen whole cells), and may also be seen by light microscopy under appropriate optical conditions (Binnerts *et al.*, 1982).

The central nucleoid contains a major part of the bacterial genome, which is present as highly condensed strands of supercoiled DNA. Autoradiographic studies on *Escherichia coli* and *Bacillus subtilis* (Ryter & Chang, 1975) have suggested that most of this central DNA is not involved in transcription. This takes place in peripheral regions of the cell on lateral loops of DNA within coralline, ribosome-free extensions of the central nucleoid (Bohrmann *et al.*, 1991).

Peripheral ribosomal area

In ultrathin section, the peripheral region of bacterial cells (surrounding the central nucleoid) appears electron dense and is packed with ribosomes. This part of the cell may contain a range of other fine structural features – including extensions of the central nucleoid, granular inclusions, mesosomes, refractive bodies, polar bodies and bacteriophage particles. Plasmids are also frequently present, but can only be visualised after isolation (see Fig. 8.12).

Granular inclusions

These may show some resemblance to ribosomes, but are characteristically 4–5 times larger in size, and are of various types.

In the case of some non-fluorescent pseudomonads and certain other phytopathogenic bacteria (see Table 3.1), the cell contains granules composed of the polymer poly-β-hydroxybutyrate (PHB). The presence of PHB granules can be detected at the level of the light microscope in living cells (as highly refractive bodies under phase contrast) and in heat-fixed smears after staining with safranin, dilute carbolfuchsin or using the lipid stain Sudan Black B (Hayward, 1990). Under the electron microscope, PHB granules appear electron-transparent in ultrathin section and may either be localised in

Fig. 2.3 Electron microscope details of plant pathogenic bacteria. (*a*) *Pseudomonas syringae*. Ultrathin section of a resin-embedded cell that has been fixed in glutaraldehyde/osmium tetroxide and dehydrated in ethanol. E, cell envelope; N, central nucleoid; g, electron-dense granule. Bar scale: 0.25 μm. (*b*) Negative-stain preparation of *Pseudomonas syringae*, showing two polar flagella. Bar scale: 1.0 μm. (*c*) Ultrathin section of degenerating cells of *Pseudomonas syringae* pv. *phaseolicola* in bean leaf tissue, showing numerous lipoprotein surface vesicles (V). Bar scale: 0.25 μm. (*d*) Negative-stain preparation of *Agrobacterium tumefaciens*. Surface fimbriae (fi) radiate out from the cell and are much thinner than flagella (F). Bar scale: 0.5 μm. (Photographs taken in collaboration with M. H. El-Masry, H. A. Epton and D. Howitt.)

occurrence (often near to the plasmalemma) or distributed throughout the peripheral protoplasm. PHB granules are particularly prominent in cells grown in media with a high carbon to nitrogen ratio.

Mesosomes

These appear in chemically fixed cells as complex insertions of the plasma-lemma, providing a connection between the cell surface and the central nucleoid (Rogers *et al.*, 1980). Visualisation of these structures in bacterial sections depends on the physiological state of the cell (particularly characteristic of log phase cells) and also the fixation procedure (Silva *et al.*, 1976). Although the presence of mesosomes has not been widely reported in plant pathogenic bacteria, they have been noted in certain cases, e.g. the xylem-limited bacterium causing Sumatra disease of clove trees (Bennett *et al.*, 1987).

Refractive bodies and polar bodies

Refractive bodies (R-bodies) have been observed in a limited number of non-pathogenic bacteria, and in a single pathogenic species: *Pseudomonas avenae* (Bird & Gibson, 1987). In this species, R-bodies have been observed (one or two per cell) in cells as a proteinaceous ribbon. In the same species, spherical electron-transparent polar bodies have also been observed in a small proportion of cultured cells. The role of R-bodies and polar bodies is unknown and no link with pathogenicity has so far been established.

The cell surface complex

The cell surface of bacteria generally has been the subject of intensive study in recent years and a considerable amount of information is now known about this part of the cell (Hammond *et al.*, 1984; Hancock & Poxton, 1988). The cell surface of plant pathogenic bacteria is particularly important in their interactions with higher plant cells, and consists of three main aspects: cell envelope, capsule and appendages.

Cell envelope

The cell envelope comprises the combined membrane/cell wall complex of the bacterial cell and is of two main types, depending on whether the bacterium is Gram-negative or Gram-positive. In the classical Gram test, Gram-positive cells are able to retain Gram stain (a complex of crystal violet and iodine) against elution with alcohol, while Gram-negative cells do not.

The cell envelope of Gram-negative cells is shown diagrammatically in Fig. 2.4, and consists of (i) an inner cytoplasmic membrane, involved in respi-

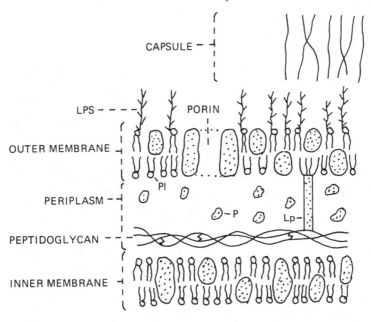

CAPSULE

LPS

PORIN

OUTER MEMBRANE

PERIPLASM

PEPTIDOGLYCAN

INNER MEMBRANE

Fig. 2.4 General diagram of Gram-negative bacterial cell envelope. LPS, lipopolysaccharide; Lp, lipoprotein linker molecule; Pl, phospholipid; P, protein (stippled). (Adapted from Hancock & Poxton, 1988.)

ration, electron transport, and nutrient uptake; (ii) a periplasmic space, a metabolic reservoir and site of cell wall material; and (iii) an outer membrane, involved in nutrient uptake, lipid metabolism and maintenance of cell wall integrity. The cell envelope of Gram-positive bacteria is also highly complex (Fig. 2.5), but differs in having a much thicker cell wall (peptidoglycan) layer and no outer lipoprotein membrane.

Plant pathogenic bacteria are almost entirely within the Gram-negative group (with only the coryneform bacteria being Gram-positive), and **molecular aspects of the Gram-negative envelope** will now be considered in more detail.

Inner and outer membrane The outer membrane of Gram-negative bacteria resembles the inner (cytoplasmic) membrane in having a clear bilayer (unit membrane) structure when examined in ultrathin section, but differs in the possession of lipopolysaccharide (LPS) molecules and also in its protein composition.

Lipopolysaccharide molecules The structure of a typical LPS molecule is shown in Fig. 2.6, and consists of three main regions: an inner lipid region (which attaches the molecule to the membrane), a core region of polysacchar-

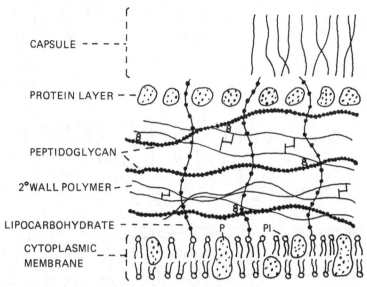

CAPSULE

PROTEIN LAYER

PEPTIDOGLYCAN

2°WALL POLYMER

LIPOCARBOHYDRATE

CYTOPLASMIC
MEMBRANE

Fig. 2.5 General diagram of Gram-positive bacterial cell envelope. Pl, phospholipid; P, protein (stippled). (Adapted from Hancock & Poxton, 1988.)

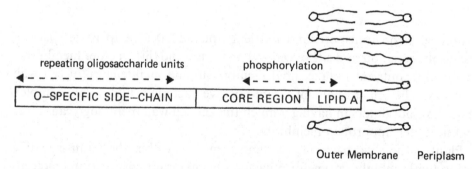

repeating oligosaccharide units   phosphorylation

O–SPECIFIC SIDE–CHAIN   CORE REGION   LIPID A

Outer Membrane   Periplasm

Fig. 2.6 Diagrammatic representation of lipopolysaccharide (LPS) molecule in the outer membrane of a Gram-negative bacterium. The lipid component of the molecule (lipid A) is typically a glucosamine disaccharide with associated fatty acids. The polysaccharide component comprises an inner core region and an outer region (O-specific side-chain) which is composed of repeating units of short and mostly branched oligosaccharides.

ide (next to the membrane) and an outer region of polysaccharide (the 'O-specific chain'). The lipid region and inner part of the polysaccharide core are highly phosphorylated, while the O-specific side-chain is largely composed of repeating units of short, branched oligosaccharides. This outer part of the LPS molecule has a major role in determining the surface characteristics of the bacterial cell, and is largely responsible for the antigenic properties of the bacterial strain. It is also thought to be important in the initial interaction that takes place between bacterium and plant cell, where it may determine the difference between compatible and incompatible reactions.

Membrane lipids The membrane lipid composition of plant pathogenic bacteria shows considerable variation in relation to growth phase, type of substrate and other environmental conditions. Under standard culture conditions, however, it can provide an important taxonomic feature – particularly in relation to fatty acid composition (Chapter 3). The phospholipid composition has recently been shown to be of major importance in ice nucleating activity (Chapter 4).

Membrane proteins In Gram-negative bacteria, approximately 50 different proteins are present in the outer membrane, compared with about 300 species in the cytoplasmic membrane (Pugsley and Schwartz, 1985).

The occurrence and function of outer membrane proteins has been examined in particular detail in *Escherichia coli* (Hancock & Poxton, 1988), which serves as a useful model for other Gram-negative bacteria. In this organism, the outer membrane proteins have been individually named with reference to the genes that encode them, and include:

1. Proteins that form diffusion (porin) channels (Omp C, Omp F and Pho E) and show close homologies with each other.
2. Omp A protein, which spans the outer membrane and acts as an external phage receptor.
3. Some proteins (Pho E and Lam B) which are inducible and become major components under certain growth conditions. Receptor proteins of siderophores fall into this category and are only present as functional components under iron-limiting conditions (see Fig. 4.5).

In plant pathogenic bacteria, the presence of receptor molecules in the inner and outer surface membranes is important for bacterial activities both inside and outside the plant. These molecules include receptors to secretory proteins (e.g. *Erwinia chrysanthemi*, see Fig. 2.11), siderophore complexes (mainly outer membrane proteins, see Fig. 4.5 and Table 8.14) and chemo-attractants (e.g. VirA inner membrane receptor of *Agrobacterium tumefaciens*, see Fig. 8.14). The location of ice nucleation proteins is also thought to be membrane-based, with at least one type of nucleation complex in the outer membrane (see Chapter 4).

The periplasmic space The periplasmic space in Gram-negative cells is an important reservoir of secretory and other proteins and contains the cell wall material. It is clearly visible in ultrathin sections of osmium-fixed bacterial cells as a space (of variable thickness) between inner and outer membranes, and has been shown to be iso-osmotic with the general bacterial matrix (Stock *et al.*, 1977).

Periplasmic proteins comprise about 100 different species, and are mainly involved in nutrient uptake and catabolism. They can be selectively released from the periplasm by osmotic shock or treatment with organic solvents.

The cell wall material contained within the periplasm is important in determining the rigidity of the bacterial cell, and is composed largely of a layer of peptidoglycan, a polymer composed of disaccharide tetrapeptide subunits. The disaccharide (*N*-acetylglucosamine-β1,4-*N*-acetylmuramic acid) components are linked in a linear sequence to form a glycan backbone, with the tetrapeptide groups linked to the muramic acid residues. Rigidity is conferred by crosslinking between the peptides of adjacent peptidoglycan chains. In addition to peptidoglycans, other cell wall polymers are also present, including techoic acids, polysaccharides and proteins.

Bacterial capsule

Most plant pathogenic bacteria have a layer of surface slime, which is partly present as a well-defined capsule and partly diffuses into the surrounding environment. The presence of a capsule may be shown under the light microscope using negative-stain preparations, and under the transmission electron microscope using a stain such as ruthenium red, which specifically attaches to acidic sugar residues.

The surface slime is composed principally of a matrix of polysaccharides, but polypeptides and glycopeptides may also occur. The polysaccharides that form the matrix are either simple polymers (homopolysaccharides), which may be neutral (e.g. levan), or acidic (e.g. alginate) molecules, or they may be a more complex assembly (heteropolysaccharide) of different oligosaccharide subunits. Fluorescent pseudomonads appear to produce mainly homopolysaccharides, while erwinias (e.g. *Erwinia stewartii*) and xanthomonads have largely heteropolysaccharide surface layers, some details of which are given below. The polysaccharide backbone may also have associated non-sugar groups such as acyl residues and ketal-linked pyruvic acid, and cell surface peptidoglycans have been reported from corynebacteria.

Fluorescent pseudomonads Pathovars of *Pseudomonas syringae* frequently produce substantial amounts of extracellular polysaccharide, in which alginate (a linear polysaccharide of C-1–C-4-linked mannuronic and galuronic acids) and levan (a branched polysaccharide with a C-2–C-6 backbone) have recently been identified as the major components. The production of alginate is a characteristic of many fluorescent pseudomonads (Fett *et al.*, 1986), but depends to some extent on culture conditions. In *Pseudomonas syringae* pv. *glycinea*, for example, Osman *et al.* (1986) have shown that levan is the major polysaccharide when sucrose is used as the primary carbon source, while alginate is produced with glucose. Bacteria growing *in planta* produce mainly alginate.

Erwinia stewartii In this bacterium, which causes wilt of maize, the capsule is composed of a very large (*c.* 45 megadaltons) and viscous heteropolysaccharide containing glucose, galactose and glucuronic acid units (Huang, 1980). The synthesis of this extracellular polysaccharide appears to be mediated by galactosyl transferase enzymes, at least one of which has been detected in the soluble fraction of sonicated cells (Huang, 1980).

The production of extracellular polysaccharides by plant pathogenic bacteria is considered important in a number of host–pathogen interactions, where the polysaccharide is either directly involved as a virulence factor (see Chapter 7) or is indirectly important as a barrier to cell contact and recognition.

Surface appendages – flagella, fimbriae and protuberances

Bacterial flagella Bacterial flagella consist of a basal body (embedded in the cell surface), a universal joint to allow articulation and a projecting filament which is 50–100 μm long and 20 nm in diameter. Although the limited thickness of flagella means that they are not directly visible at the level of the light microscope, they can be visualised indirectly using video-enhanced imaging of live cells (Block *et al.*, 1991) or specific staining of air-dried (non-living) preparations. Light microscope observation of hanging drop preparations may also be useful in distinguishing between polar and peritrichous flagella (see below), since individual cells show abrupt changes in direction in the former case but a more irregular, sluggish motility in the latter. Under the transmission electron microscope, they are readily observable in negatively stained (Fig. 2.3*b*) or shadowed preparations, where the frequency and characteristics of flagellar attachment in a population of cells can be investigated.

Considerable variation may occur within a single population of bacterial cells in terms of flagellar occurrence, with some cells being completely non-flagellate while others have varying numbers of flagella. Recent electron microscope studies by Sigee and El-Masry (1989) on *Pseudomonas syringae* pv. *tabaci* cultured *in vitro*, for example, have shown a major increase in the proportion of cells with flagella during the exponential growth phase (rising from 25 to 75% of the total population) followed by a decline in stationary phase. Analysis of bacterial cells directly isolated from infiltrated tobacco leaves at various times during the progression of wildfire disease showed that bacteria were also flagellate *in planta*, and that the degree of flagellation varied according to the stage of the disease. The proportion of cells within the leaf tissue that had flagella showed a marked increase during the major phase of bacterial multiplication (see Fig. 5.5), indicating a close parallel with the *in vitro* situation and suggesting that the degree of flagellation is primarily determined by endogenous factors (relating to growth rate) rather than by the microenvironment.

Although considerable variation in flagellation may be observed within a single culture, consistent differences in the number and location of flagella do occur between different genera and species of phytopathogenic bacteria, providing an important taxonomic characteristic (Chapter 3). In general, flagella may either occur singly at one end of the cell (monotrichous: typical of coryneform bacteria and *Xanthomonas*), as a polar group (lophotrichous: *Pseudomonas*) or arranged around the periphery of the cell (peritrichous: *Agrobacterium* and *Erwinia*). A complete absence of flagella has been reported for certain coryneform species, *E. stewartii* and other vascular wilt pathogens.

Bacterial flagella can be dissociated from whole cells by sonication and then subsequently isolated as a purified preparation after removal of the cells by centrifugation (Smith & Koffler, 1971). Chemical analysis of the flagellar fraction shows it to be largely composed of a globular protein (flagellin), which forms the main filament of the flagellum.

Bacterial motility and chemotaxis   Motility of plant pathogenic bacteria is an important aspect of the infection process (Chapter 5) and is mediated by helical rotation of the flagella.

Bacterial motility and chemotaxis have been investigated in particular detail in *Escherichia coli* (Shaw, 1991) which serves as a useful model for other organisms. In this bacterium, movement in a single direction is caused by counter-clockwise (CCW) rotation of the flagella, which operate as a bundle or single unit. If any of the flagella switch to a clockwise (CW) rotation, bacterial coordination is lost and the run ends in a disorganised rotation of the cell or tumble. Re-attainment of CCW in all flagella results in a new run, in a random direction.

Bacterial response to stimuli is determined by the balance between runs and tumbles. When the bacterium is moving in a favourable direction, runs predominate. On encountering unfavourable conditions, movement changes into a tumble. This results in a change of direction, which may extend into a run if the new direction leads to a favourable microenvironment.

In the case of chemotaxis, where the bacterial cell is moving in relation to a chemical gradient, the response is mediated by binding of the chemo-attractant to surface receptor molecules. Transduction of the signal then occurs through the cytoplasm via products of specific chemotaxis genes, resulting in suppression of a gene transcript (the 'tumble factor'). In the absence of this gene product, periodic production of which normally results in a regular change of direction, the bacterium will continue to run in one direction.

In phytopathogenic bacteria, mechanisms of motility and chemotaxis have been looked at in most detail in *Agrobacterium tumefaciens* (Shaw, 1991). This bacterium differs from *Escherichia coli* in that movement typically occurs as long runs (in excess of 500 μm) with infrequent tumbles, and is achieved by unidirectional movement of flagella in a clockwise direction. Various lines of evidence suggest that although motility of *A. tumefaciens* differs from *E. coli* in a number of respects, there is a similar signal transduction pathway in the two organisms.

Fimbriae In addition to flagella, many bacteria also possess appendages referred to as fimbriae or pili. These surface structures are readily seen under the electron microscope in negative stain preparations (Fig. 2.3*d*) and are typically thin, non-sinusoidal filaments which radiate out from the organism, normally appearing smaller and more numerous than flagella. Fimbriae are composed of single proteins, and have a constant diameter for a particular cell type (normally 3–7 nm) – with variation between different strains and species.

Recent studies by Stemmer and Sequeira (1987) have demonstrated the presence of fimbriae in a range of genera and species of plant pathogenic bacteria cultured *in vitro*, including *Pseudomonas solanacearum*, *Pseudomonas savastanoi*, *Pseudomonas syringae* pathovars and *Agrobacterium tumefaciens*. The culture conditions required for most abundant production of fimbriae (shallow, static aerobic cultures) by these bacteria resembled those reported as most suitable for the production of fimbriae by mammalian bacterial pathogens.

Fimbriae can be removed from bacteria by mechanical shearing and obtained as a pure protein fraction. Young *et al.* (1985) have described the isolation and purification of fimbral protein from *Pseudomonas solanacearum* and shown it to be strongly hydrophobic (containing 41.8% hydrophobic amino acids) with no associated carbohydrate. The purified protein reassembled *in vitro* to produce long fibres which were similar in diameter to naturally occurring fimbriae and had a similar ability to cause haemagglutination. This suggests that the formation of fimbriae *in vivo* may operate by a simple self-assembly process.

Bacterial fimbriae in general have been implicated in a number of cell functions, including bacterial attachment to surfaces, sexual conjugation (involving a special class of fimbriae, the sex pili) and as specific sites for viral recognition and attachment.

Role of fimbriae in bacterial attachment   In the case of bacterial pathogens of animals, invasion of the host tissue is normally preceded and initiated by specific adhesion of bacteria to target cell surfaces. This adhesion is mediated by specific bacterial surface molecules (adhesins), which may or may not be fimbriae-associated.

Although the role of fimbriae in animal pathogenesis has been clearly established, the importance of these structures in the attachment and pathogenicity of phytopathogenic bacteria is not clear. Stemmer and Sequeira (1987) have concluded that results so far obtained do not support any universal involvement of fimbriae in specific attachment of bacteria since in some species of bacteria (e.g. *Agrobacterium tumefaciens, Pseudomonas solanacearum*) there is no consistent distinction between virulent and avirulent strains in terms of fimbrial occurrence.

In other cases, the role of fimbriae in mediating bacterial attachment to plant cells appears to be well authenticated. Romantschuk and Bamford (1985a) have shown that chromosomally-encoded fimbriae of *Pseudomonas syringae* pv. *phaseolicola* are involved in bacterial attachment to leaf surfaces and may be important during the early stages of halo blight disease. Attachment was assessed by radioassay (using C^{14}-labelled cells) of bacterial retention to leaf strips over a 60-minute time period, and showed direct correlation with the level of fimbriae present on the cells. Cells without fimbriae showed no increase in attachment (above a low base level) over the time course, while wild-type and super-fimbriate cells showed a progressive increase in association. Scanning electron microscope studies also showed that fimbriate strains had a high level of attachment to the leaf surface (particularly the stomata of the abaxial surface), while non-fimbriate strains had only a low level random association. In terms of pathogenicity, both fimbriate and non-fimbriate pathogen strains caused disease when directly infiltrated into the leaf tissue, but only fimbriate strains caused infection when sprayed onto the outer surface of the leaf – suggesting that fimbrial-mediated attachment to the leaf surface may be an important early stage in the normal infection process.

Studies by Korhonen *et al.* (1986) have suggested that fimbriae may also be important for bacterial attachment to root surfaces. These workers showed that highly purified fimbriae from *Klebsiella* and *Enterobacter* showed specific binding to roots of *Poa pratensis* (a natural site for these bacteria) and that the binding could be blocked by fimbriae-specific antibody fragments, presumably acting at plant receptor sites.

Fimbriae as viral receptor sites   Bacterial fimbriae may act as specific and primary receptor sites for viral attachment, as shown, for example, during the infection of *Pseudomonas syringae* pv. *phaseolicola* by bacteriophage φ6 (Romantschuk & Bamford, 1985b). This phage has a lipid-protein envelope, a genome comprising three pieces of double-stranded RNA, and has also been shown to infect *P. syringae* pv. *syringae* and *P. pseudoalcigenes*. Viral infection of *P. syringae* pv. *phaseolicola* is initiated by attachment to φ6-specific host fimbriae (which have also been implicated in bacterial attachment to leaf surfaces, see above). The attached virus particles subsequently appear to be pulled through the extracellular polysaccharide layer by retraction of the fimbriae, bringing them into contact with the outer cell membrane, where fusion with phage membrane occurs. The importance of the fimbriae to the infection process was shown by electron microscope examination of phage-resistant isolates, the majority of which were devoid of fimbriae, while the remainder had a different mode of fimbriation from wild-type.

Surface protrusions   Under certain circumstances, the outer membrane of plant pathogenic bacteria appears to evaginate, forming blebs which can detach from the surface as lipopolysaccharide (LPS) vesicles. Surface protrusions can be observed under the electron microscope in both whole cell preparations and sectioned material (Fig. 2.3c) and show considerable variation in size between different cell populations (within the same species or pathovar). These surface protrusions have been observed within infected plant tissue on *Pseudomonas syringae* pv. *phaseolicola* (Epton *et al.*, 1977) and *P. syringae* pv. *pisi* (Sigee & Al-Issa, 1983), where they appear to characterise a late stage in the degeneration of the bacterial cells. The release of LPS vesicles during disease development may be important in the host response, since infiltration of an LPS preparation from *P. syringae* pv. *phaseolicola* was shown to cause water-soaking in bean tissue.

Variation in prokaryote structure

Although plant pathogenic prokaryotes all have the same fundamental features of small size and simple structure, there is considerable variation in general morphology and ultrastructural detail. This variation arises in relation to taxonomic differences, experimental manipulation and endogenous changes relating to growth phase and the local environment. An example of the latter is the changes that occur in bacterial fine structure within infected plant tissue, and is discussed in Chapter 6.

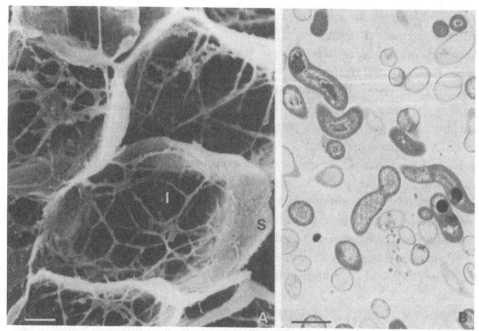

Fig. 2.7 Electron microscope examination of *Streptomyces ipomoea*: a plant pathogenic actinomycete. (*a*) Scanning electron micrograph of section of sweet potato storage root infected with the pathogen. Mycelial strands can be seen both within (I) and at the surface (S) of the parenchyma cells. Bar scale: 10 μm. (*b*) Transmission electron micrograph of *S. ipomoea* mycelial strands in agar culture, showing simple (prokaryote) internal structure. Bar scale: 1 μm. (Photographs taken by, and reproduced with the permission of, C. A. Clark and S. W. Matthews.)

Actinomycetes and Mollicutes

Two major groups of plant pathogenic prokaryotes, the Actinomycetes and Mollicutes show major differences from the structural features already described.

Actinomycetes

Although Actinomycetes do not constitute a major group of plant pathogens, two particular species: *Streptomyces ipomoea* (causing soil rot of sweet potato) and *Streptomyces scabies* (common scab on several tuber crops) are economically significant and widely distributed. These organisms have a filamentous growth habit (Fig. 2.7*a*) which is superficially similar to that of fungi (eukaryotes). Recent studies by Clark and Matthews (1987) on the growth of these organisms within host tissue have shown, however, that they do not produce the specialised penetration structures (i.e. appressoria, infection cushions, haustoria) typical of fungi, but simply penetrate host cell walls by direct growth of the main hypha or lateral branches through extensively degraded sites. Break-up of the hyphae into spore chains can be observed in host tissue at late stages of infection. The prokaryote status of these organisms is clear in ultrathin section (Fig. 2.7*b*), where there is a complete absence of membrane-bound nuclei and eukaryote cytoplasmic organelles.

Mollicutes

A number of plant diseases, initially thought to be caused by viruses due to the ability of the pathogenic agent to pass through a 0.2 μm filter, have been shown to be mediated by small prokaryotes which belong to the class Mollicutes. These organisms are characterised by their small size (diameter typically 0.2–0.5 μm), complete absence of a cell wall and various biochemical features (see Chapter 3). As a probable consequence of the lack of rigidity resulting from the absence of a cell wall they are frequently highly variable in shape (with coccoid elements, larger swollen forms and filamentous cells of varying length occurring within a single population) and are also osmotically very labile, growing only in hypertonic or isotonic media.

Although plant pathogenic Mollicutes are as yet poorly defined taxonomically, they appear to belong to two main groups: mycoplasma-like organisms (MLOs) and spiroplasmas.

Members of the genus *Spiroplasma* differ from MLOs in having the ability to produce motile helical filaments, which can be seen particularly clearly in whole-mount light and electron microscope preparations (rather than ultrathin section). The motility of *Spiroplasma* filaments is not caused by flagella or any other type of locomotory appendage but by intracellular fibrils, which generate a twitching or flexional locomotion. Plant pathogenic Mollicutes, as with these organisms generally, have very exacting nutritional requirements and only members of the genus *Spiroplasma* have as yet been cultured *in vitro* (Davis, 1979).

Studies on the fine structure of Mollicutes have been limited by a number of factors, including the inability to isolate and cultivate many of them *in vitro* and problems with examination *in planta*. The latter problem arises from the fact that they are typically localised to small regions within the plant (frequently occurring only in sieve tubes), and they are not easily extracted for electron microscope examination of whole-mount preparations.

In spite of these problems, MLOs have been observed in ultrathin section of infected plant tissue by a number of workers (Davis, 1979), providing an important means of identifying these organisms as potential agents of plant disease. The extraction of MLOs from infected plant tissue is limited both by their fragility (and resulting ease of damage during tissue processing) and by the presence of contaminating plant material. These problems have been overcome to some extent by new techniques of homogenisation and centrifugation. Recent procedures developed by Jiang and Chen (1987), for example, on aster yellow disease of lettuce have succeeded in isolating a relatively uniform preparation of aster yellow MLO, the causal agent of this disease. These procedures involve:

1. Homogenisation of infected plant tissue with rapid isolation of the pathogen in a medium containing mannitol to protect the plasmalemma, and with maintained osmolarity and pH.
2. Separation of the MLOs from contaminating plant debris by discontinuous density gradient centrifugation using Percoll which provides a non-toxic, low osmotic pressure medium.

The resulting preparation (Fig. 2.8) contains a relatively uniform population of spherical cells (less pleomorphic than cells directly observed in intact tissue) and has considerable potential for electron microscope examination and further cytological characterisation of the pathogen.

Experimental manipulation of bacterial cells: spheroplast formation

Treatment of normal bacterial cells with wall-degrading enzyme (lysozyme) leads to wall-less cells or spheroplasts and provides a good example of how experimental manipulation can lead to marked changes in the structure and activity of living cells. Under conditions of mild osmotic shock and destabilisation of the outer membrane (Witholt *et al.*, 1976), lysozyme is able to penetrate the outer membrane and degrade the peptidoglycan cell wall by breaking the $\beta(1,4)$ links of the glycan backbone. The resulting cell assumes a spherical shape and is very fragile, requiring careful stabilisation in an isosmotic medium.

Spheroplasts have been used in the production of inner and outer membrane fractions, the selective release of periplasmic constituents (Thurn & Chatterjee, 1985) and in facilitating the transfer of foreign DNA into the cell by transfection (Yeh *et al.*, 1985).

Chemical analysis of bacterial cells

A wide range of histochemical and biochemical techniques are available for the chemical analysis of plant pathogenic bacteria. Biochemical techniques, such as DNA and protein electrophoresis and fatty-acid analysis, are dis-

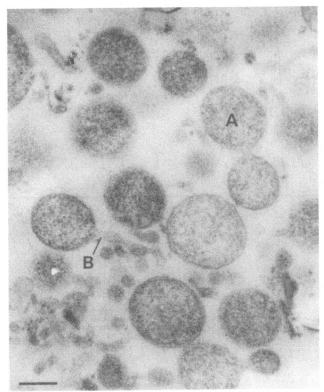

Fig. 2.8 Transmission electron micrograph (ultra-thin section) of cells of mycoplasma-like organisms associated with aster yellows disease. The cells are from a purified preparation, and show evidence of binary fission (A) and production of branched filaments (B). Note the absence of a cell wall, and the simple internal cell structure comprising pale nucleoid regions (with DNA fibrils) and peripheral ribosomes. Bar scale: 0.2 μm. (Micrograph reproduced, with permission, from Jiang & Chen, 1987.)

cussed in subsequent chapters, and are normally applied to bulk samples of cells cultured *in vitro*.

Histochemical techniques, involving light or electron microscopy, have the advantage that they can be used directly on cells growing *in planta* and are potentially able to obtain chemical information on individual bacteria during their interaction with plant cells. These techniques include electron probe X-ray microanalysis, autoradiography, specific stains and immunocyto-chemistry.

Electron probe X-ray microanalysis

This electron microscope technique provides information on the elemental composition of biological microsamples, and has been used to obtain data on the level of cations and other major constituents in various plant pathogenic

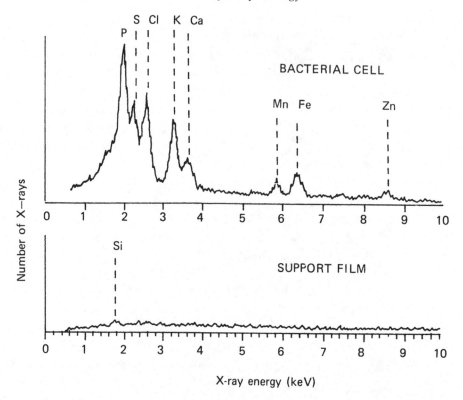

Fig. 2.9 Chemical analysis of plant pathogenic bacteria: determination of elemental composition by electron probe X-ray microanalysis. X-ray emission spectra are shown for a single air-dried cell of *Pseudomonas syringae* (on support film) and support film only. The bacterial cell shows clear peaks of monovalent (K) and divalent cations (Ca, Mn, Fe and Zn) in addition to P, S and Cl. Elements with an atomic number of 8 (oxygen) and below are not detectable with this technique. The support film spectrum shows a small (contaminant) peak of Si, but no other elements.

bacteria, including *Pseudomonas syringae* pv. *tabaci* (El-Masry & Sigee, 1989) and *Erwinia amylovora* (Sigee *et al.*, 1989). The specimen, which may be a whole bacterial cell or a cell in section, is irradiated by an electron beam (the electron probe) and the energy spectrum of emitted X-rays is analysed both qualitatively (assessment of detectable elements) and quantitatively (determination of mass fractions). A discussion of specimen preparation and microscope procedures is given by Sigee *et al.* (1985). Fig. 2.9 shows a characteristic X-ray emission spectrum taken from a whole air-dried cell of *Pseudomonas syringae* and has clear peaks of P, S, Cl, K, Ca and various transition metals.

X-ray microanalysis has been used to monitor whole cell changes in elemental composition during bacterial growth *in vitro* and *in planta*, and under conditions of metal toxicity (Sigee & Al-Rabaee, 1986). It has also provided information on the cation associations of extracted bacterial DNA (Sigee & El-Masry, 1988).

Autoradiography

In this technique, the location of radioactive tracers can be determined in bacterial cells using a photographic emulsion. Autoradiography has been used to monitor the uptake of H^3-uridine (RNA synthesis) and H^3-thymidine (DNA synthesis) in bacteria and host tissue during plant infection (reported in Sigee, 1989), and has also been used to determine the localisation of Ni^{63} during nickel toxicity (Sigee & Al-Rabaee, 1986). Fig. 2.10*a* shows the typical appearance of an electron microscope autoradiograph processed by a medium-grain chemical developer.

Histochemical stains

Histochemical staining reactions are normally carried out on sectioned material, and provide information on the chemistry of both internal and external (cell surface) components. As an example of this, ruthenium red stain has been used to detect acidic polysaccharides at the surface of pseudomonad bacteria in oat (Smith & Mansfield, 1982). In these studies, it was shown that production of extracellular polysaccharide was greater in the compatible compared with the incompatible situation, possibly preventing close contact and the induction of a resistant response in the former case.

Immunocytochemistry

This technique uses the specificity of the antibody–antigen reaction to locate particular macromolecules (including proteins and carbohydrates) in the cell. Experimental procedures are detailed by Sigee (1989), but in practice, antibodies are raised against the chemical component to be localised, and the location of these antibodies is then determined using secondary antibodies that have been tagged by a marker (such as immunogold). As an example of this technique, Fig. 2.10*b* shows an immunogold preparation of cells of *Erwinia amylovora* that have been probed with a primary antibody raised against cell surface polysaccharide material. Further details of the use of monoclonal and polyclonal antibodies in relation to plant pathogenic bacteria are given in Chapter 3.

Compartmentation and function of bacterial cells

Cell function and compartmentation are closely related. Although plant pathogenic bacteria, as prokaryotes, have much less compartmental com-

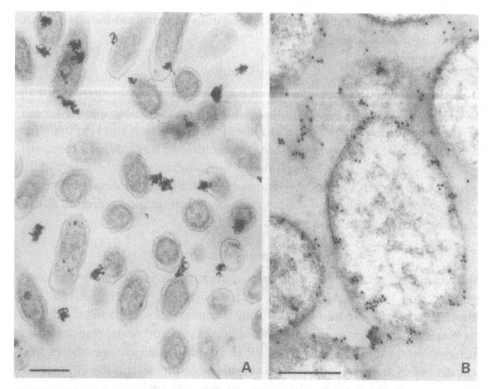

Fig. 2.10 Chemical analysis of plant pathogenic bacteria: use of autoradiography and immunocytochemistry. (*a*) Electron microscope autoradiograph of section of *Pseudomonas syringae* pv. *tabaci* cells, labelled with Ni^{63}. Uptake of radioactive cations into individual cells is shown by the presence of associated coiled filaments (photographic silver grains). Scale bar: 1 μm. (Photograph taken in collaboration with RH Al-Rabaee.) (*b*) Ultrathin section of glutaraldehyde-fixed cells of *Erwinia amylovora*, treated with rabbit antiserum raised against whole cells and labelled with anti-rabbit immunogold-IgG conjugate. The primary (rabbit) antibody reacts mainly with peripheral and cell surface epitopes (possible polysaccharides), as shown by the localised distribution of gold particles. Scale bar: 0.5 μm.

plexity than their eukaryote host cells, this is still an important aspect of their activity. Fig. 2.2 indicates some of the major compartments in a typical bacterial cell, including central nucleoid, peripheral cytoplasm, periplasmic space and capsule.

The separation of actively transcriptive and non-transcriptive DNA between the peripheral cytoplasm and the central nucleoid has already been mentioned, and is an important aspect of compartmentation of function in these cells.

In Gram-negative bacteria, the presence of inner and outer membranes in the envelope also defines compartments and provides differential barriers to the entry and exit of molecules. The outer membrane is relatively permeable due to the presence of porin channels, which allow the diffusion of hydrophilic solutes within a limited size range across the cell surface. The specific importance of these channels in solute transport has been emphasised by the fact that purified porin proteins have been reconstituted in artificial lipid bilayers (Darveau *et al.*, 1983) to give functional pores. The inner membrane does not possess such channels, and the passage of solutes through this normally involves active transport.

The inner and outer membranes also differ in the transport of macromolecules. This is particularly apparent in the process of secretion and partitioning of proteins between the peripheral cytoplasm, periplasm and extracellular environment. The secretion of enzymes by plant pathogenic bacteria, particularly those degrading the host cell wall, is an important aspect of their pathogenic activity.

Protein partitioning by the cell surface membranes

The role of the surface membranes of plant pathogenic bacteria in the transport and partitioning of internally synthesised protein molecules has been investigated in particular detail in *Erwinia chrysanthemi* (Mildenhall *et al.*, 1988; He *et al.*, 1991) as summarised in Fig. 2.11. In this bacterium, protein molecules that are transcribed and translated in the cytoplasm may either be retained in that part of the cell (e.g. β-galactosidase), pass into and remain in the periplasmic space (e.g. acid phosphatase) or be secreted to the outside of the cell. Exponentially growing cells of *E. chrysanthemi* normally secrete about 98% of their pectate lyase (PEL) and protease (PT), 60% of their cellulase (CL) and 40% of the polygalacturonase (PG).

The passage of proteins across the surface membranes of Gram-negative bacteria involves both recognition of the export protein and the activity of membrane transport molecules, and occurs by one of two mechanisms described below.

Carboxyl terminal recognition

The secretion of proteases by *E. chrysanthemi* involves membrane recognition of targeting sequences at the carboxyl end of the molecule by the products of three genes (see Table 8.10), forming protein transport complexes at the inner and outer membranes. The proteases are synthesised inside the cell as inactive precursors, secreted as such into the external medium, then activated outside the cell by autocatalytic cleavage of a short amino terminal extension, which is not a recognition (signal peptide) sequence.

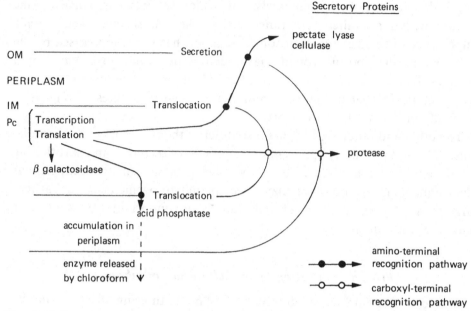

Fig. 2.11 Protein production and cell compartmentation in *Erwinia chrysanthemi*. OM, outer membrane; IM, inner membrane; Pc, peripheral (ribosomal) cytoplasm.

Amino terminal recognition

The majority of extracellular proteins in Gram-negative bacteria are secreted by a two-stage process involving separate transport steps across the inner and outer membranes.

Inner membrane translocation The passage of proteins across the inner (cytoplasmic) membrane is referred to as translocation, and involves the specific migration of molecules that are destined for either external secretion or for retention within the outer membrane, periplasmic space or the cytoplasmic membrane itself.

Translocated proteins are synthesised as precursor molecules with a specific NH_2 terminal polypeptide region. This signal peptide sequence is important for recognition at the inner membrane surface, and is absent from proteins such as β-galactosidase that are not exported. Signal peptide sequences in bacterial export proteins are structurally and functionally similar to those found in eukaryotes (Pugsley & Schwartz, 1985), and in *Erwinia chrysanthemi* have so far been demonstrated in two of the secretory proteins: pelC and pelE (Collmer & Keen, 1986).

Outer membrane secretion The release of secretory proteins from the periplasmic space to the outside of the cell occurs across the outer membrane. In *Erwinia chrysanthemi* the secretion of proteins such as pel enzymes has been investigated in relation to selective inhibition and genetic control (Mildenhall *et al.*, 1988).

Effect of inhibitors The secretion of pel enzymes is not inhibited by fluoracetate or KCN (which block active membrane transport), suggesting that the passage of these proteins across the outer membrane is a passive process. Secretion is, however, inhibited by treatment with salts LiCl and NaCl, which appear to block the release of the proteins from the periplasm, leading to a localised accumulation of secretory protein in this part of the cell. This is not an osmotic effect, since there is no inhibition with sucrose.

Genetic control Using transposon Tn_5 insertion mutagenesis (see Chapter 8), Thurn and Chatterjee (1985) have demonstrated the occurrence of three loci ('*out*' genes) in the chromosome of *Erwinia chrysanthemi* which control the final passage of secretory proteins to the outside of the cell.

Tn_5 mutagenesis leads to a block in the secretion of the wall-degrading enzymes pel, PG and CL (but not protease or phospholipid C), indicating a common pathway for at least some of the secretory products. Periplasmic accumulation of the wall-degrading enzymes in this situation indicates that the *out* genes only control transport across the outer membrane. These TN_5 insertion mutants have no apparent defect in the composition of their outer membrane, but do lack a 35 kDa periplasmic protein, suggesting that control of the final stage of secretion may reside in the periplasmic space.

The role of the *out* genes in secretion is also shown where *pel* genes have been cloned and inserted into cells of *Escherichia coli*. In this situation, the gene product is able to pass into the periplasmic space (translocation of the signal peptide protein) but is not secreted out of the periplasmic space since the bacterium does not possess *out* genes (Collmer & Keen, 1986).

Enzymes such as acid phosphatase are normally retained in the periplasmic space, and are not secreted by the cell. These periplasmic proteins may be experimentally released by degrading the outer membrane (using chloroform treatment) or by extraction of cell wall material (leading to spheroplast formation).

References

Bennett C. P. A., Jones P. & Hunt P. (1987). Isolation, culture and ultrastructure of a xylem-limited bacterium associated with Sumatra disease of cloves. *Plant Pathol.* **36**, 45–52.

Binnerts J. S., Woldringh C. L. & Brakenhoff G. J. (1982). Visualisation of the nucleoid in living bacteria on polylysine-coated surfaces by the immersion technique. *J. Microsc.* **125**, 359–63.

Bird B. & Gibson I. (1987). Studies on R-bodies of *Pseudomonas avenae*. *Micron Micros. Acta* **18**, 187–91.

Block S. M., Fahrner K. A. & Berg H. C. (1991). Visualisation of bacterial flagella by video-enhanced light microscopy. *J. Bacteriol.* **173**, 933–6.

Bohrmann B., Villiger W., Johansen R. & Kellenberger E. (1991). Coralline shape of the bacterial nucleoid after cryofixation. *J. Bacteriol.*, **173**, 3149–58.

Clark C. A. & Matthews S. W. (1987). Histopathology of sweet potato root infection by *Streptomyces ipomoea*. *Phytopathology* **77**, 1418–23.

Collmer A. & Keen N. T. (1986). The role of pectic enzymes in plant pathogenesis. *Ann. Rev. Phytopathol.* **24**, 383–409.

Darveau R. P., MacIntyre S., Buckley J. T. & Hancock R. E. W. (1983). Purification and reconstitution in lipid bilayer membranes of an outer membrane, pore-forming protein of *Aeromonas salmonicida*. *J. Bacteriol.* **156**, 1006–11.

Davis R. E. (1979) Spiroplasmas: Newly recognised arthropod-borne pathogens. In *Leafhopper Vectors and Plant Disease Agents*, ed. K. Maramorosch & K. F. Harris, pp. 451–84. New York: Academic Press.

El-Masry M. H. & Sigee D. C. (1989). Electron probe X-ray microanalysis of *Pseudomonas syringae* pv. *tabaci* isolated from inoculated tobacco leaves. *Physiol. Mol. Plant Pathol.* **34**, 557–73.

Epton H. A. S., Sigee D. C. & Passmoor M. (1977). The influence on pathogenicity of ultrastructural changes in *Pseudomonas phaseolicola* during lesion development. In *Current Topics in Plant Pathology*, ed. Z. Kiraly. Budapest: Akademiai Kiado.

Fett W. F., Osman S. F., Fishman M. L. & Siebles T. S. (1986). Alginate production by plant-pathogenic pseudomonads. *Appl. Environ. Microbiol.* **52**, 466–73.

Hammond S. M., Lambert P. A. & Rycroft A. N. (1984). *The Bacterial Cell Surface*. Beckenham: Croom Helm.

Hancock I. C. & Poxton I. R. (1988). *Bacterial Cell Surface Techniques*. Chichester: Wiley.

Hayward A. C. (1990). Proposals for a quick practical identification. In *Methods in Phytobacteriology*, ed. Z. Klement, K. Rudolph, D. C. Sands, pp. 269–74. Budapest: Akademiai Kiado.

He S. Y., Schoedel C., Chatterjee A. K. & Collmer A. (1991). Extracellular secretion of pectate lyase by the *Erwinia chrysanthemi* Out pathway is dependent upon Sec-mediated export across the inner membrane. *J. Bacteriol.* **173**, 4310–17.

Huang J. S. (1980). Galactosyltransferase activity in *Erwinia stewartii* and its role in biosynthesis of extracellular polysaccharide. *Physiol. Plant Pathol.* **17**, 73–80.

Jiang Y. P. & Chen T. A. (1987). Purification of mycoplasma-like organisms from lettuce with aster yellow disease. *Phytopathology* **77**, 949–53.

Korhonen T. K., Haahtela K., Romantschuk M. & Bamford D. H. (1986). Role of fimbriae and pili in the attachment of *Klebsiella, Enterobacter* and *Pseudomonas* to plant surfaces. In *Recognition in Microbe–Plant Symbiotic and Pathogenic Interactions*, ed. NATO ASI series, B. Lugtenberg. Berlin: Springer-Verlag.

Mildenhall J. P., Lindner W. H., Prior B. A. & Tutt K. (1988). Elevation and release of cell-associated pectate lyase in *Erwinia chrysanthemi* by lithium and sodium chloride. *Phytopathology* **78**, 213–217.

Murray R. G. E. (1978). Form and function. II. Bacteria. In *Essays in Microbiology*, ed. J. R. Norris, M. H. Richmond. Chichester: Wiley.

Osman S. F., Fett W. F. & Fishman M. L. (1986). Exopolysaccharides of the phytopathogen *Pseudomonas syringae* pv. *glycinea*. *J. Bacteriol.* **166**, 66–71.

Pugsley A. P. & Schwartz M. (1985). Export and secretion of proteins by bacteria. *FEMS Microbiol. Rev.* **32**, 3–38.

Romantschuk M. & Bamford D. H. (1985a). Chromosomally coded pili of *Pseudomonas syringae* pv. *phaseolicola* function as adhesins in early steps of halo blight of bean. In *Plant Pathogenic Bacteria*, Proceedings 6th International Conference on Plant Pathology and Bacteria. ed. E. L. Civerolo, A. Collmer, R. E. Davis & A. G. Gillaspie, pp. 573–7. Dordrecht: Martinus Nijhoff.

Romantschuk M. & Bamford D. H. (1985b). Function of pili in bacteriophage φ6 penetration. *J. Gen. Virol.* **66**, 2461–9.

Rogers H. J., Perkins H. R. & Ward J. B. (1980). Microbial Cell Walls and Membranes. I. Ultrastructure of Bacterial Envelopes, pp. 1–71. London: Chapman & Hall.

Ryter A. & Chang A. (1975). Localisation of transcribing genes in the bacterial cell by means of high resolution autoradiography. *J. Mol. Biol.* **98**, 797–810.

Shaw C. H. (1991). Swimming against the tide: Chemotaxis in *Agrobacterium*. *BioEssays* **13**, 25–9.

Sigee D. C. (1989). Microscopical techniques for bacteria. In *Methods in Phytobacteriology*, ed. Z. Klement, K. Rudolph, D. Sands, pp. 24–41. Budapest: Akademiai Kiado.

Sigee D. C. & Al-Issa A. N. (1983). The hypersensitive reaction in tobacco leaf tissue infiltrated with *Pseudomonas pisi*. 4. Scanning electron microscope studies on fractured leaf tissue. *J. Phytopathol.* **106**, 1–15.

Sigee D. C. & Al-Rabaee R. H. (1986). Nickel toxicity in *Pseudomonas tabaci*: single cell and bulk sample analysis of bacteria cultured at high cation levels. *Protoplasma* **130**, 171–85.

Sigee D. C. & El-Masry M. H. (1988). Electron probe X-ray microanalysis of bacterial DNA. *J. Biochem. Biophys. Methods* **15**, 215–28.

Sigee D. C. & El-Masry M. H. (1989). Changes in cell size and flagellation in the phytopathogen *Pseudomonas syringae* pv. *tabaci* cultured *in vitro* and *in planta*: A comparative electron microscope study. *J. Phytopathol.* **125**, 217–230.

Sigee D. C., El-Masry M. H. & Al-Rabaee R. H. (1985). The electron microscope detection and quantitation of cations in bacterial cells. *Scanning Electron Microsc.* 1985/III, 1151–63.

Sigee D. C., Hodson N. & El-Masry M. H. (1989). X-ray microanalytical determination of the ionic composition of plant pathogenic bacteria. In *Plant Pathogenic Bacteria*, Proceedings 7th International Conference on Plant Pathology and Bacteria, ed. Z. Klement, pp. 509–14. Budapest: Akademiai Kiado.

Silva M. T., Sousa J. C. F., Polonia J. J., Macedo M. A. E. & Parente A. M. (1976). Bacterial mesosomes real structures or artefacts? *Biochim. Biophys. Acta* **443**, 92–105.

Smith A. R. & Koffler H. (1971). Production and isolation of flagella. In *Methods in Microbiology*, ed. R. R. Norris & W. D. Ribbons, pp. 165–72. London: Academic Press.

Smith J. J. & Mansfield J. W. (1982). Ultrastructure of interactions between pseudomonads and oat leaves. *Physiol. Plant Pathol.* **21**, 259–66.

Stemmer W. P. C. & Sequeira L. (1987). Fimbriae of phytopathogenic and symbiotic bacteria. *Phytopathology* **77**, 1633–9.

Stock J. B., Rauch B. & Roseman S. (1977). Periplasmic space in *Salmonella typhimurium* and *Escherichia coli*. *J. Biol. Chem.* **252**, 7850–7861.

Thurn K. K. & Chatterjee A. K. (1985). Single-site chromosomal Tn5 insertions affect the export of pectolytic and cellulolytic enzymes in *Erwinia chrysanthemi* EC16. *Appl. Environ. Microbiol.* **50**, 894–8.

Witholt B., Boekhout M., Brock M., van Heerikhuizen H. & de Leij L. (1976). An efficient and reproducible procedure for the formation of spheroplasts from variously grown *Escherichia coli*. *Anal. Chem.* **74**, 160–70.

Yeh P., Oreglia J. & Sicard A. M. (1985). Transfection of *Corynebacterium lilium* protoplasts. *J. Gen. Microbiol.* **131**, 3179–83.

Young D. H., Stemmer W. P. & Sequeira L. (1985). Reassembly of a fimbrial haemagglutinin from *Pseudomonas solanacearum* after purification of the subunit by preparative sodium dodecyl sulphate polyacrylamide gel electrophoresis. *Appl. Environ. Microbiol.* **50**, 605–10.

3

Taxonomy of plant pathogenic bacteria: classification, nomenclature and identification

The taxonomy of plant pathogenic bacteria, with its three interrelated aspects of classification, nomenclature and identification, is a central aspect of bacterial plant pathology. It is clearly important to be able to establish the identity of an isolated plant pathogenic bacterium, so that the agent causing a particular disease – or with the potential to cause disease – can be clearly defined. Phytopathogenic bacteria may be isolated from a wide range of sources, including infected plant tissue, seed surfaces, soil and water environments, and the identification of bacteria from such sites has implications not only for disease diagnosis and pathogenicity but also for studies on disease epidemiology and aetiology. In addition to defining bacteria as agents of disease, taxonomic studies can also provide useful insights into phylogenetic relationships between the phytopathogens.

The taxonomy of plant pathogenic bacteria will be considered in relation to two major aspects:

1. The establishment of a clearly defined and internationally accepted system of classification and nomenclature.
2. Bacterial identification, including: isolation and identification from different sites, pathogenicity testing, *in vitro* diagnosis, and computer identification by numerical analysis. There is now a wide range of features on which bacterial classification can be based, and which can be used for identification.

General principles of bacterial taxonomy, with details of the various characteristics that can be used for classification, are discussed in *Bergey's Manual of Systematic Bacteriology* (Kreig & Holt, 1984).

Bacterial classification and nomenclature

Classification can be defined as the arrangement of organisms into taxonomic groups on the basis of similarity, while nomenclature involves the assignment of names to the taxonomic groups in accordance with the International Code of Nomenclature of Bacteria (Lapage *et al.*, 1975).

Classification of plant pathogenic bacteria has always been a matter for discussion and controversy, with many changes in nomenclature. The fire-blight pathogen, for example, has been variously referred to as *Micrococcus amylovorus* (1882), *Bacillus amylovorus* (1889), *Bacterium amylovorus* (1915) and finally *Erwinia amylovora* (1920).

Traditionally, classification of plant pathogenic bacteria was based on a few biochemical characters plus pathogenicity. The latter became increasingly important, with less attention being paid to biochemical and other features, leading to the situation where a bacterium was given species rank if it was isolated from a particular host and shown to be pathogenic. With the need for a fuller and more broadly based description of plant pathogenic bacteria, the International Committee on Systematic Bacteriology (ICSB) instituted a new nomenclature (with 1980 as the new starting date) in which only those names of bacteria which were adequately described were placed on the approved list, the lowest ranks of which were species and subspecies. Any previously designated species simply based on pathogen status would be reduced to the rank of pathovar.

All official names and taxonomic criteria relating to plant pathogenic bacteria are determined by the International Society for Plant Pathology. In general, the classification of plant pathogenic bacteria can be considered within three major categories:

1. Standard taxonomic ranks. Particularly families, genera, species and subspecies, where the taxonomic unit is fully defined by classical microbiological features such as biochemistry, growth characteristics and cell morphology.
2. Pathovar and race. These are taxonomic units below the level of subspecies (i.e. infra-subspecific), where the taxon is determined with reference to the ability of the pathogen to cause disease in particular hosts.
3. Biotypes. These are further infra-subspecific ranks, which are defined by diagnostic features such as special biochemical or physiological properties (biovar), distinctive bacteriophage lysis (lysotype), serology (serotype), protein analysis and DNA homology.

A pure bacterial culture which is defined simply in terms of its single isolation from a particular source (e.g. an infected plant or a specific soil sample) is referred to as a strain or isolate. Different isolates may be biologically identical.

Standard taxonomic ranks

Plant pathogenic bacteria are classified within the standard taxonomic ranks of kingdom, division, class, order, family, genus and species, as described by Sneath (1984).

The basic unit of bacterial classification is the species, which may be defined as a collection of strains that have many features in common, differing in these features from other groups of strains. Unlike higher organisms, sexual compatibility is not used to define bacterial species, since relatively few bacteria undergo conjugation. All valid names of bacterial species have been published in the *International Journal of Systematic Bacteriology* (IJSB) from 1980 (Skerman *et al.*, 1980) onwards. Subspecies represent major divisions within the species taxon. Ranks below the level of subspecies have no official standing in nomenclature, but have considerable practical usage.

Within the prokaryote kingdom, four divisions are recognised, distinguished largely on the basis of cell wall characteristics (Murray, 1984). These are:

Division I	Gracilicutes	Gram-negative type cell wall (contains Gram-negative bacteria)
Division II	Firmicutes	Gram-positive type cell wall (contains Gram-positive bacteria and actinomycetes)
Division III	Tenericutes	No cell wall (contains class Mollicutes)
Division IV	Mendosicutes	Give evidence of earlier phylogenetic origin (contains class Archaeobacteria)

Plant pathogenic prokaryotes are present in divisions I to III, and can thus be divided into organisms with a cell wall (true bacteria plus actinomycetes) and those without.

Gram-positive and Gram-negative bacteria

The Gram staining reaction reflects fundamental differences in cell wall structure of plant pathogenic bacteria (see Chapter 2), and shows a high correlation with two other tests, solubility in 3% KOH and aminopeptidase activity, both of which are characteristic of Gram-negative cells. A simplified flowchart showing this major dichotomy based on the Gram reaction, with other features used for the identification of the major groups of plant pathogenic bacteria, is shown in Fig. 3.1. Identification of plant pathogenic bacteria at genus and species level is based largely on major biochemical

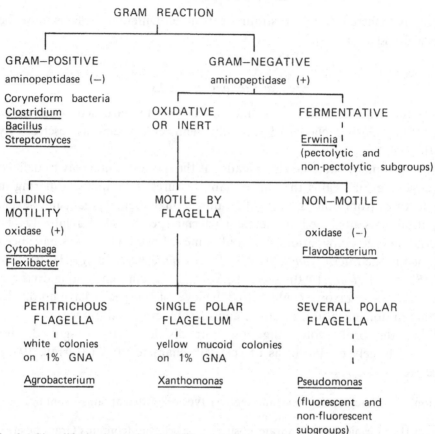

Fig. 3.1 Simplified scheme for the differentiation of major groups of plant pathogenic bacteria. In the Gram-positive group, the coryneform bacteria comprise the genera *Clavibacter*, *Corynebacterium*, *Arthrobacter*, *Curtobacterium* and *Rhodococcus*. Mollicutes (mycoplasma-like organisms and spiroplasmas) are not included in this scheme. GNA, glucose nutrient agar. (Adapted from Hayward, 1990.)

properties, growth characteristics in culture and cell morphology (Stead, 1990; Hayward, 1990). The use of taxonomic features based on cultured cells presents particular problems for the classification and identification of fastidious prokaryotes, including both walled and naked cell types (see below), where many pathogens have not yet been obtained in pure culture due to their exacting growth requirements.

Within the Gram-negative bacteria, glucose metabolism provides a key differentiation of *Erwinia* (fermentative) from other pathogens (oxidative or inert). In the latter group, the ability to demonstrate a clear oxidase reaction (Kovacs' oxidase test) varies considerably. This reaction is positive, for example, in some fluorescent and some non-fluorescent pseudomonads (but not others), and with certain oxidase-positive species a clear demonstration of

Table 3.1. *Presence of PHB inclusions*

	Present	Absent
Erwinia sp.	−	+
Pseudomonas		
fluorescent sp.	−	+
non-fluorescent sp.	*P. solanacearum*	*P. amygdali*
	P. andropogonis	
Agrobacterium	*A. rhizogenes*	*A. tumefaciens*
		A. rubi

Data from Hayward (1990).

oxidase activity only occurs if the level of glucose in the medium is below a critical level (Hayward, 1990).

The microscopical appearance of bacteria (Chapter 2) is also of great importance in classification, with features such as cell size and shape, presence and characteristics of flagella and presence of polyhydroxybutyrate inclusions providing useful taxonomic characteristics.

Bacterial flagellation Although bacterial flagellation provides a useful criterion for classification, some variation does occur within particular groups. Almost all members of the genus *Xanthomonas*, for example, have a single polar flagellum, but some non-pathogenic bacteria with more than one flagellum (e.g. *Xanthomonas maltophilia*, a common bacterium on plant surfaces) are now included in this taxon. In groups of bacteria that are multiflagellate, the number of flagella varies between bacteria within a single population and at different stages of the growth cycle (Chapter 2).

Polyhydroxybutyrate (PHB) inclusions The presence or absence of PHB inclusions can be determined at the level of the light or electron microscope (see Chapter 2), and represents a clearly defined characteristic which is very useful in bacterial classification (Table 3.1). As noted previously, the occurrence of these inclusions is enhanced by growth in media with a high ratio of utilisable carbon to nitrogen.

Naked prokaryotes

Phytopathogenic mollicutes are divided into two groups: mycoplasma-like organisms (MLOs) and spiroplasmas, and have largely been identified on the basis of their small size, general morphology and fine-structural features (see Chapter 2, Fig. 2.8). Difficulties encountered in isolating these organisms from infected plants and growing them in culture present a major limitation for their taxonomy and identification, and many phytopathogenic MLOs are characterised in terms of their ability to induce a particular disease in a specific host plant (see Chapter 7). The problems encountered in culturing these organisms arise due to their exacting nutritional requirements, with both cholesterol and long chain fatty acids being necessary in the complex growth medium. Physical parameters must also be exact, including the need for a high osmotic pressure in the external medium.

In common with other mollicutes, plant pathogenic MLOs and spiroplasmas are not able to synthesise precursors of peptidoglycan and are naked at all stages of their life cycle. These organisms generally have a smaller genome size ($0.5–1.0 \times 10^9$ daltons) than other prokaryotes and are characterised by a low guanine and cytosine content of both ribosomal RNA (43–48 mol%, compared with 50–54 mol% in most bacteria) and DNA (23–46 mol%). Mollicutes also differ from other phytopathogenic prokaryotes in being resistant to beta-lactam antibiotics such as penicillin.

Pathovars and races

Some species of plant pathogenic bacteria, such as *Erwinia amylovora* and *Pseudomonas solanacearum*, have a broad host range in which there is no clear differentiation within the species in reference to particular hosts and, therefore, no subdivision into pathovars and races. In other cases, clear distinctions within the species in terms of pathovars and races can be demonstrated.

Pathovars

For the plant pathologist, the pathovar is a taxonomic unit of major importance, since it distinguishes those isolates which can cause disease on a particular plant from those which cannot. Pathovars are defined simply in terms of distinctive pathogenicity on one or more hosts, with considerable variation between pathovars in terms of host range. With *Pseudomonas syringae*, for example, pathogenicity for pv. *tabaci* has been demonstrated only on two host plants: *Glycine* and *Nicotiana*, while pv. *syringae* has a host range which includes cereals, legumes, woody ornamentals and fruit trees.

Pathovars are present in five genera of plant pathogenic bacteria, within which they are limited to the species shown in Table 3.2. No pathovars are recognised at present in the genera *Agrobacterium*, *Bacillus*, *Mycoplasma*, *Rhodococcus*, *Spiroplasma* or *Streptomyces*.

The use of pathogenicity as the definitive criterion of taxonomic status does not preclude the subsidiary use of other criteria (e.g. biochemical or serological), but does imply that pathogenicity is of prime taxonomic significance at

Table 3.2. *Species of bacterial plant pathogens that contain pathovars*

Genus	Species	Number of pathovars
Curtobacterium[a]	*flaccumfaciens*	4
Clavibacter[a]	*michiganensis*	7[b]
Erwinia	*chrysanthemi*	6
	herbicola	1
Pseudomonas	*gladioli*	2
	marginalis	3
	syringae	41
Xanthomonas	*campestris*	123

[a] Originally placed in the genus *Corynebacterium*. [b] Now ranked as subspecies.
Data taken mainly from Dye *et al.* (1980).

subspecies level. One drawback with the use of a single taxonomic criterion is that it provides no information on taxonomic relationships between pathovars. A further problem arises in genera such as *Agrobacterium*, where pathogenicity is transmissable (via plasmids) and does not represent a stable taxonomic character.

Pathovar status is determined by experimental inoculation of potential host plants, followed by positive or negative confirmation of disease development to determine host specificity. A number of practical problems can arise:

1. In determining the pathovar status of a particular isolate, it is practically impossible to test every potential host plant for the development of disease. Because of this, the host range of many bacterial plant pathogens is not completely known. In some cases, pathovars have simply been assigned to the host plant from which they were originally isolated, with no appreciation of alternative hosts, some of which may be of primary importance.
2. Infiltration of a phytopathogenic bacterium into a susceptible host plant does not invariably result in clear disease symptoms, since the development of infection may be dependent on environmental factors, type and maturity of the host material and the state of the bacterial culture.
3. In some cases, pathovars are closely related and appear to have intermediate forms, making clear determination of pathovar status difficult. *Pseudomonas syringae* pathovars *syringae* and *mors-prunorum*, for example, both infect stone fruit but have slightly different host ranges and environmental requirements. In South Africa (Roos & Hattingh, 1987a) pv. *mors-prunorum* predominates on cherries in the summer rainfall region, while pv. *syringae* is favoured in South-West Cape province, a winter rainfall region. Mixed populations, with intermediate forms, are very difficult to resolve on a pathovar basis.

Table 3.3. *Pathovar races and host cultivars in the interaction between*
Pseudomonas syringae *pv.* phaseolicola *and bean*

Race	Canadian Wonder	Red Mexican	Tendergreen
1	S	HR	S
2	S	S	S
3	S	S	HR

S susceptible interaction; HR, hypersensitive (resistant) interaction.

Table 3.4. *Differentiation of pathovar races on host lines*

Species	Pathovar	No. of races	Host	Reference
Pseudomonas syringae	*glycinea*	9	Soybean	Fett & Sequeira, 1981
	mors-prunorum	2	Cherry	Lelliott & Stead, 1987
	pisi	6	Pea	Taylor *et al.*, 1989
	phaseolicola	3[a]	Bean	Harper *et al.*, 1987
	tabaci	2	Tobacco	Knoche *et al.*, 1987
Xanthomonas campestris	*malvacearum*	18	Cotton	Brinkerhoff, 1970
	vesicatoria	3	Pepper	Hibberd *et al.*, 1988

[a] The number of races in pv. *phaseolicola* is now updated to nine (unpublished report).

Races

Although plant pathogenic bacteria may typically be defined in terms of a clear-cut ability to cause disease in particular species of higher plants, some variation within this may occur. The importance of environment, bacterial state and host maturity to disease development have already been noted.

The variety of host plant (cultivar) and type of bacterium (race) may also be important, with many pathovar taxa being further subdivided into distinct races on the basis of a differential reaction to a range of cultivars. This is illustrated in Table 3.3, which shows the interaction of different races of *Pseudomonas syringae* pv. *phaseolicola* with cultivars of bean. Race 1 of this pathovar, for example, will produce halo blight disease after infiltration into cultivars Canadian Wonder and Tendergreen, but not Red Mexican. In some cases the cultivar and pathogen can be genetically defined and the host–pathogen interaction shown to conform to a gene-for-gene relationship (see Chapter 8). In the above example, two resistant genes are present respectively in cv. Red Mexican and Canadian Wonder.

Examples of pathovars where races have been identified and distinguished on sets of differential (cultivar) host lines are shown in Table 3.4. The numbers of races given in this table for particular pathovars are subject to revision, and increase as more studies are carried out with further ranges of bacterial isolates and host cultivars. In the case of *Pseudomonas syringae* pv. *phaseolicola*, for example, the number of races is now reported to be nine, based on five pairs of resistance and avirulence genes.

Although bacterial races are typically defined at the subpathovar level in relation to host cultivars, the term has also been used in a less specific context to define differences between strains over a wide host range. In *Pseudomonas solanacearum*, for example, three races have been distinguished (Buddenhagen and Kelman, 1964) as follows:

Race 1. Infects most members of the Solanaceae plus many other plants
Race 2. Infects *Heliconia* spp. and triploid banana (causing Moko disease)
Race 3. Infects potato and tomato

Biotypes

In the previous section, plant pathogenic bacterial species were divided into pathovars or races on a strictly pathogenic basis. Species and pathovars can also be divided into subgroups using non-pathogenic characteristics, including geographic distribution, biochemical characters (biovars), serology (serotypes) and sensitivity to phages (lysotypes).

The differentiation of biotypes is most useful where a species or pathovar is widespread, has a broad host range and is heterogeneous in a range of characteristics. A good example of this is *Xanthomonas campestris* pv. *citri*, which causes citrus canker in many countries and has been divided into four main groups on the basis of geographic distribution, host range, symptoms, genetic characteristics and phage sensitivity (Gottwald *et al.*, 1988a). These groups are:

Group A (causes Asiatic citrus canker). Most widely distributed group (present in Asia, Africa, Oceania and S. America) with broadest host range.
Group B (causes cancrosis type B). Primarily affects lemon in Argentina, Uruguay and probably Paraguay.
Group C (causes Mexican lime cancrosis). Infects only *Citrus aurantifolia*.
Group E (present only in Florida). Causes flat, water-soaked lesions rather than the raised lesions found in Groups A–C.

The genetic characteristics of this bacterium parallel the geographic distribution in demonstrating three different clonal groups: A, B/C and E (Gabriel *et al.*, 1987).

Bacterial isolation and identification from different sites

The occurrence of plant pathogenic bacteria at different locations and in different situations leads to a variety of procedures for isolation and identification. Bacterial isolates may originate from infected tissue (i.e. disease conditions), plant aerial or subterranean surfaces, or the physical environment.

Identification from infected tissue

The identification of plant pathogenic bacteria within infected plant tissue is closely tied to diagnosis of the disease itself and the use of Koch's postulates to establish causality (see Chapter 1, Fig. 1.1).

Experimental procedures involved in the identification of phytopathogenic bacteria within infected plant tissue, and the diagnosis of disease, have been considered in detail in a number of recent publications including Fahy and Persley (1983) and Lelliott and Stead (1987). In general, the diagnosis of bacterial disease follows a three-stage process.

Microscopical examination of diseased tissue

Light microscope examination of small pieces of detached diseased tissue, mounted in a drop of sterile water, can provide useful information on the main cause of disease. Where this is caused by bacteria as the primary pathogen, large numbers of bacteria are normally released into the droplet. There are some types of diseased tissue, however, where bacteria are not present in large numbers, including hyperplastic tissue (plant galls and tumours) and chlorotic tissue. In the first case, bacteria are simply few in number. In the second case, particularly where regions of chlorosis are occurring around necrotic areas, the symptoms are caused by bacterial toxins which have moved out of regions of pathogen occupation. In cases where fungal hyphae or nematodes are seen, it is more likely that these are the primary cause of disease, and that any bacteria observed are present secondarily.

Preliminary diagnosis

With a known host plant, once it has been established that the disease has been caused by bacteria as the primary agent, preliminary identification of disease (and pathogen) can normally be made on the basis of observed symptoms. Both Fahy and Persley (1983) and Lelliott and Stead (1987) give detailed listings of these.

Diagnostic testing

Once the probable cause of disease has been narrowed down to a few potential pathogens, isolation and growth of bacteria, with *in vitro* diagnostic testing using biochemical and physiological characteristics can then be carried out.

Following the report of disease in the field, diagnosis should be carried out as quickly as possible. Lelliott and Stead (1987) distinguish between rapid (presumptive) diagnosis, based on observations of symptom expression and a minimal number of tests on isolates, and subsequent confirmatory diagnosis involving inoculation of the host plant.

Identification of phytopathogenic epiphytes

The detection and identification of plant pathogenic bacteria on plant surfaces presents a different situation from bacterial identification in the diseased plant, since numbers are frequently very small and symptoms are often not present. In this situation, bacteria either have to be identified *in situ* using very sensitive histochemical tests or they have to be isolated and grown for *in vitro* diagnosis.

Seedborne pathogens

Identification of epiphytic pathogens is particularly important in control programmes where phytopathogen-free plants or seeds require certification. With over 45 plant pathogenic bacteria known to be transmitted by seed, often at very low level, the development of sensitive techniques for the detection of seedborne pathogens has become an area of major interest (Schaad, 1982).

The traditional method for detecting seedborne pathogenic bacteria was simply to grow plants from seed in the field or greenhouse and note the development of disease. The use of such 'growing-on tests' had a number of drawbacks, not least of which was the time required to obtain a result. The alternative host plant procedure of injecting seed washings into susceptible plants and monitoring the incidence of disease (pathogenicity testing, see below) was also time consuming, and results were often difficult to interpret.

The above constraints have led to the development of *in vitro* laboratory assays for seed bacteria. These have the advantages of being faster, easier to perform, less expensive and are more sensitive than previous *in planta* procedures. The development of a successful laboratory assay for soilborne bacteria involves three major steps (Schaad, 1982):

1. Extraction of the bacterium from the seed, followed by *in vitro* culture.
2. Development of sensitive and specific *in vitro* tests, including the use of selective media and serological testing.
3. Determination of assay sensitivity and tolerance limits.

Identification of bacteria in the physical environment

Identification of bacteria present in the atmospheric (e.g. aerosol), water or soil environments involves isolation and culture prior to *in vitro* diagnosis. Selective media are particularly useful where there is a high microbial diversity, such as occurs in a typical soil environment.

Identification by pathogenicity testing

Pathogenicity testing can be carried out on bacterial isolates from any of the sites described above and typically involves infiltration of a suspension of cultured bacteria into plant tissue and subsequent determination of the response. This can provide information on two related aspects of bacterial identity:

1. Bacterial status as a general plant pathogen. The development of a localised necrosis in a nonspecific test plant (such as tobacco, see Fig. 6.1) demonstrates an incompatible interaction and shows that the bacterial isolate has general phyto-pathogenic capability. Further details on the use and limitations of this experimental procedure (HR test for pathogenicity) are given in Chapter 6.
2. Bacterial identity as the causal agent of a specific disease. The ability of a bacterial isolate to cause disease in a particular plant species or cultivar provides an important criterion for determining its species and pathovar status. This is particularly important where *in vitro* tests are difficult to apply or are of limited value. An example of this is provided by *Clavibacter michiganensis* subsp. *michiganensis* which causes bacterial canker of tomato and is difficult to differentiate from other Gram-positive bacteria on semi-selective media (Gitaitis & Beaver, 1990).

Although pathogenicity testing has been widely used for phytopathogen diagnosis it has a number of limitations including the need for a constant supply of plants, the latent period of disease (before symptoms occur) and the danger of introduction of other pathogenic contaminants. Because of these limitations, many laboratories are now turning towards *in vitro* tests which have the potential for more rapid but highly specific diagnosis.

In vitro diagnosis of plant pathogenic bacteria

A wide range of *in vitro* diagnostic tests (carried out on bacteria growing in culture) are available to assist in the identification of plant pathogenic bacteria. These include the use of specific growth media, metabolic and physiological characteristics, identification of specific biochemical products, phage tests, serology and molecular analytical procedures.

Specific growth media

Specific growth media fall into two major categories:

1. Differential media. Where the bacterium being identified is distinguished in terms of colony characters.
2. Selective media. Where the bacterium being identified grows on a medium which does not support the growth of other micro-organisms.

In practice these two aspects are frequently combined into a single growth medium which supports the growth of a limited range of bacteria (semi-selective) and permits identification on the basis of growth characteristics.

Differential media

Non-selective media allow good growth of bacteria with the development of a distinctive colony morphology. Although different plant pathogenic bacteria have different growth requirements, many grow well on maintenance media such as nutrient agar and yeast peptone agar.

These simple media have been widely used in the isolation and differentiation of a variety of pathogens. In the case of *Erwinia amylovora*, for example, the pathogen is clearly distinguishable from other epiphytic bacteria (such as *Pseudomonas syringae*) in terms of colony morphology:

1. Nutrient agar containing 50 g/l sucrose (NSA). *Erwinia amylovora* colonies are typically domed, mucoid, cream-coloured and uniform in structure.
2. Yeast peptone agar (YPA). *Erwinia amylovora* colonies are typically circular, with an entire edge, white or grey-cream and slightly irridescent.

A range of differential media are available to promote particular colony characteristics, including King's medium B, which typically leads to pigment production by fluorescent pseudomonads (Fig. 3.2) and pectate gel media, on which pectolytic bacteria produce characteristic pits (Lelliott & Stead, 1987).

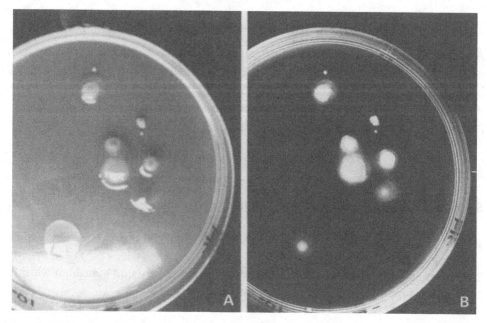

Fig. 3.2 Use of semi-selective medium to monitor fluorescent pseudomonads on pear leaves. Leaf washings have been plated onto Kings medium B (selective for sidero-phore-producing bacteria) and incubated for three weeks at 20°C. The plate has been photographed under ordinary fluorescent light (*a*) and ultraviolet light (*b*), where bacterial colonies show the characteristic fluorescence of pyoverdine pigment production.

Selective media

Selective or semi-selective media are generally designed to allow growth of the pathogen, but inhibit other micro-organisms. These two objectives are rarely compatible, and many truly selective media have a low plating efficiency for the pathogen (colony count on the selective medium compared with a non-selective medium).

Many of the selective media that were initially developed combined high nutrient status to promote growth of the pathogen and high levels of inhibitor to suppress other micro-organisms. Some of these complex media were limited by the build-up of toxic substances, while others were moderately successful. An early semi-selective medium developed for use with *Erwinia amylovora*, for example, comprised YPA plus 50 g/l sucrose, and contained 2 mg/ml crystal violet (to inhibit Gram-positive bacteria) and 100 mg/ml cycloheximide (to inhibit fungi). This medium was semi-selective for *Erwinia* spp., and was used largely as a differential medium for the fireblight pathogen.

In more recent times, the strategy for design of selective media has changed,

Table 3.5. *Some recently developed selective media*

Bacterial genus or species	Pathovar	Medium	Reference
Clavibacter michiganensis	—	SCM	Fatmi & Schaad, 1988
Pseudomonas syringae	*syringae*	KBC	Mohan & Schaad, 1987
	phaseolicola	MSP	
Pseudomonas savastanoi	*savastanoi*[a]	PVF-1	Surico & Lavermicocca, 1989
Xanthomonas			
Range of species		XOS	Ming *et al.*, 1991
campestris	*phaseoli*	MXP	Clafkin *et al.*, 1987
	pruni	XPSM	Civerolo *et al.*, 1982
	vesicatoria	Tween	McGuire *et al.*, 1986
Fastidious prokaryotes			
Xylella fastidiosa		PD3/P10	Hopkins, 1988
Xylem-limited bacteria		SC/PW	Davis, 1990
Spiroplasmas		C-3G	Liao & Chen, 1977

[a] Subspecies.

with nutrients being added at the lowest level to support good growth of the target pathogen, and inhibitors being added at the highest level that will still permit good growth of this organism. These media are often well defined, containing one carbon source, one nitrogen source, inorganic salts, a buffer and agar, and have only been developed after extensive testing in reference to the particular pathogen concerned. Over 350 selective or semi-selective media have now been developed for use with plant pathogenic bacteria. Some recent examples are given in Table 3.5, and a more extensive list is presented by Roy & Sasser (1990). Many of these media are able to distinguish plant pathogenic bacteria at the subspecies level, with differentiation of both pathovars and biovars (e.g. of *Agrobacterium tumefaciens*, Lelliott & Stead, 1987).

The major advantages of using selective media are that they are relatively inexpensive, give rapid results, are easy to use and result in a culture of the suspected pathogen. They have particular potential where a large amount of routine testing is involved, such as screening for seedborne pathogens. Fatmi and Schaad (1988), for example, used a semi-selective medium to monitor *Clavibacter michiganensis* in tomato seed, and were able to detect a single seed containing as few as 50 CFU in a sample of 10^4 seeds.

The major disadvantages of selective media in terms of poor selectivity and plating efficiency have already been referred to, and may lead to the use of differential rather than selective media in studies on bacterial populations (e.g. Gitaitis *et al.*, 1988). A further limitation in their general applicability is that although many selective media have been developed, these are for relatively few pathogenic bacteria.

Selective media and fastidious pathogens

Although the term 'selective growth medium' normally implies the selective exclusion of other micro-organisms, this is not the primary consideration with fastidious phytopathogens, where the object is to develop a medium that will support their very stringent growth requirements (no growth on simple nutrient agar). The use of selective growth media with these organisms therefore represents a special situation, and is an important aspect of their *in vitro* culture and study.

The causal agent of Pierce's disease of grapevine (*Xylella fastidiosa*) was the first fastidious xylem-limited bacterium to be isolated in axenic culture. Some recently developed media (e.g. PD3) will only support the growth of specific pathogens, while others (e.g. SC and PW media) support the primary isolation and continuous culture of xylem-limited bacteria generally (Table 3.5). Culture media have also been developed for the isolation and growth of spiroplasmas (Table 3.5), but two major categories of fastidious phytopathogen: phloem-limited bacteria and mycoplasma-like organisms have not yet been grown *in vitro*.

In those cases where fastidious prokaryotes can be cultured, growth rate is typically low and is outstripped by contaminating micro-organisms. As with non-fastidious prokaryotes, details of growth rate, colony morphology and culture requirements may help differentiate these pathogens.

Metabolic and physiological characteristics

A considerable variety of metabolic and physiological characteristics are routinely used in the diagnosis of plant pathogenic bacteria, including the ability to use particular substrates, production of reducing substances from sucrose, oxidative or fermentative metabolism, temperature of optimum growth, and ability to cause rot of potato slices and hypersensitive reaction in tobacco.

These features may be used to determine major taxonomic divisions (see Fig. 3.1), define species and biovars, and in some cases distinguish between pathovars. An example of major taxonomic division is provided by the LOPAT scheme of Lelliott *et al.* (1966), which separates green-fluorescent pseudomonads into eight major groups on the basis of presumptive characteristics (levan production, oxidase reaction, potato rot, arginine dihydrolase and hypersensitivity) plus confirmatory features.

Table 3.6. *Substrate utilisation in* Pseudomonas solanacearum

Substrate Utilised	Biovar			
	I	II	III	IV
Maltose	−	+	+	−
Lactose	−	+	+	−
Cellobiose	−	+	+	−
Mannitol	−	−	+	+
Sorbitol	−	−	+	+
Dulcitol	−	−	+	+

Data from Lelliott & Stead (1983)

Table 3.7. *Biochemical differentiation of pathovars*

Test	pv. *phaseolicola*	pv. *syringae*
Aesculin or arbutin hydrolysis	−	+
Gelatin liquefaction	−	+
Grease spots on detached bean pods	+	±
Utilisation of sorbitol, inositol and erythritol	−	+

Data from Lelliott & Stead (1983).

At the subspecies level, nutritional characteristics have been particularly useful in *Pseudomonas solanacearum*, with individual biovars having different patterns of substrate utilisation (Table 3.6). Metabolic characteristics may also be useful in the separation of pathovars, where symptoms on the host are difficult to distinguish. An example of this is given by *Pseudomonas syringae* pathovars *phaseolicola* and *syringae*, both of which can cause necrotic spots on bean. These pathovars can be distinguished by a range of biochemical tests (Table 3.7).

Further information on the use of metabolic and physiological characteristics in the diagnosis of plant pathogenic bacteria is provided by Lelliott and Stead (1983).

Biochemical products

A variety of specific biochemical products can be readily identified and used as potential taxonomic markers, including toxins (see Chapter 7), bacteriocins (Chapter 9) and pigments.

Table 3.8. *Pigments from phytopathogenic bacteria*

Pigment	Colour	Producing organism
Membrane-bound pigments *(extractable by organic solvents)*		
Carotinoids		
Lycopene	Yellow-pink	*C. flaccumfaciens* pv. *poinsettiae*
Xanthomonadins		
Various partly characterised xanthomonadins	Yellow	*X. campestris pvs* *X. albilineans* *X. fragariae*
Diffusible extracellular *pigments (water-soluble)*		
Pyridine-derivatives		
Indigoiodine	Blue	*Cl.m.* pv. *insidiosus* *E. chrysanthemi*
Ferrorosamine A	Red	*E. rhapontici*
Rubrifacine	Red	*E. rubrifaciens*
Pyoverdines		
Pyoverdine ps	Yellow-green fluorescent pigments	*P.s.* pv. *phaseolicola*
Pyoverdine pa	″	*P. aeruginosa*
Pyoverdine ps	″	*P. fluorescens*
Catechol-Siderophores		
Agrobactin	Red ferric complex	*A. tumefaciens*
2,3-dihydroxybenzoic acid	Blue-violet ferric complex	*E. carotovora*

Adapted from Feistner, 1990

Isolation and identification of bacterial pigments

The importance of pigment production as a taxonomic characteristic for plant pathogenic bacteria has already been noted in Fig. 3.1, where pseudomonad species are separated into fluorescent (producing a yellow-green fluorescent pigment) and non-fluorescent groups. The use of pigments as taxonomic markers is enhanced by the fact that pigment production under appropriate growth conditions can be readily observed, isolation and identification of pigment molecules can be carried out according to standard procedures and there is a wide range of pigments which are characteristic of particular phytopathogenic bacteria (Feistner, 1990).

General characteristics of some of the major phytopathogenic bacterial pigments are shown in Table 3.8, and the molecular structures of representa-

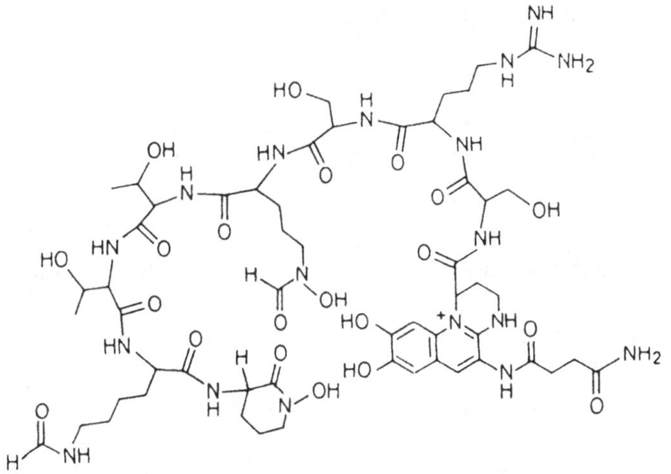

Pyoverdine pa

Fig. 3.3 Structure of pyoverdine pigment from a fluorescent pseudomonad (*Pseudomonas aeruginosa*).

tive examples in Figs. 3.3 and 3.4. In terms of their location and extractability, the pigments of phytopathogenic and saprophytic bacteria can be separated into three groups:

1. Membrane-bound pigments (comprising the polyene compounds carotinoids and xanthomonadins) which can only be extracted from bacterial cells by organic solvents such as acetone and methanol.
2. Pigments which diffuse out of the bacterial cells into the surrounding culture fluid, from which they can be extracted, and are referred to as 'extracellular'. This group contains a wide range of phytopathogenic bacterial pigments.
3. Pigments which are retained within the cell and do not diffuse out (intracellular). Relatively few pigments fall into this category, but the blue pigment lemonnierin produced by the saprophytic bacterium *Pseudomonas lemonnieri* is one example.

Pigment separation from organic or aqueous solvents may be achieved by thin layer chromatography (TLC) or high performance liquid chromatography (HPLC) and subsequent identification carried out by ultraviolet and visual spectroscopy. Elemental composition and molecular structure may be determined by techniques such as nuclear magnetic resonance (NMR) and mass spectometry (MS).

Although the data presented in Table 3.8 might suggest that particular pigments are restricted to particular taxonomic groups, it should be noted that the pigment composition of many phytopathogenic bacteria has not yet

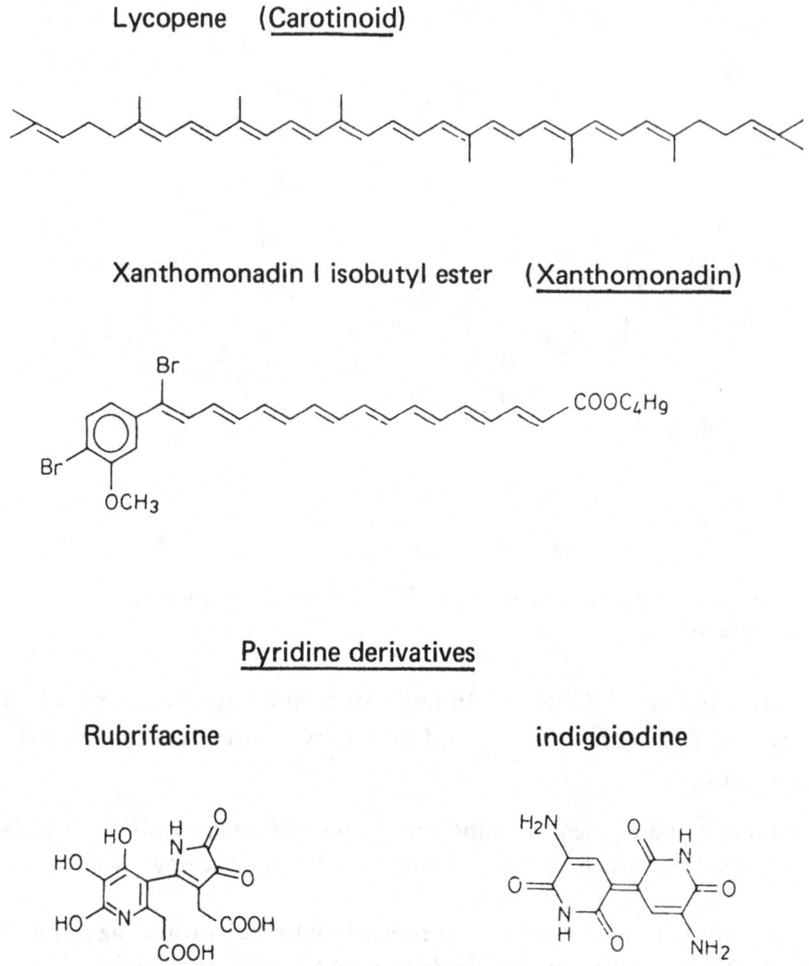

Lycopene   (Carotinoid)

Xanthomonadin I isobutyl ester   (Xanthomonadin)

Pyridine derivatives

Rubrifacine   indigoiodine

Fig. 3.4 Structure of carotinoid, xanthomonadin and pyridine-derivative pigments from plant pathogenic bacteria.

been investigated and that a general understanding in this area is very incomplete. Some pigments are clearly not restricted to particular genera and species of bacteria and future studies will probably show that specific pigments are more widespread than currently appears to be the case. It is also clear that many of these pigments are not restricted to pathogenic bacteria. Pyoverdines, for example, are not exclusively secreted by pathogenic pseudomonads but are also produced by saprophytes such as *Pseudomonas fluorescens* and *P. aeruginosa*. Similarly ferrorosamine A from *Erwinia rhapontici* is also synthesised by saprophytic pseudomonad species.

The production of pigments by phytopathogenic bacteria shows a close dependence on environmental conditions such as temperature, aeration and type of growth medium. This variability limits the use of pigments as taxonomic markers, as also does the fact that ability to produce pigments may be easily lost in culture due to mutation.

Bacterial pigments have a variety of functions, including iron chelation (siderophores) and prevention of damage by oxygen radicals. Pigment production *in vitro* is generally stimulated by high oxygen tension, and all of the pigments appear to have the potential to either scavenge or prevent the generation of oxygen radicals, thus protecting the bacterial cell from being oxidatively degraded. This may be particularly important as a protection against plant defence mechanisms (Feistner, 1990).

Bacteriophage testing

Phages are normally isolated from natural habitats of the host bacterium (including diseased plant tissue, irrigation water and soil) and show considerable diversity in terms of their host specificity. In general, phages isolated from diseased tissues are more specific than those isolated from soil. Phages with a broad spectrum of activity against several bacterial hosts are potentially useful in providing information on bacterial strain relationships, while phages with a high degree of host specificity may be used for more precise diagnostic purposes.

The presence of a virulent phage is usually determined experimentally by the resulting host cell lysis, causing plaque formation in bacterial cultures on semi-solid medium or clearing of cultures in liquid medium. In the first case, the phage sample may either be applied as spots to a bacterial lawn (surface plating method) or mixed with the host bacterial suspension in molten agar prior to plating (pour plate method).

A good example of the use of bacteriophages in diagnosis is provided by the work in Japan of Wakimoto (1960), who grouped strains of *Xanthomonas campestris* pv. *oryzae* (Table 3.9) into five main lysotypes (A–E) on the basis of their sensitivity to bacteriophages (designated OP1, OP1h, OP1h2 and OP2). Phages have also been used to differentiate subgroups in other pathovars, such as *Xanthomonas campestris* pv. *citri* (Gottwald *et al.*, 1988b).

Although bacteriophages provide a simple and rapid method of identification, there are a number of problems:

1. The specificity and stability of bacteriophages do not appear to be absolute since there is evidence that phages may change by repeated passage through certain bacterial isolates.

Table 3.9. *Classification of strains of* Xanthomonas campestris *pv.* oryzae
by phage sensitivity

Lysotype	Bacteriophage			
	OP1	OP1h	OP1h2	OP2
A	+	−	+	+
B	−	+	+	+
C	−	−	−	−
D	−	−	+	+
E	−	−	−	+

Data from Ou (1985).

2. Phage-resistant mutants and strains may arise.
3. There is lowered sensitivity to phage when large numbers of other bacteria are present.

Studies have generally shown that the sensitivity of bacterial strains to phages do not parallel their serological grouping, and bear no relation to the virulence of the strains.

Serology

Serology is a very useful technique for the identification of plant pathogenic bacteria down to subgroups (serovars) within species, using (by definition) the serum from immunised animals. The technique depends on the specific recognition and attachment that occurs between bacterial antigens and vertebrate antibodies. The antibodies are raised within the organism in response to the injection of bacterial components, and may be used to label and identify similar bacteria growing *in planta* or *in vitro*.

Serology has wide applications within bacterial plant pathology, including ecological studies, identification of pathogens for the diagnosis of disease, detection of pre-infective bacteria for plant and seed certification and identification of gene products in molecular genetics.

Under ideal circumstances, serology is a rapid, relatively low-cost technique with high specifity and sensitivity. The procedure may be used, for example, in association with light or electron microscopy, to detect the presence of a single phytopathogenic bacterial cell within a region of infected plant tissue. The general success of the technique depends on the specificity and detection of the antigen–antibody interaction, which relates to the nature of the bacterial antigens used, the type of antibodies produced and the

Table 3.10. *Serological diagnosis of plant pathogenic bacteria*

Bacterium	Antigen	Test	Reference
POLYCLONAL ANTIBODIES			
Agrobacterium tumefaciens	Lipopolysaccharide	Gel immunodiffusion	Bouzar *et al.*, 1988
Erwinia chrysanthemi	Whole cells or protein extract	Gel immunodiffusion	Dickey *et al.*, 1987
Erwinia amylovora	Lipo- & extracellular polysaccharide	Gel immunodiffusion	Laroche & Verhoyen, 1986
Xanthomonas campestris pv. *citri*	Whole cells	Immunofluorescence	Gottwald *et al.*, 1988b
Xanthomonas campestris pv. *translucens*	Membrane proteins	Gel immunodiffusion	Azad & Schaad, 1988
Xanthomonas campestris pv. *pelargonii*	Whole cells	Dot-immunobinding assay	Anderson & Nameth, 1990
MONOCLONAL ANTIBODIES			
Clavibacter michiganensis subsp. *sepedonicum*	Extracellular polysaccharide	Agglutination ELISA	De Boer *et al.*, 1988
Erwinia amylovora	Whole cells	Immunofluorescence & ELISA	Lin *et al.*, 1987
Xanthomonas campestris pv. *oryzae*	Surface antigens	Immunofluorescence, ELISA & Immunogold	Benedict *et al.*, 1989

procedures used to monitor the antibody–antigen interaction. The use of different approaches in recent studies on plant pathogenic bacteria is summarised in Table 3.10.

Bacterial antigens

The simplest and most direct procedure is to inject whole bacterial cells into the experimental animal. The immune system is then challenged by a multiplicity of antigens, located mainly on the bacterial cell surface, and the resulting antiserum will therefore contain a variety of antibodies. Some of these will not be specific to that particular type of bacterium, leading to cross-reaction with other bacteria, so the antiserum will have limited value for bacterial diagnosis. The situation may be improved either by using specific cell components as antigen, by removal of cross-reacting antigens using absorption techniques or by selecting particular cells producing specific (monoclonal) antibodies.

Various cell components have been used as antigens in an attempt to produce specific antibodies (Table 3.10), including extracellular polysacchar-

ides, surface lipopolysaccharides, membrane proteins, flagella preparations
and enzyme preparations. Some of these extracts appear to be highly specific
and potentially very useful in bacterial diagnosis. The LiCl-extracted membrane protein complex isolated by Azad and Schaad (1988), for example, was
shown to comprise five distinct antigens, use of which led to a clear separation
of *Xanthomonas campestris* pv. *translucens* into two distinct serovars.

Polyclonal and monoclonal antibodies

The majority of serological studies have used antiserum directly obtained
from the experimental animal. Even where this is produced in response to a
specific antigen, it will contain a mixture of antibodies derived from a number
of different cell lines (polyclonal). Some of these may not be specific, resulting
in a degree of cross-reaction with other types of bacteria during testing. The
polyclonal antibodies raised by Anderson and Nameth (1990) to *Xanthomonas campestris* pv. *pelargonii*, for example, gave a high reactivity with this
pathovar, but also moderate reactions with seven other pathovars tested.

The generation of monoclonal antibodies derived from a single cell line
(Kohler and Milstein, 1975) provides the opportunity for complete specificity,
and is being increasingly used in phytopathogen diagnosis (Table 3.10). This
approach typically results in the production of a range of hybridoma cell
lines, each producing an antibody which relates to a particular antigen or
antigen component (epitope). Only some of these antibodies will be specific to
that particular bacterial type. In the studies of Lin *et al.* (1987), for example,
out of 48 hybridoma clones producing monoclonal antibodies to *Erwinia
amylovora*, only 10 monoclonals were completely specific to this bacterium.

Antibody–antigen interaction

Once antibodies have been raised to particular bacterial antigens, positive
diagnosis may be achieved by demonstrating specific coupling between the
two. A range of procedures is available for this (Table 3.10), including gel
immunodiffusion, enzyme-linked immunosorbent assay and microscopical
techniques.

Gel immunodiffusion This technique had been widely used in serological
studies on bacterial plant pathogens, and involves the diffusion of soluble
antibodies and antigens through agar gel, with the formation of a white line of
precipitation (precipitin line) where homologous molecules interact. The most
commonly used procedure is the Ouchterlony double diffusion test (see, for
example, Azad & Schaad, 1988), and may result in the formation of multiple

precipitin bands where there are multiple antigens or antibodies within the preparation. The presence of minor levels of cross-reacting antibodies does not present a problem with this technique, since they are simply observed as secondary precipitin lines.

Enzyme-linked immunosorbent assay (ELISA) In this standard serological procedure, antibody molecules are conjugated to a suitable enzyme marker (e.g. alkaline phosphatase) and allowed to react with bacteria. The amount of antibody that remains attached to the bacterial cells (after washing) can be assessed by addition of enzyme substrate and subsequent colorimetric determination of the level of reaction product that is formed. This technique has the advantage of being rapid, highly sensitive and can be performed with tissue homogenates. For example, De Boer *et al.* (1988) used the procedure to monitor the presence of *Clavibacter michiganensis* subsp. *sepedonicum* in bulked samples of potato tubers, and were able to detect levels of infection as low as one contaminated tuber per 100.

Microscopical techniques Light microscope tests can provide a very rapid serological diagnosis of plant pathogenic bacteria. The simplest procedure is slide agglutination, where a droplet of the bacterial cells under test is mixed with specific antigen at appropriate dilution and the resulting positive reaction observed as an aggregation of crosslinked bacterial cells. A recent development of this approach uses antibody-coated latex beads (latex agglutination test) which can give a positive agglutination reaction in the presence of low numbers of the homologous bacterium (e.g. De Boer *et al.*, 1988).

Other microscopical procedures involve the use of labelled antibody, conjugated with fluorescein (for light microscope immunofluorescence) or with gold particles (for electron microscope immunogold labelling). Both of these procedures can be used for the specific identification and location of plant pathogenic bacteria in infected plant tissue and for the diagnosis of bacteria cultured *in vitro*. Immunofluorescence has proved particularly useful for the detection of pathogens *in planta*, where it may be carried out either directly on tissue samples (Fig. 3.5*b*) or on cells that have been rapidly filtered from the infected tissue (Gottwald *et al.*, 1988a). Fig. 3.5*a* shows the appearance of individual cells of *Erwinia amylovora* labelled by immunofluorescence. At the level of the electron microscope, immunogold labelling provides a high resolution technique which enables the identification of individual bacterial cells and provides information on the location of antigens at the ultrastructural level (see Fig. 2.10*b*).

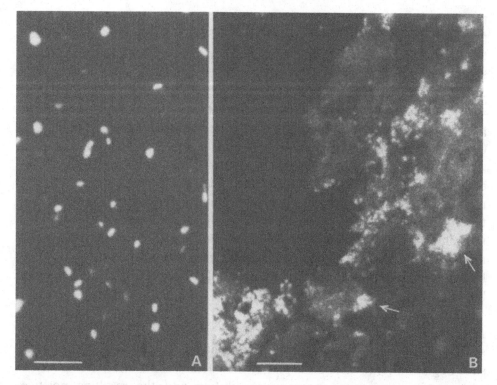

Fig. 3.5 Immunofluorescent labelling of plant pathogenic bacteria. (*a*) Light micrograph of air-dried smear of *Erwinia amylovora* (Ea) on a glass slide, treated with rabbit anti-Ea serum (primary antibody) and subsequently stained with fluorescein isothiocyanate (FITC) conjugated anti-rabbit IgG (secondary antibody). Individual bacterial cells have a bright yellow fluorescence. Bar scale: 15 μm. (*b*) Section of secretory surface of hawthorn nectary, 72 hours after surface inoculation with a suspension of *Erwinia amylovora*, treated with primary and secondary antibodies as above. Bacteria have invaded the secretory tissues of the nectary and appear as brightly fluorescing groups (arrows). Note the slight background fluorescence (autofluorescence) of the nectary tissue. Bar scale: 25 μm. (Photographs taken in collaboration with H. A. Epton and M. Wilson.)

Molecular analytical procedures

With the advent of molecular biology, a number of new techniques are becoming routinely available for rapid and specific diagnosis of plant pathogenic bacteria; these include DNA restriction fragment profile analysis, use of DNA probes and analysis of whole cell polypeptides and fatty acids.

DNA restriction fragment profile analysis

Bacterial genomic (chromosomal) or plasmid DNA can be readily isolated and the nucleotide sequence characterised by restriction endonuclease analysis. This involves digestion of the sample to completion with a specific endonuclease, separation of the fragments on an agarose or polyacrylamide gel, then observation of ethidium bromide-stained bands by ultraviolet light. The pattern of bands is determined by the distribution of restriction endonuclease cleavage sites in the DNA sample, and is unique and stable for a particular clone of cells.

Analysis of the chromosomal DNA (genomic fingerprinting) provides useful information on the genetic diversity within a particular population or taxonomic group. For example, Malvick and Moore (1988) found 21 different DNA restriction profiles in epiphytic populations of *Pseudomonas syringae* in Oregon State (USA). Restriction analysis of *Xanthomonas campestris* pv. *citri* by Hartung and Civerolo (1987) demonstrated five major groupings within this pathovar, corresponding to the five major biovars defined by geographic distribution, host range, phage typing and serology.

The use of restriction analysis with plasmid DNA has considerable potential, but can obviously only be used where plasmids are present in the bacterial cells. This may be a major limitation, and in a recent study on *Xanthomonas campestris*, Lazo and Gabriel (1987) showed that out of 26 pathovars analysed, 10 had stable plasmids, 3 were variable in plasmid content and 13 had no detectable plasmids. Where plasmids were present, these studies on DNA homology permitted clear differentiation at pathovar level, with close similarity in plasmid fragment pattern within but not between pathovars.

DNA probe

The genetic constitution of an organism is the fundamental basis for classification and is reflected in the DNA sequence that it possesses. Since specific parts of the sequence define the specific phenotype of the organism, the presence of a unique nucleotide sequence (or sequences) within a particular DNA sample may provide a useful diagnostic characteristic. This can be exploited by using a labelled DNA fragment with a complementary sequence (the DNA probe) which hybridises with and identifies the unique DNA. In addition to this DNA–DNA interaction, DNA probes may also be used to identify specific RNA sequences (DNA–RNA interaction).

Procedures involved in the preparation of DNA probes and the detection of homologous DNA sequences in the target bacterial DNA are described in detail by Gillis *et al.* (1990). In general, the DNA probe can be labelled radioactively or non-radioactively (using biotin), and hybridisation can either be carried out against purified test bacterial DNA or against DNA liberated *in situ* by lysed colonies on nitrocellulose or nylon filters. Unlike other types of

Table 3.11. *DNA probes for specific nucleotide sequences*

Bacterium	Probe	Diagnostic characteristics	Reference
DNA–DNA probes			
Erwinia carotovora	Isolated from *Erwinia* genomic library	Species-specific. Can detect 1/1000 colonies of pathogen/soil bacteria	Ward & de Boer (1990)
Pseudomonas syringae pv. *tomato*	3.5 and 3.6 kb *Eco*R1 restriction fragments	Probe hybridises with pv. *tomato* but not pv. *syringae* DNA	Denny (1988)
Pseudomonas syringae pv. *phaseolicola*	2.6 kb fragment from phaseolo-toxin gene	Pathovar-specific	Schaad *et al.* (1989)
Western X-disease MLO	Obtained from MLO in infected leaf-hopper	Used for identification of MLO *in planta*	Kirkpatrick *et al.* (1987)
DNA–RNA probe			
Xanthomonas sp.	Probe for 16S RNA	Used to identify xanthomonads in bean seed extract	DeParasis & Roth (1990)

diagnosis, the use of DNA probes does not depend on the physiological state of the organism and does not require specific gene expression.

DNA–DNA interactions Specific DNA probes have recently been developed for a number of phytopathogenic bacteria, some examples of which are summarised in Table 3.11. This technique has been used, for example, in the identification of *Pseudomonas syringae* pv. *tomato*, where the ability to distinguish this phytopathogen from pv. *syringae* is important for crop certification procedures. Denny (1988) has developed a DNA probe that will hybridise strongly with both purified and crude pv. *tomato* DNA, but not with that of pv. *syringae*.

In other cases, DNA probes have been developed to diagnose pathogens in soil samples (e.g. *Erwinia carotovora*) and also directly *in planta*. As far as the latter is concerned, DNA probes have considerable potential for identification of the phytopathogen directly within host tissue, since hybridisation of a specific probe will only occur with pathogen and not host DNA. This is particularly important for mycoplasma-like organisms where identification from *in vitro* culture is not possible, and has been used by Kirkpatrick *et al.* (1987) to identify 'western X-disease MLO' in infected plants and insects.

Although the ideal DNA probe would involve known DNA sequences (of large copy number) that define a particular taxonomic feature, in the majority of cases the DNA sequence that is being probed occurs as a single copy within the genome and is of unknown function. Notable exceptions to this are the use of a specific toxin gene DNA fragment to identify *Pseudomonas syringae* pv. *phaseolicola* (Schaad *et al.*, 1989), and the use of DNA–RNA hybridisations.

DNA–RNA interactions Recent studies by DeParasis and Roth (1990) have suggested that the use of DNA probes specific for 16s ribosomal RNA may have particular diagnostic potential. Although this molecule has been highly conserved during evolution, at least one region (bases 1057 to 1090) was identified as being genus-specific. Apart from this specificity, probing for 16s RNA also has the advantage that sequence information is easy to obtain and that the molecule is abundant (more than 10 000 copies in an actively growing cell). The approach adopted by these workers was to identify short-sequence regions of 16s RNA that were specific to the target organism, then raise complementary oligodeoxynucleotides (using reverse transcriptase) which were subsequently labelled and used as identification probes.

Polypeptide analysis

When grown under standard conditions, a particular isolate of bacterium will always produce the same set of proteins. This is a characteristic of the bacterium, and provides a diagnostic tool which shows good correlation with results obtained from DNA–DNA hybridisation (Kersters, 1985).

Although the analysis of whole cell proteins by gel electrophoresis has been widely used in bacterial identification generally, there are relatively few published accounts for plant pathogenic bacteria.

Protein fractions of cell envelopes have been used, however, to characterise certain species of *Xanthomonas* and *Pseudomonas* (Schnaitman, 1970) and also to distinguish pathogenic xanthomonads at the pathovar level (Santos & Dianese, 1985). More recently, Dristig and Dianese (1990) have examined the surface membrane proteins of strains of *Pseudomonas solanacearum*, identifying over 29 distinct bands in polyacrylamide gels and comparing the three biovars of this pathogen in terms of a PAGE similarity index. The band patterns obtained showed clear characterisation at the biovar level, with biovars 1 and 3 having a 35–37 kDa protein which was absent from biovar 2. Close similarity occurred between the surface protein pattern of this pathogen and other non-fluorescent pseudomonads, but not with pathovars of *Xanthomonas campestris*, which had a distinct common band at 44 kDa. SDS PAGE analysis of *Pseudomonas solanacearum* surface envelope proteins did not reveal any differences between high virulence and low virulence strains.

Two-dimensional gel electrophoresis has particular potential in resolving the large number of proteins involved in this type of analysis, and the use of computer-enhanced image analysis facilitates the interpretation and comparison of data. Using this approach, Moline and Hruschka (1987) analysed the acidic ribosomal proteins of *Erwinia* and showed that polypeptide patterns followed taxonomic differences based on other criteria. In this study, the polypeptide maps of *Erwinia amylovora* and *E. herbicola* differed completely from the soft-rot erwinias, and *E. carotovora, E. chrysanthemi* and *E. rhapontici* had 18 matching polypeptide clusters, indicating a close evolutionary homology.

Gel electrophoresis can also be used to analyse variations in genetically determined polymorphic enzymes (isozymes) within bacterial populations and comparison of isozyme patterns can assist in the determination of taxonomic relationships. In *Xanthomonas campestris* pv. *citri*, for example, Kubicek *et al.* (1989) analysed the isozyme variation at structural genes encoding 14 enzymes, 10 of which were polymorphic and 4 monomorphic. Little polymorphic variation occurred in Asiatic citrus canker strains but, in line with other evidence, Florida citrus spot strains of this pathogen represented a very diverse assemblage.

Fatty acid composition

Fatty acids of plant pathogenic bacteria are located in the surface membranes of the cell, and are typically between 9 to 20 carbon atoms in length. These fatty acids are chemically diverse and include hydroxy acids, saturated and unsaturated molecules, comprising either straight or branched chains. The types and proportions of these acids is often unique and constant for particular taxonomic groups, and the analysis of fatty acid composition therefore provides a very useful taxonomic approach. This technique involves:

1. Cell culture under standard conditions.
2. Release of fatty acids from the cell surface (inner and outer membranes) by saponification.
3. Methylation of fatty acids to increase volatility.
4. Analysis by high resolution gas chromatography.
5. Comparison of the fatty acid profile with standard profiles in a microbial library to obtain the best fit in a genus, species and subspecies match.

Identification from fatty acid profiles has become a rapid and precise diagnostic procedure and is particularly useful where pathogenicity testing and general *in vitro* tests are problematic – as in the case of *Clavibacter michiganensis* subsp. *michiganensis*. Gitaitis & Beaver (1990) showed that different strains of this pathogen have a standard profile of C12–C17 fatty acids, and that the presence of unsaturated branched chain a15:1, anteiso-pentadeconoic acid was highly specific for this bacterium. Successful application of this procedure requires that fatty acid profiles should not only be standard over the range of strains within the taxon, but should also be standard in bacteria from different sites. In the case of *Clavibacter*, similar profiles were obtained from cultured isolates derived from irrigation pools, seed surfaces, soil, transplants and weeds.

Computer identification by numerical analysis

In some cases, bacterial identification on the basis of observed phenotypic characteristics may be complex and problematical. This is particularly the case where two or more closely related pathogens occupy the same micro-environment and are probably involved in genetic exchange. Resolution of the taxonomic grouping is not necessarily made easier by the increasing number and diversity of phenotypic characteristics which are available for use. In this situation, the most objective approach is to grade individual phenotypic characteristics on a numerical basis and assess all of the data by computer, thus providing a consistent methodology for both classification and identification of the bacteria concerned.

Useful accounts of numerical taxonomy in relation to bacterial classification are provided by Sneath (1984) and Goor *et al.* (1990). In practice, this approach involves:

1. Selecting a wide range of entities (referred to as operational taxonomic units or OTUs) to be classified. In most microbial studies these will be bacterial strains, and should represent as broad an assemblage as possible, including type strains, representative strains from culture collections and recently isolated strains of different ecological and geographic origins.
2. Selecting a wide range of phenotypic characters, where a character is defined as any property that varies between OTUs and can assume two or more possible values. It is generally accepted that at least 50 characters and preferably more (up to 300) are required to give enough discrimination to resolve the different taxonomic groupings and to arrive at a stable classification.
3. Computer analysis: phenotypic characters are coded on a numerical basis then arranged in a data matrix which can be input for numerical analysis by computer. Similarities and dissimilarities between OTUs can be analysed in two main ways: hierarchical cluster analysis or non-hierarchical ordination techniques. In the

former case, similar OTUs are arranged in a branching sequence of groups, producing a dendrogram of affinities. In the second (non-hierarchical) case, OTUs are arranged in two- or three-dimensional space in the form of a scatter diagram (Goor *et al.*, 1990).

Numerical analysis has been used by Roos & Hattingh (1987b) to characterise and identify South African strains of *Pseudomonas syringae*, which occur as a heterogeneous assemblage clustered around *P.s.* pv. *syringae* and *P.s.* pv. *mors-prunorum*. Both of these pathovars are pathogenic on stone fruits, but phenotypic characteristics – including colony morphology, enzyme activities, antibiotic sensitivity and utilisation of amino acids, amines, organic acids and carbohydrates – are highly variable, and intermediate forms exist over a wide spectrum. In this situation, the whole population can be assessed on a numerical basis, with individual phenotypic characters being graded as positive (score 2), negative (score 1) or neutral (score 0). Roos & Hattingh (1987b) computer-analysed a variety of isolates for over 200 phenotypic characters to determine the degree of matching between strains and to draw up a dendrogram based on affinities. The results obtained provided an objective assessment of the phenotypes encountered, and allowed most of the isolates to be clearly assigned to one of the two major pathovars.

References

Anderson M. J. & Nameth S. T. (1990). Development of a polyclonal antibody-based serodiagnostic assay for the detection of *Xanthomonas campestris* pv. *pelargonii* in geranium plants. *Phytopathology* **80**, 357–60.

Azad H. & Schaad N. W. (1988). Serological relationships among membrane proteins of strains of *Xanthomonas campestris* pv. *translucens. Phytopathology* **78**, 272–7.

Benedict A. A., Alvarez A. M., Berestecky J. & Imanak W. (1989). Pathovar-specific monoclonal antibodies for *Xanthomonas campestris* pv. *oryzae* and *X.c.* pv. *oryzicola. Phytopathology* **79**, 322–8.

Bouzar H., Moore L. W. & Schaup H. W. (1988). Lipopolysaccharide from *Agrobacterium tumefaciens* B6 induces the production of strain-specific antibodies. *Phytopathology* **78**, 1237–41.

Brinkerhoff L. A. (1970). Variation in *Xanthomonas malvacearum* and its relation to control. *Ann. Rev. Phytopathol.* **8**, 85–110.

Buddenhagen I. & Kelman A. (1964). Biological and physiological aspects of bacterial wilt caused by *Pseudomonas solanacearum. Ann. Rev. Phytopathol.* **2**, 203–30.

Civerolo E. L. & Fan F. (1982). *Xanthomonas campestris* pv. *citri* detection and identification by enzyme-linked immunosorbent assay. *Plant Dis.* **66**, 231–6.

Clafkin L. E., Vidaver A. K. & Sasser M. (1987). MXP, a semi-selective medium for *Xanthomonas campestris* pv. *phaseoli. Phytopathology* **77**, 730–4.

Davis M. J. (1990). Fastidious prokaryotes. In *Methods in Phytobacteriology*, ed. Z. Klement, K. Rudolph & D. C. Sands. Budapest: Akademiai Kiado, pp. 75–84.

De Boer S. H., Wieczorek A. & Kummer A. (1988). An ELISA test for bacterial ring rot of potato with a new monoclonal antibody. *Plant Dis.* **72**, 874–8.

Denny T. P. (1988). Differentiation of *Pseudomonas syringae* pv. *tomato* from *P.s.* pv. *syringae* with a DNA hybridisation probe. *Phytopathology* **78**, 1186–93.

DeParasis J. & Roth D. A. (1990). Nucleic acid probes for identification of phytobacteria: Identification of genus-specific 16s RNA sequences. *Phytopathology* **80**, 618–21.

Dickey R. S., Clafkin L. E. & Zumoff C. H. (1987). *Erwinia chrysanthemi*: Serological comparisons of strains from *Zea mays* and other hosts. *Phytopathology* **77**, 426–30.

Dristig M. C. & Dianese J. C. (1990). Characterisation of *Pseudomonas solanacearum* biovars based on membrane protein patterns. *Phytopathology* **80**, 641–6.

Dye D. W., Bradbury J. F., Goto M., Hayward A. C., Lelliott R. A. & Schroth M. N. (1980). International standards for naming pathovars of phytopathogenic bacteria and a list of pathovar names and pathogenic strains. *Rev. Plant Pathol.* **59**, 153–68.

Fahy P. C. & Persley G. J. (1983). *Plant Bacterial Diseases*. Sydney: Academic Press.

Fatmi M. & Schaad N. W. (1988). Semiselective agar medium for isolation of *Clavibacter michiganense* subsp. *michiganense* from tomato seed. *Phytopathology* **78**, 121–6.

Feistner G. (1990). Pigments. In *Methods in Phytobacteriology*, ed. Z. Klement, K. Rudolph, D. C. Sands. Budapest: Akademiai Kiado, pp. 233–44.

Fett W. F. & Sequeira L. (1981). Further characterisation of the physiologic races of *Pseudomonas glycinea*. *Can. J. Bot.* **59**, 283–7.

Gabriel D. W., Hunter J. E., Kingsley M. T., Miller J. W. & Lazo G. R. (1987). Clonal population structure of *Xanthomonas campestris* and genetic diversity among citrus canker strains. *Mol. Plant Microbe Interact.* **1**, 59–65.

Gillis M., Roth D. A., Johnson J. & Rudolph K. (1990). Characterisation by nucleic acids. In *Methods in Phytobacteriology*, ed. Z. Klement, K. Rudolph, D. C. Sands. Budapest: Akademiai Kiado, pp. 216–30.

Gitaitis R. D., Hamm J. D. & Bertrand P. F. (1988). Differentiation of *Xanthomonas campestris* pv. *pruni* from other yellow-pigmented bacteria by the refractive quality of bacterial colonies on agar medium. *Plant Dis.* **72**, 416–17.

Gitaitis R. D. & Beaver R. W. (1990). Characterisation of fatty acid methyl ester content of *Clavibacter michiganensis* subsp. *michiganensis*. *Phytopathology* **80**, 318–21.

Goor M., Kersters K., Mergaert J., *et al.* (1990). Numerical analysis of phenotypic features. In *Methods in Phytobacteriology*, ed. Z. Klement, K. Rudolph, D. C. Sands. Budapest: Akademiai Kiado, pp. 145–52.

Gottwald T. R., Civerolo E. L., Garnsey S. M., BrLansky R. H., Graham J. H. & Gabriel D. W. (1988a). Dynamics and spatial distribution of *Xanthomonas campestris* pv. *citri* group E strains in simulated nursery and new grove situations. *Plant Dis.* **72**, 781–7.

Gottwald T. R., McGuire R. G. & Garran S. (1988b). Asiatic citrus canker:

Spatial and temporal spread in simulated new planting situations in Argentina. *Phytopathology* **78**, 739–45.

Harper S., Zewdie N., Brown I. R. & Mansfield J. W. (1987). Histological, physiological and genetical studies of the responses of leaves and pods of *Phaseolus vulgaris* to three races of *Pseudomonas syringae* pv. *phaseolicola* and to *P.s.* pv. *coronafaciens*. *Physiol. Mol. Plant Pathol.* **31**, 153–72.

Hartung J. S. & Civerolo E. L. (1987). Genomic fingerprints of *Xanthomonas campestris* pv. *citri* strains from Asia, South America and Florida. *Phytopathology* **77**, 282–5.

Hayward A. C. (1990). Proposals for a quick practical identification. In *Methods in Phytobacteriology*, ed. Z. Klement, K. Rudolph, D. C. Sands. Budapest: Akademiai Kiado, pp. 269–74.

Hibberd A. M., Stall R. E. & Bassett M. J. (1988). Quantitatively assessed resistance to bacterial leaf spot in pepper that is simply inherited. *Phytopathology* **78**, 607–12.

Hopkins D. L. (1988). Natural hosts of *Xylella fastidiosa* in Florida. *Plant Dis.* **72**, 429–31.

Kirkpatrick B. C., Stenger D. C., Morris T. J. & Purcell A. J. (1987). Cloning and detection of DNA from a nonculturable plant pathogenic mycoplasma-like organism. *Science* **238**, 197–200.

Kersters K. (1985). Numerical methods in the classification of bacteria by protein electrophoresis. In *Computer-Assisted Bacterial Systematics*, ed. M. Goodfellow, R. G. Board. London: Academic Press.

Knoche K. K., Clayton M. L. & Fulton R. W. (1987). Comparison of resistance in tobacco to *Pseudomonas syringae* pv. *tabaci* races 0 & 1 by infectivity titration and bacterial multiplication. *Phytopathology* **77**, 1364–8.

Kohler G. & Milstein C. (1975). Continuous cultures of fused cells secreting antibody of predefined specificity. *Nature* **256**, 495–7.

Kreig N. R. & Holt J. G. (1984). *Bergey's Manual of Systematic Bacteriology*, 8th edn. Baltimore: Williams & Wilkins.

Kubicek Q. B., Civerolo E. L., Bonde M. R., Hartung J. S. & Peterson G. L. (1989). Isozyme analysis of *Xanthomonas campestris* pv. *citri*. *Phytopathology* **79**, 297–300.

Lapage S. P., Sneath P. H. A., Lessel E. F., Skerman V. B. D., Seeliger H. P. R. & Clark W. A. (1975). International code of nomenclature of bacteria. Revision. Washington, D.C.: American Society of Microbiology.

Laroche M. & Verhoyen M. (1986). The search for a specific antigen for *Erwinia amylovora*. *Phytopathol. Z.* **116**, 269–77.

Lazo G. R. & Gabriel D. W. (1987). Conservation of plasmid DNA sequences and pathovar identification of strains of *Xanthomonas campestris*. *Phytopathology* **77**, 448–53.

Lelliott R. A., Billing E. & Hayward A. C. (1966). A determinative scheme for the fluorescent plant pathogenic pseudomonads. *J. Appl. Bacteriol.* **29**, 470–89.

Lelliott R. A. & Stead D. E. (1983). *Methods for the Diagnosis of Bacterial Diseases of Plants*. Oxford: Blackwell Scientific.

Liao C. H. & Chen T. A. (1977). Culture of corn stunt *Spiroplasma* in a simple medium. *Phytopathology* **67**, 802–7.

Lin C. P., Chen T. A., Wells J. M. & van der Zwet T. (1987). Identification and detection of *Erwinia amylovora* with monoclonal antibodies. *Phytopathology* **77**, 376–80.

Malvick D. K. & Moore L. W. (1988). Population dynamics and diversity of *Pseudomonas syringae* on maple and pear trees and associated grasses. *Phytopathology* **78**, 1366–70.

McGuire R. G., Jones J. B. & Sasser M. (1986). Tween media for semiselective isolation of *Xanthomonas campestris* pv. *vesicatoria* from soil and plant material. *Plant Dis.* **70**, 887–91.

Ming D., Huazhi Y., Schaad N. W. & Roth R. A. (1991). Selective recovery of *Xanthomonas* sp. from rice seed. *Phytopathology* **81**, 1358–63.

Mohan S. K. & Schaad N. W. (1987). An improved agar plating assay for detecting *Pseudomonas syringae* pv. *syringae* and *P.s.* pv. *phaseolicola* in contaminated seed. *Phytopathology* **77**, 1390–5.

Moline H. E. & Hruschka W. R. (1987). Computer-enhanced image analysis of bacterial polypeptide patterns on two-dimensional polyacrylamide gels. *Phytopathology* **77**, 745–7.

Murray R. G. E. (1984). The higher taxa, or, a place for everything? In *Bergey's Manual of Systematic Bacteriology*, ed. N. R. Krieg, J. G. Holt. Baltimore: Williams & Wilkins.

Ou S. H. (1985). *Rice Diseases*. Slough: Commonwealth Agricultural Bureau.

Roos I. M. & Hattingh M. J. (1987a). Systemic inversion of cherry leaves and petioles by *Pseudomonas syringae* pv. *mors-prunorum*. *Phytopathology* **77**, 1246.

Roos I. M. & Hattingh M. J. (1987b). Pathogenicity and numerical analysis of phenotypic features of *Pseudomonas syringae* strains isolated from deciduous fruit trees. *Phytopathology* **77**, 900–8.

Roy M. A. & Sasser M. (1990). Selective media: Principles of design and use. In *Methods in Phytobacteriology*, ed. Z. Klement, K. Rudolph & D. C. Sands. Budapest: Akademiai Kiado, pp. 61–5.

Santos R. M. & Dianese J. C. (1985). Comparative membrane characterisation of *Xanthomonas campestris* pv. *cassavae* and *X.c.* pv. *manihotis*. *Phytopathology* **75**, 581–7.

Schaad N. W. (1982). Detection of seedborne bacterial plant pathogens. *Plant Dis.* **66**, 885–90.

Schaad N. W., Azad H., Peet R. C. & Panopoulos N. J. (1989). Identification of *Pseudomonas syringae* pv. *phaseolicola* by a DNA hybridisation probe. *Phytopathology* **79**, 903–7.

Schnaitman C. A. (1970). Comparison of envelope protein compositions of several Gram-negative bacteria. *J. Bacteriol.* **104**, 1404–5.

Skerman V. B. D., McGowan V. & Sneath P. H. A. (1980). Approved lists of bacterial names. *Int. J. Syst. Bacteriol.* **30**, 225–420.

Sneath P. H. A. (1984). Bacterial nomenclature. In *Bergey's Manual of Systematic Bacteriology*, ed. N. R. Krieg, J. G. Holt, Vol. 2, Baltimore: Williams & Wilkins, pp. 19–23.

Stead D. E. (1990). Differentiation of commonly isolated pathogenic species. In *Methods in Phytobacteriology*, ed. Z. Klement, K. Rudolph & D. C. Sands. Budapest: Academiai Kiado, pp. 65–74.

Surico G. & Lavermicocca P. (1989). A semiselective medium for the isolation of *Pseudomonas syringae* pv. *savastanoi*. *Phytopathology* **79**, 185–90.

Taylor J. D., Bevan J. R., Crute I. R. & Reader S. L. (1989). Genetic relationship between races of *Pseudomonas syringae* pv. *pisi* and cultivars of *Pisum sativum*. *Plant Pathol.* **38**, 364–75.

Wakimoto S. (1960). Classification of strains of *Xanthomonas oryzae* on the basis of their susceptibility against bacteriophages. *Ann. Phytopathol. Soc. Japan* **25**, 193–8.

Ward L. J. & De Boer S. H. (1990). A DNA probe specific for serologically diverse strains of *Erwinia carotovora*. *Phytopathology* **80**, 665–9.

4

Plant pathogenic bacteria in the environment

Plant pathogenic bacteria are not restricted in their occurrence to infected plant tissue, but are widely dispersed throughout the external environment. This chapter will consider the general occurrence of plant pathogenic bacteria in the aerial and soil/water environments, environmental interactions at the micro-level and the association of these bacteria with invertebrates (vectors).

The aerial environment

The aerial occurrence of plant pathogenic bacteria is clearly of particular relevance to those pathogens that infect aerial parts of plants, including leaves, flowers and fruit. The aerial environment includes both physical aspects (e.g. occurrence of bacteria in rain and aerosols) and biotic aspects (occurrence of bacteria on plant surfaces and aerial dispersal by vectors).

Occurrence of bacteria in rain and aerosols

The aerial environment presents a potentially important medium for both survival and transmission of plant pathogenic bacteria, particularly where cells are contained in rain water or fine water droplet dispersions (aerosols).

Rain-water from infected foliage may contain high levels of phytopathogenic bacteria, and may be important in the spread of bacteria both within and between plants. A good example of rain dispersal of pathogen within single plants is provided by the studies of Miller on *Erwinia amylovora* (reported in Van der Zwet & Keil, 1979), who showed that if a source of inoculum was present in the upper part of a tree, the region of secondary infection below was cone-shaped due to downward dispersal of bacteria by rain-splash. Dispersal of pathogen between plants has been shown for citrus bacterial canker (caused by *Xanthomonas campestris* pv. *citri*), spread of which is directly related to windblown rain (Gottwald *et al.*, 1988), particularly at windspeeds greater than 8 m/sec. Rain-water derived from infected plants contained up to 10^4 cfu/ml.

Aerosol dispersal has been demonstrated for a number of pathogens, including *Erwinia carotovora* and pathovars of *Pseudomonas syringae* and *Xanthomonas campestris*. The presence of bacteria within aerosols can be detected using an air sampler, in which the impaction of droplets onto agar plates can be graded in terms of particle size. A recent example of where this has been carried out is provided by the work of McInnes *et al.* (1988), where air samplers were placed at canopy height in tomato fields which had been inoculated with *Xanthomonas campestris* pv. *vesicatoria* and *Pseudomonas syringae* pv. *tomato*.

In general, the efficiency of aerosol dispersal and long-distance survival of bacteria depends on a number of interrelated factors such as aerosol droplet size, external conditions and cultural practices.

Droplet size

The mean particle size of aerosols appears to be in the range 2–7 μm diameter, and is probably of considerable importance in terms of both the mechanics of dispersal and the ultimate survival of the pathogen. Although there is little experimental information on this, it seems likely that small droplets will tend to be dispersed longer distances, while bacterial survival and establishment of epiphytic populations will be greater with large droplets. In the latter case, droplets may contain more than one bacterium, and may support cell division during the dispersal process.

External conditions

Monitoring of aerial populations suggests that warm, sunny periods produce the highest levels of aerosol bacteria. Radiant (solar) energy represents the main factor for aerosol generation, with upward convection and dispersal of droplets in the resulting air currents.

High rainfall and irrigation appear to have variable effects. In some cases, a clear increase in the aerosol population has been demonstrated, probably resulting from rain-splash of the epiphytic population, while in other cases there has been a decline. The aerosol dispersal of *Xanthomonas campestris* pv. *vesicatoria* and *Pseudomonas syringae* p.v *tomato* on tomato plants in the field, for example, showed a clear decrease under these conditions (McInnes *et al.*, 1988), probably due to a strong downward flux of bacteria during rain and a large scale loss of epiphytic populations as they are washed into the soil.

Cultural practices

Although most studies on aerosol dispersion of bacteria have dealt with natural processes of generation, it is clear that cultural practices may also be important. Studies by McInnes *et al.* (1988) have shown, for example, that clipping and harvesting of tomato crops lead to increased levels of airborne bacteria, and that chemical control sprays may also promote production of aerosols. The importance of cultural practices in the spread of plant pathogenic bacteria is discussed more fully in Chapter 9.

Aerial plant surfaces

In recent years, it has become widely recognised that many phytopathogenic and non-pathogenic bacteria are residents of aerial plant surfaces (referred to collectively as the phyllosphere) along with other micro-organisms such as small invertebrates, fungi and yeasts (Fokkema & Van den Heuvel, 1986). These epiphytic bacteria are capable of growth and multiplication, and are naturally present on both healthy and diseased plants. They have been studied particularly in relation to their presence on the leaf surface (phylloplane) which is regarded as the major aerial habitat. The best known and most widely investigated of the epiphytic bacteria belong to the *Pseudomonas syringae* group.

In general, epiphytic bacteria have been studied either as naturally occurring populations, or after artificial inoculation of plant surfaces.

Monitoring epiphytic populations

Epiphytic bacteria can be detected and quantified either by standard microbiological techniques or by direct observation using electron microscopy. In the first case, the bacteria are removed by washing, then plated out on standard growth media (see Fig. 3.2) to retrieve as many cells as possible in a viable condition for colony counts and subsequent identification. Bacterial populations are subsequently expressed as colony forming units (CFU) per gram fresh or dry weight of leaf tissue, per unit area or per leaf.

Counts of viable bacteria on leaves show that these epiphytic organisms do not occur either uniformly or randomly on plant surfaces, but follow a log-normal distribution (Hirano *et al.*, 1982). This type of distribution arises where localised exponential multiplication is superimposed on a random distribution of cells and has important implications for the statistical evaluation of bacterial populations:

1. Estimates of population size based on the common practice of using

bulked samples will overestimate the population median by a factor of approximately 1.15 σ^2 (where σ^2 is the population variance).

2. From this known probability distribution, the frequency with which a particular population of bacteria on individual leaves is equalled or exceeded can be calculated. Since the induction of disease relates to population levels on individual leaves rather than the whole canopy, this value is important in epidemiology.

The localised occurrence of bacteria on plant surfaces has been demonstrated in a number of scanning electron microscope studies and often appears to indicate their limitation to microenvironments within the surface area which support their survival and multiplication. This is shown, for example, during floral colonisation and infection by *Erwinia amylovora*, where multiplication of the pathogen on the nectarial surface is initially restricted to nectarthodal depressions (Fig. 4.2*b*; Wilson *et al.*, 1990) and the occurrence of bacteria on the anther and stigmatic surfaces is limited to grooves between cells. Surface structures such as trichomes may also support populations of bacteria, and studies by Bashan *et al.* (1981) on *Pseudomonas syringae* pv. *tomato* have shown that dead tomato leaf trichomes provide important long-term survival and multiplication sites for this pathogen.

The general phytopathogenic nature of epiphytic bacteria can be investigated in terms of the ability of plant surface isolates to induce a hypersensitive reaction (HR) in a non-host plant such as tobacco (HR test for pathogenicity: see Chapter 6). This has been investigated particularly in relation to strains of *Pseudomonas syringae* occurring on foliage of fruit trees, and shows a widespread presence of potentially pathogenic cells. Studies by Olive and McCarter (1988), for example, on the occurrence of strains of *Pseudomonas syringae* on apple and peach trees in Georgia (USA), showed that all strains tested were phytopathogenic (HR test), but considerable variation occurred in the ability of bacteria to cause disease on their associated plant.

Specificity of bacteria–plant associations

A substantial amount of information has now been obtained from studies of naturally occurring and surface-inoculated populations on the epiphytic occurrence and specificity of association of plant pathogenic bacteria.

Naturally occurring populations Detailed studies carried out by Roos & Hattingh (1987) in South Africa, and Malvick & Moore (1988) in the North-west USA, on the occurrence of strains and pathovars of *Pseudomonas syringae* on fruit trees, provide good examples where specificity of bacterial association has been investigated. Although in some cases (e.g. pv. *mors-*

prunorum) the pathogen is epiphytically restricted to its (compatible) host plant, this is normally not the case, and other pathovars (e.g. pv. *syringae*) occur over a wide range of both host and non-host plants. Malvick and Moore showed that epiphytic populations of *Pseudomonas syringae* in pear and red maple orchards were highly heterogeneous in terms of pathogenicity, and that the bacteria were widespread on both host trees and non-host plants, including orchard grasses such as perennial rye, field brome and red fescue.

Surface-inoculated bacteria Monitoring of epiphytic populations following spray inoculation of experimental plants has provided useful information on the relationship between pathogenic compatibility and epiphyte development. In general, the results show that incompatible or non-pathogenic bacteria are just as able to grow epiphytically as compatible ones.

Timmer *et al.* (1987) investigated the survival of compatible and incompatible pathovars of *Xanthomonas campestris* on leaves of tomato plants under a variety of conditions. The most important factor determining epiphytic survival was humidity. The compatible pathovar *X.c.* pv. *vesicatoria* grew well on detached and attached leaves of tomato at high (90–95% RH) but not low (10–25% RH) humidity. At high humidity, the incompatible pathovars *X.c.* pv. *alfalfae*, pv. *campestris* and pv. *translucens* multiplied on tomato leaf surfaces at the same rate as pv. *vesicatoria*, showing that epiphytic growth is not host-specific.

This lack of epiphytic specificity is further demonstrated by studies within particular race–cultivar systems and by work on non-pathogenic Tn5 mutants. Cooksey (1988) has shown, for example, that races of *Pseudomonas syringae* pv. *tomato* can survive on both resistant as well as susceptible tomato cultivars, and that non-pathogenic Tn5 mutants are also able to survive as epiphytes.

Epiphytic populations on host plants: sites of multiplication and seasonal variation

Where phytopathogenic bacteria do occur epiphytically on a susceptible host, they clearly provide a source of inoculum for disease development (see Chapter 5) and also for transmission between plants.

In some cases, the bacteria occur on different parts of the host during the growth season, relating to different phases of transmission and infection as the year progresses. An example of this is shown in Fig. 4.1, which illustrates the rather complex annual cycle of *Erwinia amylovora* on a rosaceous host (hawthorn). This pathogen appears to occupy key surface sites for limited duration at different times of year, including buds and cankers (early Spring),

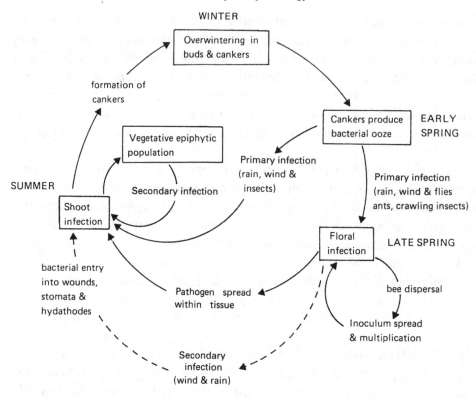

Fig. 4.1 Annual growth and infection cycle of *Erwinia amylovora* on a rosaceous host plant.

flowers (late Spring) and vegetative foliage (on heavily infected plants in Summer). It does not occur as a general phylloplane resident, either on host or non-host plants.

The presence of *Erwinia amylovora* as an epiphyte on flowers is important for floral infection, the principal route of bacterial entry with fireblight disease (Van der Zwet & Keil, 1979), and is discussed in detail in Chapter 5. The major sites of epiphytic multiplication are shown diagrammatically in Fig. 5.4 and include the stigmatic surface, pollen grains (Fig. 4.2*a*) and the surface of the nectary (Fig. 4.2*b*).

In situations where a mixture of pathogenic bacteria occurs on a particular plant surface, seasonal changes may occur in the proportion of the different pathogens and the resulting incidence of disease. This has been shown by Legard and Schwartz (1987) for phylloplane pathogens of snap bean, including *Pseudomonas syringae* pv. *syringae* (causal agent of brown spot disease) and *P. s.* pv. *phaseolicola* (causing halo blight). The level of *P. s.* pv. *syringae* on leaflets of this host showed a consistent increase throughout the

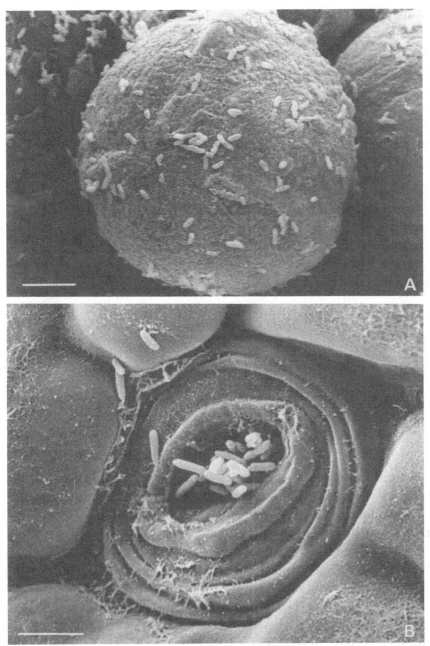

Fig. 4.2 Scanning electron micrographs of epiphytic bacteria on hawthorn flower surfaces. (*a*) Cells of *Erwinia amylovora* on an exposed pollen grain in the region of the anther dehiscence zone, 48 hours after artificial inoculation of the anther surface. Bar scale: 5 μm. (*b*) Detail from nectary surface, showing localised groups of unidentified bacteria (probably saprophytic). These naturally occurring bacteria are mainly apparent within the nectarthode pore, but scattered cells are also present in the surrounding nectarthodal depression. Bar scale: 4 μm. (Photographs taken in collaboration with H. A. Epton and M. Wilson.)

growing period, with a mid-season discontinuity which was thought to relate to physiological changes in the host at the time of flower initiation. In the same study, these authors demonstrated a change in the balance of epiphytic populations over the growing period, with increased proportions of *P. s.* pv. *phaseolicola* (compared with pv. *syringae*) late in the season. These changes were reflected in decreased levels of brown spot, and increased levels of halo blight in late July/early August, demonstrating a correlation between the composition of epiphytic populations and the pattern of disease.

The soil–water environment

The soil environment provides an important medium for the survival and dissemination of bacterial phytopathogens that cause disease of roots and ground storage organs. The widespread occurrence of these bacteria in the soil, where they survive as saprophytes, has been investigated both in the limited confines of the rhizosphere and in the general soil–water environment.

The rhizosphere

The rhizosphere comprises the root surface and its immediate environment. This surface is a major interface for the spread of bacteria through soil, and the rhizosphere region as a whole is an important site for the build-up of bacterial populations due to the localised accumulation of high levels of organic nutrients derived from the host plant.

Colonisation of the rhizosphere

The bacterial colonisation of the rhizosphere is important to plant pathology in terms of the spread and establishment of phytopathogenic bacteria and also in relation to the development of populations of bacteria involved in disease suppression (i.e. bacterial antagonists).

Bacterial colonisation has been investigated with reference to both established populations and populations of bacteria inoculated under glasshouse or field conditions. An example of the former is provided by the work of Van Vuurde and Schippers (1980) who investigated the microbial colonisation of wheat roots grown in soil. They showed a linear sequence in the occurrence of rhizosphere bacteria in relation to daily formed regions of the root (diurnal segments), with two maxima in the level of bacterial colonisation. These appeared to relate to two peaks in the release of organic material (due to root cell lysis) during the root ontogeny (Table 4.1).

Studies on the colonisation of root systems by inoculated bacterial strains provide useful information on the ability of newly introduced cells to migrate

Table 4.1 *Colonisation of wheat roots by rhizosphere bacteria*

Diurnal segment	Root activity	Bacterial population
1–2	Root tip. Cell lysis commences.	Low level
3–4	Emergence of lateral roots and increasing lysis of epidermal and cortex cells	Rapid increase to 10^5 CFU/cm of root
5–6	Lysis reduced	Fall in bacterial growth
7–8	Second increase in cell lysis	Second increase in bacterial growth

Adapted from Van Vuurde & Schippers (1980)

throughout the rhizosphere and multiply, and have enabled an assessment of colonisation ability in terms of motility, chemotaxis and growth rate, all of which might be expected to be important. Previous studies have shown that there is a chemical attraction of rhizobacteria to exudates and mucilage from both roots and germinating seeds (De Weger *et al.*, 1988), suggesting that chemotactic movement of bacteria in the rhizosphere may be an important factor in root colonisation.

In general, the results obtained from inoculation experiments have been somewhat contradictory, and no general principles have emerged. For example, studies by Scher *et al.* (1987) on the colonisation of soybean roots by *Pseudomonas putida* and *Pseudomonas fluorescens* have demonstrated no positive correlation between the extent of rhizosphere colonisation and motility, chemotaxis (to root exudates) or generation time. Tn5 non-motile mutants of *P. putida* colonised roots as well as the motile parent, indicating that motility in this case was not essential for colonisation. In contrast to these results, De Weger *et al.* (1988) reported that Tn5 mutants of *P. fluorescens* without flagella were impaired in their ability to colonise roots of potato, indicating that motility was important in this particular instance.

Accumulation of *Agrobacterium tumefaciens* in the rhizosphere of host plants is an important precursor to wound invasion, and depends both on bacterial motility and chemotactic behaviour (Shaw, 1991). In this bacterium, non-motile mutants are impaired in root colonising ability. Chemotaxis within the general rhizosphere probably involves attraction to plant saccharides, to which all agrobacteria show a highly sensitive response.

Specificity of rhizosphere associations

The size and composition of the rhizosphere microflora is to a large extent plant-dependent, with various authors reporting qualitative and quantitative differences in microbial populations in relation to particular plant species and varieties.

A recent example of this is provided by the studies of Glandorf *et al.* (1990) on the fluorescent pseudomonad populations of various crop plants, including potato, wheat, tomato and carnation. These authors demonstrated the presence of distinct bacterial populations in reference to lipopolysaccharide (LPS) and cell envelope protein (CEP) characteristics. The majority of LPS and CEP patterns detected for a particular plant species were not observed with any other species, demonstrating a clear crop specificity for certain pseudomonad populations. The reason for these differences may lie in the specificity of the root exudates, which are important in the establishment and maintenance of rhizosphere populations in young plants. Some of these exudates may act as agglutinins, leading to specific root surface adhesion of certain bacteria.

Pathogen populations and disease occurrence

Populations of rhizosphere bacteria for a single plant may be determined by gently shaking the complete root system to remove soil, placing it in a known volume of buffer solution or sterile water then shaking vigorously for a given period of time prior to dilution plating on appropriate agar media. Bacterial levels are generally expressed as colony forming units (CFU) per cm of root length or per gram of soil.

Loper *et al.* (1984) have shown that considerable variation exists in rhizosphere populations in both glasshouse and field crops, and that the distribution of bacteria in the rhizosphere follows a log-normal pattern. In this respect the rhizosphere shows a close similarity to the phyllosphere.

In some cases, the presence of phytopathogenic bacteria in soil shows a close correlation with disease occurrence. This has been shown with corky root disease of lettuce in California (Van Bruggen *et al.*, 1988), where the causal agent could be isolated from the root surface of diseased but not healthy plants.

The general soil–water environment

Survival of pathogenic bacteria away from the plant surface

Although the long-term survival of soil pathogens is thought to occur normally in association with the host, some bacteria – such as *Agrobacterium tumefaciens*, *Pseudomonas solanacearum* and certain species of *Erwinia* – have been shown to occur throughout the general soil–water environment. Soft rot erwinias, in particular, have been recovered from irrigation water, rivers and oceans, where they constitute a continuous source of inoculum for future infection of host plants.

Erwinia carotovora subsp. *carotovora* and *E.c.* subsp. *atroseptica*, causal agents of aerial stem rot and blackleg of potatoes respectively, commonly occur in irrigated fields of temperate zones, including parts of the USA (Cappaert *et al.*, 1988). The presence of *E.c.* subsp. *carotovora* in soil water is particularly interesting, since this pathogen is serologically diverse, and the relative importance of waterborne bacteria as a potential inoculum source could be determined by tracing individual strains using serological techniques. In Oregon and Colorado, Cappaert *et al.* (1988) demonstrated seven different serological strains in irrigation water early in the season. These particular strains were subsequently detected on plant leaflets and later in diseased stems, suggesting a clear time course of bacterial migration and infection, and a direct role for irrigation water in the development of aerial stem rot of potato.

Influence of competition and predation on population dynamics of soil bacteria

Competition with other soil bacteria, and predation by bacteria and protozoa, may have important effects on the growth and spread of both naturally occurring and introduced bacterial species. The roles of competition and predation in controlling soil bacterial populations have been investigated from two main aspects:

1. Biocontrol of soil pathogenic bacteria, where introduction of predatory or antibiotic-secreting antagonists may be used to control soilborne disease. This situation leads to consideration of the effects of introduced bacteria on indigenous (pathogen) populations (discussed in Chapter 9).

2. Factors affecting the survival of bacteria that have been added to improve crop yields, including nitrogen-fixing bacteria and plant growth-promoting rhizobacteria. In contrast to the previous situation, the effect of indigenous or already established microbial populations on recently inoculated bacteria should be considered.

Recent studies by Postma and van Veen (1990) on the population dynamics of *Rhizobium leguminosarum* under a range of biotic soil conditions provide useful insights into the general effects of competition and predation on introduced bacteria.

The general inhibitory effect of other soil micro-organisms was shown by the fact that inoculation of the bacterium into non-sterilised soils resulted in a drastic decrease in numbers, while in sterilised soils the bacterial population rose to high levels. Pre-inoculation of sterilised soil with various bacterial isolates, or with flagellate protozoa that had been fed on *Rhizobium*, led to a significant reduction in the *Rhizobium* population only when these factors were combined, suggesting that a multiplicity of factors may be involved. The synergistic effect of competition and predation may arise due to the fact that fewer rhizobial bacteria are particle-associated in a soil that is already colonised by other bacteria, thus making the *Rhizobium* cells more accessible to predation by flagellates. Under natural conditions, predation also occurs by other protozoa (particularly amoebae), bacteria in the genus *Bdellovibrio* (see Chapter 9) and bacteriophages.

Environmental interactions at the micro-level

Previous sections have considered the general occurrence of plant pathogenic bacteria on plant surfaces in the aerial and subterranean environments. In both of these situations the bacteria exhibit complex biological, physical and chemical interactions at the level of the microenvironment. Some of the biological interactions will be considered later in the chapter on biological control, where competition between pathogenic and antagonistic bacteria is discussed in the context of disease reduction. Relatively little is known about the physical and chemical interactions between plant pathogenic bacteria and their microenvironments, but two aspects which have been studied in some detail are the effects of bacteria on the freezing characteristics of plant foliage (ice nucleation activity) and the effect of bacteria on the surrounding level of Fe^{3+} ions (siderophore production).

Ice nucleation activity

Some strains of bacteria on aerial plant surfaces, including both pathogenic and non-pathogenic organisms, have the capacity to act as sites of ice nucleation or rapid freezing at temperatures above $-10°C$. These ice nucleation-active (INA^+) bacteria prevent supercooling, and may lead to:

1. Ice damage to the plant (i.e. they increase plant sensitivity to subzero temperatures).
2. Entry and multiplication of pathogenic bacteria in ice-damaged tissues.

Laboratory tests for ice nucleation

Various laboratory tests have been developed to screen for bacterial ice nucleation activity. In general, these may either be carried out on bacterial isolates in suspension (droplet test) or on phylloplane bacteria *in situ* (tube nucleation test).

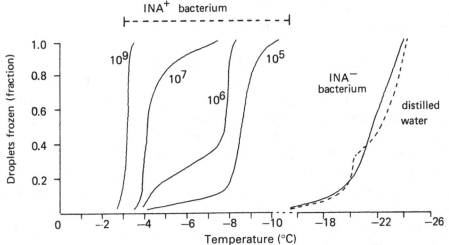

Fig. 4.3 Freezing spectra of cell suspensions of an ice nucleating (INA⁺) strain of *Erwinia herbicola*, a non-ice nucleating (INA⁻) strain and distilled water. Population densities of the INA⁺ bacterium range from 10^5 to 10^9 cells per ml. (Adapted from Lindow *et al.*, 1982.)

In the droplet test, initially developed by Lindow *et al.* (1982), 10 droplets of bacterial suspension are placed on a hydrophobic surface (e.g. paraffin film) at a specified freezing temperature and the number of droplets freezing within a three-minute period noted. The proportion of droplets frozen at decreasing temperatures can then be plotted (freezing spectrum) to reveal whether the bacterial strain is ice nucleation-active (INA⁺) or non-ice nucleating (INA⁻) at temperatures below −2°C. In the examples of freezing spectra illustrated in Fig. 4.3, the INA⁻ strain showed a closely similar spectrum to distilled water (freezing over the range −18 to −24°C), while the INA⁺ strain showed freezing activity over the range −2 to −10°C. This figure also demonstrates that the freezing activity of a suspension of INA⁺ bacteria depends on the population level, both in terms of the 'average freezing temperature' (i.e. where 50% of the droplets are frozen) and the shape of the freezing spectrum.

At a particular temperature, the number of active ice nuclei per unit volume of bacterial suspension (N_t) can be calculated using the equation of Vali (1971), subsequently modified by Lindow (1990), where

$$N_t = [\ln\frac{N_0}{N_0 - N_f}].V^{-1}$$

and

N_0 = total number of droplets tested from a given dilution of bacterial suspension

N_f = number of droplets that had frozen
V = volume of individual droplets used

Within a population of bacteria, the number of cells that directly cause freezing through ice nucleation is very low. In the case of an INA$^+$ strain of *Erwinia herbicola*, for example, Lindow *et al.* (1982) obtained a nucleation frequency at $-2.6°C$ of 10^{-8} nuclei per bacterial cell, which is equivalent to one active cell in 10^8 if nucleation sites occur singly.

The presence of INA$^+$ bacteria on leaf surfaces, and their effect on the temperature at which freezing occurs, can be monitored *in situ* using the tube nucleation test of Hirano *et al.* (1985); this is a rapid procedure for use with detached whole leaves. In this test, a number of sterile test tubes containing sterile phosphate buffer are initially screened for absence of non-specific nuclei by cooling at $-10°C$ for 2 to 3 hours. Tubes that do not freeze are then warmed to room temperature, and individual leaves to be tested are placed in a tube, chilled to $0°C$, then tested for freezing at -2, -4, -6 and $-10°C$ for periods of 90 minutes. Leaves which show ice nucleation activity may then be shaken with the buffer, and the frequency of INA$^+$ bacteria determined by the droplet test.

Bacterial species with INA$^+$ activity

Out of the many thousands of bacterial isolates that have now been tested for ice nucleation activity, only six species of bacteria have been shown to contain ice nucleating strains: *Erwinia herbicola*, *Erwinia ananas*, *Pseudomonas syringae*, *Pseudomonas viridiflava*, *Pseudomonas fluorescens* and *Xanthomonas campestris*. The occurrence of the INA$^+$ phenotype within these species varies considerably, and shows clear distinctions in relation to pathovar status. In *P. syringae*, for example, while many pathovars (e.g. pv. *syringae*, *pisi*, *lachrymans* and *coronafaciens*) have INA$^+$ activity, pv. *tomato* and subsp. *savastanoi* do not appear to have any ice nucleating strains. In *X. campestris*, only pv. *translucens* has so far been demonstrated to have any ice nucleating ability. These differences between pathovars can be readily determined by simple ice nucleation tests, and have considerable taxonomic potential.

Although ice nucleating strains within the above species comprise both pathogenic and non-pathogenic bacteria, they all occur predominantly as epiphytic (phylloplane) cells, and it is this feature which has been pre-eminent in the evolution of the INA$^+$ phenotype. Recent studies on phylloplane bacteria by Lindow *et al.* (1982) and Azad & Schaad (1988) have shown both nucleating and non-nucleating (INA$^-$) strains within the major leaf surface bacteria *P. syringae* and *E. herbicola*.

The INA$^+$ phenotype arises primarily due to the production of specific ice nucleating proteins, and is determined by the single genes *ina*Z (*Pseudomonas syringae*), *ina*W (*Pseudomonas fluorescens*) and *ice*E (*Erwinia herbicola*) as described in Chapter 9. Recent studies (Turner *et al.*, 1991) indicate that this protein is complexed with other molecules, forming various types of ice nucleation site. In decreasing order of ice nucleation activity, these are:

1. Protein–mannan–phosphatidyl inositol complex. The phospholipid phosphatidyl inositol (PI) and PI synthase occur at consistently higher levels in INA$^+$ compared to INA$^-$ strains, where the PI appears to link the ice nucleating protein to the cell membrane.
2. Protein–sugar complexes. In these the ice nucleating protein is complexed with mannose or glucosamine, but not phospholipid.
3. Ice nucleation protein linked to a few mannan residues, partly embedded in the outer cell membrane.

Other factors, such as ionic composition, might also be important, though recent X-ray microanalytical studies on the elemental composition of INA$^+$ and INA$^-$ bacteria indicate that no consistent differences occur in relation to major cation levels (particularly K$^+$), suggesting that differences in freezing point are not related to differences in internal molarity (Sigee & Hodson, 1993).

Factors affecting expression of ice nucleating activity

The ability to promote ice nucleation is not an inherent characteristic of every bacterial cell within a species or pathovar that has the ice nucleation phenotype. Considerable variation exists between populations of different strains within these taxonomic units, and, as noted previously, also between individual cells within a single population.

Ice nucleation ability also varies with environmental conditions and the physiological state of the bacteria (Lindow, 1990). The culture media used to grow cells for *in vitro* assay are important. Bacteria grown in liquid culture do not generally express ice nucleation as clearly as cells grown on solid media, and expression is also more efficient on media containing polyalcohols such as mannitol, sorbitol and glycerol. The temperature at which bacteria were grown prior to testing is also of major importance, with many INA$^+$ strains showing a marked decrease in nucleation activity if grown above 24°C. All of these environmental aspects suggest that the metabolic state of the bacteria is fundamental in modifying the expression of this phenotype. This is also indicated by the fact that log phase cells in both cultures have a much lower nucleation frequency than cells in early and late stationary phase.

Ice damage to plants

The ability of INA$^+$ bacteria to nucleate ice above $-10°C$ (and in some cases as high as $-1.5°C$) means that they are among the most important naturally occurring ice nucleation factors in the environment.

Ice damage to plant tissue during freezing is caused by the formation and growth of ice crystals around nucleation sites, leading to extensive physical and biochemical disruption of the cells. Most frost-sensitive plants can withstand mild frost, since they are able to supercool, and do not have active ice nuclei above $-5°C$. Epiphytic populations of INA$^+$ bacteria may limit supercooling by promoting ice formation at temperatures of -2 to $-5°C$.

Much of the evidence implicating epiphytic INA$^+$ bacteria as causal agents of frost injury comes from experimental studies on surface-inoculated bacteria. For example, Lindow *et al.* (1982) showed that separate application of a wide range of INA$^-$ bacterial strains to corn leaves did not result in any significant frost injury (compared with controls) at $-5°C$, while application of a range of INA$^+$ bacteria led to significant frost damage at $-5°C$ in a high proportion of the test plants.

The potential role of INA$^+$ bacteria in the initiation of ice crystal formation and the development of frost damage is also shown by the fact that:

(i) In experimental systems, frost damage is directly related to the number of ice crystal nuclei or the log of the number of INA$^+$ bacteria on the plant parts at the moment of freezing. Azad and Schaad (1988) demonstrated a direct relationship between the level of INA$^+$ bacteria (*Xanthomonas campestris* pv. *translucens* and *Pseudomonas syringae* pv. *syringae*) applied to leaves of wheat and bean and the level of frost damage, when exposed to $-3°C$ for 15 minutes (Fig. 4.4).

(ii) Chemical or biological treatments which reduce the number of epiphytic INA$^+$ bacteria also reduce the number of ice nuclei at the plant surface, and reduce the level of frost injury. Spray inoculation of potato plants in the field with non-nucleating (Ice$^-$) deletion mutants of *Pseudomonas syringae* leads to an 8-fold decrease in the level of INA$^+$ bacteria, and an 80% reduction in frost injury caused by radiation frosts (Lindow *et al.*, 1988).

In spite of evidence that INA$^+$ bacteria may be involved in causing frost injury to crops in certain experimental situations, their relative importance under natural conditions is not always clear. Other factors, such as the physiological ability of the plant to withstand supercooling and the occurrence of intrinsic ice nucleation sites may also be important. However, in some cases, the presence of INA$^+$ bacteria does appear to be significant. For example, studies in the USA by Marshall (1988) on different cultivars of oat

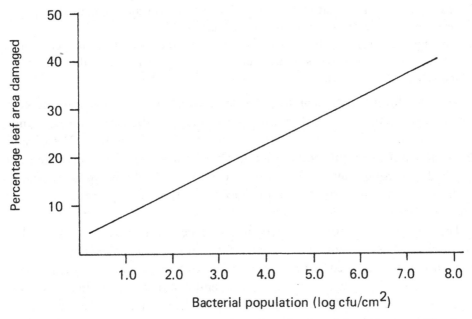

Fig. 4.4 Relationship between bacterial inoculum level and frost damage. The data have been obtained following the application of suspensions of ice nucleating *Xanthomonas campestris* and *Pseudomonas syringae* to the surface of bean leaves at different population levels. (Adapted from Azad & Schaad, 1988.)

have shown a direct correlation between frost damage and the level of ice nucleating bacteria, with the winter-tender cultivar Florida 501 supporting higher populations of INA+ bacteria than hardier cultivars. In other situations, the presence of INA+ bacteria is not considered of prime importance. In the south-east USA, for example, where INA+ bacteria have been detected in apple and peach orchards during spring, it seems unlikely that they are a major factor in causing frost damage to these crops since they occur at levels that are too low to be effective, and other ice nucleation sites – such as woody tissue – may be more important (Olive & McCarter, 1988).

Siderophore production

Siderophores are iron-chelating compounds produced by bacteria, and are important in the uptake of iron from the surrounding microenvironment.

General features of siderophores

The activity of siderophores is characterised by:

1. Synthesis and secretion only under iron-limiting conditions.

2. A specific and high affinity for the ferric ion. This is typically complexed with the siderophore via catechol or hydroxamate ligands.
3. An important role in the uptake of Fe^{3+} into the bacterial cell, permitting siderophore-producing cells to grow under conditions which limit the growth of non-siderophore bacteria.

The need for a specific iron acquisition system partly arises due to the key requirement of this micronutrient for the general metabolism of the cell, and partly due to the low levels that are present in the surroundings. These may occur due to the precipitation (under aerobic conditions) of free ferric ions as oxy-hydroxy complexes, and also due to removal by high affinity organic chelating compounds that are produced by the host plant and other micro-organisms.

The role of siderophores in the complex process of iron retrieval from the environment has been intensively investigated in *Escherichia coli* (Ecker *et al.*, 1986) and is summarised in Fig. 4.5. In addition to siderophores, this process also involves the activity of surface receptor molecules, reductases and digestive enzymes, and is tightly regulated. Although the production and control of siderophore production has not so far been investigated in such detail in phytopathogenic bacteria, it is clear that a similar system operates. In *Erwinia chrysanthemi*, for example, iron retrieval is mediated by the catechol-type siderophore chrysobactin and outer membrane transport proteins. A cluster of genes encoding chrysobactin transport and biosynthetic functions has been cloned and expressed in *Escherichia coli* (Franza and Expert, 1991). The genetic control of siderophore production is discussed further in Chapter 8.

The production of siderophores is important not only for the nutrition of the producing bacterium, but also has wider implications in terms of host–pathogen interactions and biocontrol. The production of these compounds may be important for bacterial survival *in planta*, particularly during early stages of infection, enabling the pathogen to multiply and interact within the host. Their potential role in biocontrol arises at the plant surface, where siderophore production may be important in competition with other naturally occurring pathogens, saprophytes and artificially applied antagonists (described in Chapter 9).

The ability of particular strains of bacteria to produce siderophores can be demonstrated *in vitro* by their ability to grow on an iron deficient medium, such as medium B of King *et al.* (1954). This is shown in Fig. 3.2, where phylloplane isolates are growing under iron-limiting conditions (see Fig. 3.2*a*) and are producing a yellow-green fluorescent siderophore (Fig. 3.2*b*). The structure of this pigment – which is typical of a wide range of fluorescent

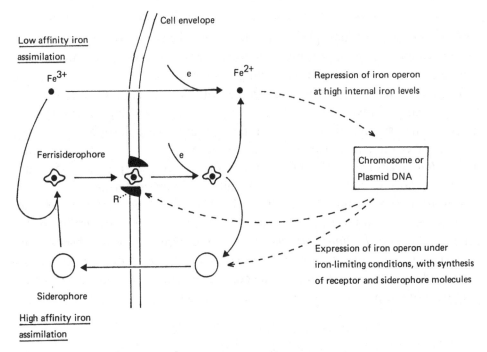

Fig. 4.5 General diagram to show the role of siderophores in iron assimilation by Gram-negative bacteria. Expression of an 'iron operon' under iron-limiting conditions results in the intracellular synthesis of both siderophore and surface receptor (R) molecules. The siderophore is secreted to the outside of the cell, where there is high-affinity binding to ferric ions and subsequent receptor-mediated uptake of the conjugate (ferrisiderophore) into the cell. Subsequent reduction of the ferric ion results in dissociation of the ferrisiderophore complex (due to low affinity of the siderophore for Fe^{2+}), with recycling of the siderophore and uptake of the Fe^{2+} into the metabolic pool. Under conditions of high (non-limiting) external Fe^{3+}, the cation is taken directly into the cell (low affinity assimilation), where supra-threshold levels lead to repression of the iron operon and resulting inhibition of siderophore and receptor synthesis. (Adapted from Leong, 1986.)

pseudomonads (including *Pseudomonas fluorescens* and *P. syringae*) – is illustrated in Fig. 3.3. Non-fluorescent pigments produced by plant pathogenic bacteria such as *Agrobacterium tumefaciens* and *Erwinia carotovora* (see Chapter 3) also have siderophore activity.

Production of siderophores by a particular strain of bacterium *in vitro* does not imply that the bacterium normally produces these molecules while present on the plant surface. It is generally considered, however, that siderophore production may be important in both the rhizosphere and phylloplane environment.

Siderophore production in the rhizosphere

The production of siderophores in the rhizosphere has been investigated mainly in reference to saprophytic bacteria, particularly where the bacterium is antagonistic to fungal and bacterial pathogens.

The importance of siderophore production by wild-type strains of these bacteria can be investigated by comparison with non-fluorescent (Flu⁻) mutants that specifically lack the ability to produce siderophores. Loper (1988), for example, has isolated a siderophore-producing strain of *Pseudomonas fluorescens* from cotton rhizosphere soil, raised transposon (Tn5) Flu⁻ mutants, and compared parental and mutant strains in terms of both soil colonisation and antagonistic ability. The results showed that the parental strain differed from the Flu⁻ mutants in being an effective antagonist to the fungal pathogen *Pythium ultimum*, but no differences were observed in bacterial growth rates or population sizes in soil. This suggests that even where siderophores are being produced, factors other than iron deficiency may be more important in limiting bacterial growth.

Siderophore production by phylloplane bacteria

A number of phylloplane bacteria, including phytopathogenic strains of *Pseudomonas syringae*, are able to produce fluorescent siderophore pigment. There are three main speculative roles for this pigment:

1. Under conditions of iron deficiency on the leaf surface, the siderophore may be important in the uptake of Fe^{3+} ions, and determine the growth and virulence of the phytopathogen.
2. In line with their *in vitro* antagonistic activity, the production of siderophores may be important in the antagonistic activity of the phytopathogenic bacteria on the leaf surface.
3. As a chromophore, the pigment may protect bacterial cells exposed to incident radiation by the absorption of both ultraviolet and visible light.

Critical experiments, comparing siderophore and non-siderophore producing bacteria on leaf surfaces, have so far generally failed to substantiate the above suggestions.

The importance of siderophore production to bacterial growth on the phylloplane has been studied by Loper & Lindow (1987), who compared the epiphytic characteristics of a siderophore-producing (wild-type) strain and a non-siderophore producing (mutant) strain of *Pseudomonas syringae* pv. *syringae* on bean leaves. This pathogenic bacterium is frequently present as an epiphyte on bean, causing brown spot disease. *In vitro* studies showed that the wild-type strain produced a fluorescent pigment and was able to grow on iron-deficient medium (King's B, supplemented with EDDA), while the non-fluorescent (Flu$^-$) strain was not able to grow on the deficient medium. The Flu$^-$ strain was also more sensitive to ultra-violet irradiation in culture compared to the wild-type strain. Application of the two strains to bean leaves, followed by monitoring of populations and disease incidence, showed that epiphytic growth rates, stationary epiphytic populations and the development of disease (brown spot) symptoms did not significantly differ between the two strains. These results provide no evidence for fluorescent siderophore production *in situ*, and suggest that the ability to produce siderophores does not contribute to the survival, growth or pathogenicity of the phytopathogen on the leaf surface.

The role of siderophore production by epiphytic bacteria in relation to their interactions with other micro-organisms has also been investigated, and is described more fully in Chapter 9. Although it is clear from *in vitro* experiments that siderophores may be important in interactions with fungi (McCracken & Swinburne, 1979) and other bacteria under laboratory conditions, there is no direct evidence that these compounds are involved in bacterial interactions on the leaf surface.

Studies by Lindow (1988) on competition between ice nucleation active (INA$^+$) and non-active (INA$^-$) bacteria have shown that there is no correlation between *in vitro* antibiosis (including siderophore activity) and antagonistic activity on the leaf surface. Only 58% of epiphytes antagonistic to INA$^+$ strains of *Pseudomonas syringae* and *Erwinia herbicola* on the leaf surface produced inhibitory compounds *in vitro*. Furthermore, elimination of the ability to produce siderophores by those strains showing this activity *in vitro* had no effect on their antagonistic activity on the leaf surface. This was demonstrated by the fact that frost injury to corn plants (which is an index of the level of naturally occurring INA$^+$ epiphytes) treated with mutant INA$^-$ antagonists deficient in siderophore production did not differ significantly from corn treated with parental strains from which the mutants were derived.

Bacterial associations with invertebrates and their importance in disease transmission

Many examples have been reported of bacterial plant pathogens being found in association with invertebrates which appear to be potential agents (vectors) for the transmission of disease. Of these, at least 60 bacterial diseases have been reported as being transmitted by insects (Harrison *et al.*, 1980), which comprise the major group of invertebrates involved.

There have been relatively few reports of plant pathogenic bacterial associations with other invertebrates though undoubtedly this must occur. One example of this is the association of bacteria in the genus *Clavibacter* with the nematode *Anguina* (Bird, 1985). In this relationship there appears to be a specific recognition between nematode species causing gall formation in particular grasses and the corresponding *Clavibacter* pathogen for that plant host.

With many plant diseases, a number of different insect species have been implicated as potential vectors, and it is often difficult to establish which organisms, if any, are important in disease epidemiology. In the case of fireblight, for example, over a hundred insect species have been implicated in the dissemination of bacteria (Van der Zwet & Keil, 1979), and the specific importance of particular insects in the spread of *Erwinia amylovora* is difficult to assess. Examples of some important bacterial diseases and their insect vectors are shown in Table 4.2.

General roles of vector in disease transmission

Vectors are involved in the transmission of disease in a number of specific ways, including pick-up of pathogen, dissemination, pathogen survival, inoculation onto the host plant and provision of wounds for bacterial entry and invasion.

All of these factors may be important in a particular disease, as shown, for example, by the role of insect vectors in the spread of fireblight (Fig. 4.1). *Erwinia amylovora* overwinters in cankers (on trunks and branches of infected trees) which produce ooze (primary inoculum) in early spring. Various insects, including aphids and Hemiptera, transmit bacteria to flowers and shoots, infection of which leads to secondary inoculum. Flowers represent the major site of inoculum increase in spring and also the major site of infection. Floral transmission is largely carried out by bees, which carry bacteria from the major epiphytic population on the stigma to both the nectary (main site of infection) and the anther (localised region of inoculum increase and contamination of pollen) of other blossoms (see Fig. 5.4). Infected flowers may lead to shoot infection either by invasion through the plant or by secondary insect transmission from flowers to shoots. Infection of shoots via damaged tissue caused by contaminated feeding insects – such as the bug *Lygus lineolaris* or the aphid *Aphis pomi* – also contributes to vegetative infection. Insect transmission of *Erwinia amylovora* occurs by attachment of the bacterium to the vector's surface, where it appears to be capable of survival for a number of days.

Table 4.2 *Role of insect vectors in the spread of some bacterial diseases*

Bacterium	Disease	Insect vector	Insect–plant relationship	Insect–pathogen relationship
A. NON-SPECIFIC TRANSMISSION				
Pseudomonas syringae pv. *phaseolicola*	Halo blight of bean	Thrips (*Heliothrips femoralis*)	Greenhouse pest	Casual association
Clavibacter michiganensis subsp. *sepedonicum*	Ring rot of potatoes	Grasshopper, (*Melonoplus*)	Feed on diseased plants	Casual association
Erwinia amylovora	Fireblight of Rosaceae	Many insects e.g. Bug (*Lygus lineolaris*)	Members of phyllosphere fauna	Casual association
B. SPECIFIC TRANSMISSION				
(1) Plant–vector specificity				
Xanthomonas juglandis	Walnut blight	Walnut mite (*Eriophyes tristriatus*)	Obligate parasite	Casual association
Erwinia amylovora	Fireblight of Rosaceae	Bee (*Apis mellifera*)	Between flowers of one species	Casual association
(2) Pathogen–vector specificity				
Pseudomonas melophthora	Bacterial rot of apples	Apple fly (*Rhagoletis pomonella*)	Egg deposition and larval food source	Specific transmission
Erwinia tracheiphila	Bacterial wilt of cucurbits	Cucumber beetles (*Diabrotica* sp.)	Not restricted to Cucurbita	Intestinal resident (overwinter)
Erwinia stewartii	Bacterial wilt of corn	Corn beetles (*Chaetocneme* sp.)	Feed on a range of grasses	Intestinal resident (overwinter)
Erwinia carotovora	Potato blackleg	Seedcorn maggot (*Hylemyia cilicrura*)	Life cycle involves potato	Intestinal resident
Pseudomonas savastanoi	Olive-knot disease	Olive fly (*Dacus oleae*)	Larval damage	Intestinal resident

Information mainly from Carter (1973).

Specificity of disease transmission

Disease transmission by a particular vector may be either non-specific or specific, depending on the nature of the interrelationships between the vector, pathogen and plant. In some cases, spread of a particular disease may involve both types of transmission (e.g. fireblight, see Table 4.2).

In non-specific transmission, the vector is not restricted in its association with the plant, and there is no interdependence between vector and pathogen. Dissemination of disease depends on a random encounter between the bacterium on the plant surface and the casual invertebrate visitor to the infected plant, and must be regarded as relatively non-specialised. Examples of non-specific transmission of bacterial disease, which is believed to be widely occurring, are given in Table 4.2.

In specific disease transmission, vector relationships are more defined, either with the host plant (plant–vector association) or with the pathogen (bacteria–vector interdependence).

Plant–vector association

The specificity and efficiency with which a vector spreads disease frequently relates to the specificity of association between vector and food plant, and the mechanism of feeding.

Where vectors have a restricted food plant range, the pick-up and transmission of pathogen will be limited. In the case of *Erwinia amylovora*, for example, the restriction of nectar collection by bees to flowers of a particular species means that dissemination from a particular host (e.g. hawthorn) will be limited to certain bees in the area, and will only occur to other flowers of the same species.

The close proximity of alternative food plants may also be important in the infection of crop plants by vectors. In California, for example, the incidence of Pierce's disease of grapevine is particularly high near to permanent water sources where weed hosts of the pathogen (*Xylella fastidiosa*) and its sharpshooter leafhopper vectors occur (Hopkins, 1988).

The mechanism of vector feeding is particularly important in the insect transmission of xylem-limited or phloem-limited pathogens. The majority of leafhoppers feed in the phloem and are ideally suited to the transmission of the spiroplasmas that are restricted to this tissue. The precise feeding habits on a particular host appear to determine the spread of disease, with efficient transmission of pathogen only occurring when the phloem is reached in a high proportion of feeds and when there is minimal damage to the phloem cells (Markham & Townsend, 1979).

Bacteria–vector relationships

In some situations, there is an intimate association between the bacterial phytopathogen and the vector, which can be a major factor in the spread of disease. In the case of bacterial rot of apple caused by *Pseudomonas melophthora*, for example, disease transmission is closely associated with the apply fly (*Rhagoletis pomonella*). Two stages of the insect are involved, the adult fly deposits contaminated eggs below the cuticle of apple and the larvae spread the bacteria through the fruit during burrowing. Within the fruit, the presence of bacteria appears to be necessary for normal development of larvae, since sterile larvae die without increasing in size (Allen *et al.*, 1934). There is some evidence that bacteria provide essential amino acids – specifically methionine and cystine – which are not present at sufficiently high level in uncontaminated apple tissue (Miyazaki *et al.*, 1968).

Transmission of *Pseudomonas melophthora* by the apple fly occurs on the insect surface, but in other cases, transport of the pathogen occurs within the body of the vector. This is probably important both for bacterial survival and multiplication, and has been reported for a number of plant diseases (Table 4.2).

Protection of the bacterium inside the vector may be essential for overwintering in temperate climates and occurs, for example, with *Erwinia stewartii*, causal agent of bacterial wilt of corn. This pathogen survives in the alimentary canal of the corn flea beetle (*Chaetocnema pulicaria*) and the toothed flea beetle (*C. denticulata*) (Harrison *et al.*, 1980). Both insect vectors feed on a range of plants, and wild host plants may be particularly important in spring in bridging the gap between insect emergence from hibernation and the appearance of freshly growing corn. The presence of phytopathogen has been demonstrated in a range of commonly occurring grass species (with and without symptoms) at this time of year (Carter, 1973).

The presence of phytopathogenic bacteria within insect hosts shows a gradation in terms of the degree of association between the two organisms:

1. In the simplest situation, the pathogen is taken up into the gut of the adult insect and is able to multiply and survive only in the imago (e.g. *Erwinia tracheiphila* and *E. stewartii*).
2. A closer association occurs where pathogenic bacteria are taken up by the larva and survive through metamorphosis to the adult stage, from which they pass to the egg during oviposition (e.g. association of *Erwinia carotovora* with *Hylemyia cilicrura*).
3. At a more advanced level, pathogenic bacteria are associated with the vector for the whole of its life cycle, with the addition of further specialised aspects.

The association of *Pseudomonas savastanoi* subsp. *savastanoi* with the olive fly is particularly important in the spread of olive-knot disease, and shows a number of specialised features. High levels of bacterial infestation of the alimentary tract of the vector occur at both pupal and imago stages of the life cycle. In the pupa, these are present in a specialised diverticulum of the oesophagus, while in the adult insect there are foci of bacterial population in sac-like evaginations at the junction of the anal tract and the vagina, which have a common opening to the exterior (Carter, 1973). Bacteria pass out of the adult insect both with the faeces and also in association with eggs. Pathogen cells on the egg surface subsequently pass through the micropyle into the developing larva, which is contaminated prior to emergence. Transmission by *Dacus oleae* thus shows two specialisations which distinguish it from previous pathogen–vector relationships: specific migration of bacteria into the developing egg (rather than remaining as casual surface contamination) and the development of gut diverticula to provide a reservoir of bacterial multiplication.

Although close pathogen–vector associations have clearly evolved in the examples quoted above, in no case (with the apparent exception of *Pseudomonas melophthora*) is the association absolutely specific and in no case does transmission of the pathogen entirely depend on the presence of the vector. An entirely different situation occurs in the fastidious prokaryotes, where the pathogen shows a much greater dependency on the insect vector.

Vectors of fastidious prokaryotes These are of particular importance to the pathogen since they are not only involved in the transmission of disease, but are probably the only site of survival of the pathogen outside the host plant. Insect vectors of these pathogens are restricted to the order Homoptera, which includes leafhoppers (family Cicadellidae), planthoppers (Cixiidae) and psyllids (Psyllidae) and the order Hemiptera, which includes leaf bugs (family Piesmidae).

The pathogens, which include mollicutes, phloem-limited bacteria and xylem-limited bacteria, can be separated into two groups in relation to the nature of the vector association and transmission:

Non-circulative prokaryotes This group, comprising the Gram-negative xylem-limited bacteria, does not pass into the insect blood circulation, but is taken up into the insect foregut, where it remains (Chiykowski, 1987). Because of this alimentary location, transmission by insects is possible shortly after acquisition from infected plants. Insects that become infected in the adult stage remain infective for life, while those that were infected as nymphs lose their infectivity after moulting. Further evidence of non-circulatory transmission was shown by the fact that injection of vectors with a suspension of the pathogen did not lead to infective insects. Recent studies (Chiykowski, 1987) have shown a polar attachment of these bacteria to the wall of the insect foregut, possibly involving fimbriae. Extracellular material that is associated with the bacterial colonies may protect the bacteria against digestive fluid.

Circulative prokaryotes These pathogens, including the Mollicutes (MLOs and spiroplasmas) and phloem-limited bacteria, enter the blood circulation of the insect where they are able to survive in the long term. The transport of spiroplasmas by, for example, leafhoppers involves a close adaptation of plant pathogen to insect vector, with circulation and multiplication of the organism inside the body of the insect (Markham & Townsend, 1979). Acquisition of the pathogen from infected plants during feeding is followed by passage through the gut wall into the haemolymph, circulation through the vector, then entry into the salivary glands, from where they are transmitted to new plants. The time between pathogen uptake by the insect and infection of the salivary glands (normally 2–6 weeks) constitutes an incubation period, during which the organisms multiply in the vector and insects are unable to transmit the disease.

The close association between mollicutes and insect vectors has clearly evolved over a long period of time, and the specificity of transmission of particular mollicutes by particular species or groups of insect reflects a close phylogenetic relationship (Markham & Townsend, 1979).

References

Allen T. C., Pinckard J. A. & Riker A. J. (1934). Frequent association of *Phytomonas melophthora* with various stages in the life cycle of the apple maggot *Rhagoletis pomonella*. *Phytopathology* **24**, 228–38.

Azad H. & Schaad N. W. (1988). The relationship of *Xanthomonas campestris* pv. *translucens* to frost and the effect of frost on black chaff in wheat. *Phytopathology* **78**, 95–100.

Bashan Y., Okon Y. & Henis Y. (1981). Scanning electron and light microscopy of infection and symptom development in tomato leaves infected with *Pseudomonas tomato*. *Physiol. Plant Pathol.* **19**, 139–44.

Bird A. F. (1985). The nature of the adhesion of *Corynebacterium rathayi* to the cuticle of the infective larva of *Anguina agrostis*. *Int. J. Parasitol.* **15**, 301–8.

Cappaert M. R., Powelson M. L., Franc G. D. & Harrison M. D. (1988). Irrigation water as a source of inoculum of soft rot erwinias for aerial stem rot of potatoes. *Phytopathology* **78**, 1668–72.

Carter W. (1973). *Insects in Relation to Plant Disease*. New York: Wiley.

Chiykowski L. N. (1987). Vector relationships of xylem- and phloem-limited fastidious prokaryotes. In *Plant Pathogenic Bacteria*. Proceedings of the 6th International Conference on Plant Pathogenic Bacteria, ed. E. L. Civerolo, A.

Collmer, R. E. Davis & A. G. Gillaspie. Dordrecht: Martinus Nijhoff, pp. 313–20.

Cooksey D. A. (1988). Reduction of infection by *Pseudomonas syringae* pv. *tomato* using a nonpathogenic, Cu-resistant strain combined with a Cu bactericide. *Phytopathology* **78**, 601–3.

De Weger L. A., van Avendonk J. C., Recourt K., van der Hofstad G., Weisbeck P. J. & Lugtenberg B. (1988). Siderophore-mediated uptake of Fe^{3+} by the plant growth-stimulating *Pseudomonas putida* strain WC3358 and by other rhizosphere micro-organisms. *J. Bacteriol.* **170**, 4693–8.

Ecker D. J., Matzanke B. F. & Raymond K. N. (1986). Recognition and transport of ferric enterobactin in *Escherichia coli*. *J. Bacteriol.* **167**, 666–73.

Fokkema N. J. & Van den Heuvel J. (1986). *Microbiology of the Phyllosphere*. Cambridge: Cambridge University Press.

Franza T. & Expert D. (1991). The virulence-associated chrysobactin iron uptake system of *Erwinia chrysanthemi* 3937 involves an operon encoding transport and biosynthetic functions. *J. Bacteriol.* **173**, 6874–81.

Glandorf D. C., Bakker P. A. & Schippers B. (1990). Crop specificity of fluorescent pseudomonads and the involvement of root agglutinins. In *Biotic Interactions and Soilborne Diseases*, ed. A. B. Beemster, G. J. Bollen, M. Gerlagh, M. A. Ruissen, B. Schippers & A. Tempel. Amsterdam: Elsevier, pp. 365–9.

Gottwald T. R., McGuire R. G. & Garran S. (1988). Asiatic citrus canker: Spatial and temporal spread in simulated new planting situations in Argentina. *Phytopathology* **78**, 739–45.

Gross M. (1990). Siderophores and fluorescent pigments. In *Methods in Phytobacteriology*. Ed. Z. Klement, K. Rudolph & D. C. Sands. Budapest: Akademiai Kiado, pp. 434–8.

Harrison M. D., Brewer J. W. & Merrill L. D. (1980). Insect involvement in the transmission of bacterial pathogens. In *Vectors of Plant Pathogens*, ed. K. F. Harris, K. Marramorosch. New York: Academic Press, pp. 201–92.

Hirano S. S., Nordheim E. V., Arny D. C. & Upper C. D. (1982). Lognormal distribution of epiphytic bacterial populations on leaf surfaces. *Appl. Env. Microbiol.* **44**, 695–700.

Hirano S. S., Baker L. S. & Upper C. D. (1985). Ice nucleation temperature of individual leaves in relation to population sizes of ice nucleation active bacteria and frost injury. *Plant Physiol.* **77**, 259–65.

Hopkins D. L. (1988). Natural hosts of *Xylella fastidiosa* in Florida. *Plant Dis.* **72**, 429–31.

King E. O., Ward M. K. & Raney D. E. (1954). Two simple media for the demonstration of pyocyanin and fluorescein. *J. Lab. Clin. Med.* **44**, 301–7.

Legard D. E. & Schwartz H. F. (1987). Sources and management of *Pseudomonas syringae* pv. *phaseolicola* and *P.s.* pv. *syringae* epiphytes on dry beans in Colorado. *Phytopathology* **77**, 1503–9.

Leong J. (1986). Siderophores: Their biochemistry and possible role in the biocontrol of plant pathogens. *Ann. Rev. Phytopathol.* **24**, 187–209.

Lindow S. E. (1988). Lack of correlation of in vitro antibiosis with antagonism of ice nucleation active bacteria on leaf surfaces by non-ice nucleation active bacteria. *Phytopathology* **78**, 444.

Lindow S. E. (1990). Bacterial ice nucleation activity. In *Methods in*

Phytobacteriology, ed. Z. Klement, K. Rudolph & D. C. Sands. Budapest: Akademiai Kiado, pp. 428–34.

Lindow S. E., Arny D. C. & Upper C. D. (1982). Bacterial ice nucleation: A factor in frost injury to plants. *Plant Physiol.* **70**, 1084–9.

Lindow S. E., Panopoulos N. J., Andersen G., Pierce C. & Lim G. (1988). Biological control of frost injury to potato with ice-deletion mutants of *Pseudomonas syringae* constructed *in vitro*. 5th International Congress on Plant Pathology, Kyoto, p. 93.

Loper J. E. (1988). Role of fluorescent siderophore production in biological control of *Pythium ultimum* by a *Pseudomonas fluorescens* strain. *Phytopathology* **78**, 166–72.

Loper J. E. & Lindow S. E. (1987). Lack of evidence for *in situ* fluorescent pigment production by *Pseudomonas syringae* pv. *syringae* on bean leaf surfaces. *Phytopathology* **77**, 1449–54.

Loper J. E., Suslow T. V. & Schroth M. N. (1984). Lognormal distribution of bacterial populations in the rhizosphere. *Phytopathology* **74**, 1454–60.

Malvick D. K. & Moore L. W. (1988). Population dynamics and diversity of *Pseudomonas syringae* on maple and pear trees and associated grasses. *Phytopathology* **78**, 1366–70.

Markham P. G. & Townsend R. (1979). Experimental vectors of spiroplasmas. In *Leafhopper Vectors and Plant Disease Agents*, ed. K. Maramorosch, K. F. Harris. New York: Academic Press, pp. 413–45.

Marshall D. (1988). A relationship between ice nucleation-active bacteria, frost damage and genotype in oats. *Phytopathology* **78**, 952–7.

McCracken A. R. & Swinburne T. R. (1979). Siderophores produced by saprophytic bacteria as stimulants of germination of conidia of *Colletotrichum musae*. *Physiol. Plant Pathol.* **15**, 331–40.

McInnes T. B., Gitaitis R. D., McCarter S. M., Jaworski C. A. & Phatak S. C. (1988). Airborne dispersal of bacteria in tomato and pepper transplant fields. *Plant Dis.* **72**, 575–9.

Miyazaki S., Boush G. M. & Baerwald R. J. (1968). Amino acid synthesis by *Pseudomonas melophthora*, bacterial symbiote of *Rhagoletis pomonella* (Diptera). *J. Insect Physiol.* **14**, 513–18.

Olive J. W. & McCarter S. M. (1988). Occurrence and nature of ice nucleation-active strains of *Pseudomonas syringae* on apple and peach trees of Georgia. *Plant Dis.* **72**, 837–43.

Postma J. & van Veen J. A. (1990). Influence of competition and predation on the population dynamics of bacteria introduced into soil. In *Biotic Interactions and Soilborne Diseases*, ed. A. B. R. Beemster, G. J. Bollen, M. Gerlagh, M. A. Ruissen, B. Schippers & A. Tempel. Amsterdam: Elsevier, pp. 35–40.

Roos I. M. M. & Hattingh M. J. (1987). Pathogenicity and numerical analysis of phenotypic features of *Pseudomonas syringae* strains isolated from deciduous fruit trees. *Phytopathology* **77**, 900–8.

Scher F. M., Kloepper J. W., Singleton C., Zaleska I. & Laliberte M. (1987). Colonisation of soybean roots by *Pseudomonas* and *Serratia* species: Relationship to bacterial motility, chemotaxis and generation time. *Phytopathology* **78**, 1055–9.

Shaw C. H. (1991). Swimming against the tide: Chemotaxis in *Agrobacterium*. *BioEssays* **13**, 25–9.

Sigee D. C. & Hodson N. (1993). X-ray microanalytical studies on plant pathogenic bacteria. In *X-ray Microanalysis in Biology: Experimental Techniques and Applications.* Cambridge: Cambridge University Press pp. 198–215.

Timmer L. W., Marois J. J. & Achor D. (1987). Growth and survival of xanthomonads under conditions nonconducive to disease development. *Phytopathology* **77**, 1341–5.

Turner M. A., Avellano F. & Kozloff L. M. (1991). Components of ice nucleation structures of bacteria. *J. Bacteriology* **173**, 6515–27.

Vali G. (1971). Quantitative evaluation of experimental results on the heterogeneous freezing nucleation of supercooled liquids. *J. Atmos. Sci.* **28**, 402–9.

Van Bruggen A. H. C., Grogan R. G., Bogdanoff C. P. & Waters C. M. (1988). Corky root of lettuce in California caused by a Gram-negative bacterium. *Phytopathology* **78**, 1139–45.

Van der Zwet T. & Keil H. K. (1979). Fire blight: A bacterial disease of rosaceous plants. Agriculture Handbook 510, Washington D.C.: U.S.D.A. Science and Education Administration.

Van Vuurde J. W. L. & Schippers B. (1980). Bacterial colonisation of seminal wheat roots. *Soil Biol. Biochem.* **12**, 559–65.

Wilson M., Sigee D. C. & Epton H. A. S. (1990). *Erwinia amylovora* infection of hawthorn blossom. III. The nectary. *J. Phytopathology* **128**, 62–74.

5

The infection process

Infection of plants by pathogenic bacteria can generally be considered in terms of three interrelated phases:

1. Population build-up, competition and migration of bacteria at the plant surface.
2. Bacterial entry into plant tissue.
3. Migration of bacteria within the plant to and from regions of multiplication.

Build-up and activity of epiphytic populations

Population level

The presence of epiphytic pathogens on host plants does not imply that disease will necessarily develop, and many cases have been reported where quite high levels of pathogenic bacteria were present on symptomless foliage. This has been noted, for example, for *Pseudomonas syringae* pathovars on red maple (Malvick & Moore, 1988) and snap beans (Legard & Schwartz, 1987) and for *Erwinia amylovora* on apple and pear blossom (see later).

In other situations, the presence of epiphytic bacteria does lead to disease development. This was initially noted by Crosse (1957), who reported the presence of *Pseudomonas syringae* pv. *mors-prunorum* as an epiphyte on cherry foliage, leading to canker formation. The relationship between epiphytic occurrence and disease development has subsequently been investigated for a wide range of bacterial pathogens by monitoring naturally occurring populations and carrying out experimental inoculations of plant surfaces. These studies have shown that plant infection and disease development depend on a number of factors, including the particular host–pathogen combination, critical environmental conditions, physiological stress of the host plant and the attainment of minimal threshold levels of the pathogen.

Naturally occurring populations

The importance of bacterial population build-up as a prelude to infection is shown by the studies of Lindeman *et al.* (1984) and Legard and Schwartz (1987) on the development of brown spot disease of bean, caused by *Pseudomonas syringae* pv. *syringae*. Epiphytic populations of the pathogen required a minimum threshold level of 10^4 cells/g leaf tissue to produce brown spot disease. For disease development (and prediction), the important factor is not mean population size, but the frequency with which populations of the pathogen exceed the critical 10^4 CFU level on individual symptomless leaflets.

The importance of epiphytic build-up is also shown by its use as a criterion in various disease forecasting models. In the prediction of fireblight outbreaks in California, for example, Miller and Schroth (1972) monitored the presence of *Erwinia amylovora* on apple and pear to correlate population development with weather conditions.

Experimental inoculation of bacteria

With a number of diseases, the relationship between epiphytic level and disease occurrence has also been demonstrated experimentally by spray inoculation of bacteria at different levels followed by disease monitoring. Recent studies by Whitesides and Spotts (1991) have shown that the incidence of blossom blast in pear, caused by *Pseudomonas syringae* pv. *syringae*, was proportional to the logarithm of the pathogen population under conditions of optimal temperature, moisture and bloom development stage. Similarly, with another floral pathogen, Norelli and Beer (1984) have demonstrated a significant relationship between inoculum level and disease occurrence after spraying apple and pear blossoms with different levels of *Erwinia amylovora*.

Bacterial competition

Plant pathogenic bacteria exist in competition with other pathogens and with saprophytes on the plant surface, involving complex interactions in terms of siderophore production, release of antibiotics and competition for nutrients. All of these factors are important for the induction of disease since they affect the distribution, micropopulation and activity of pathogenic bacteria at sites of entry into the plant. The production and possible role of siderophores is discussed in Chapter 4, and the other two aspects will be considered in the section on biological control (Chapter 9).

Migration of bacteria

Bacterial migration is important for the movement of bacterial cells to sites of entry on the plant surface (infection courts) and their passage through these into the internal regions of the plant. In some cases at least, the movement of bacteria towards the site of entry appears to be by chemotaxis, which combines the twin aspects of motility and perception of a chemical stimulus.

Measurement of motility and chemotaxis in vitro

The motility of plant pathogenic bacteria in culture can be determined in three main ways: microscopic examination, migration in semi-solid medium and capillary assay.

Microscopic examination Light microscope observation of bacterial suspensions provides direct information on motility within a cell population, and is best carried out using dark field or phase contrast microscopy on hanging drop preparations. Although electron microscopy cannot provide direct information on motility, negative stain preparations do give related information on the occurrence of flagella within the bacterial population.

Migration in semi-solid media Semi-solid media contain low concentrations of the gelling agent, giving them a sloppy consistency through which bacteria are able to move by flagellar activity. Inoculation of bacteria onto the surface of such a medium will lead either to a localised growth (non-motile or poorly motile cells) or to spread of the bacteria from the inoculation site (motile cells). The degree of motility can be quantified in relation to the increase in colony diameter over a defined time period, and was used by Bayot and Ries (1986) as a motility assay for strains of *Erwinia amylovora*.

Capillary assay Determination of motility by capillary assay is the most widely used approach to quantifying this bacterial characteristic (see, for example, Raymundo & Ries, 1980), and uses an apparatus similar in type to that shown in Fig. 5.1. In this technique, bacteria are able to migrate from suspension culture into the test medium contained in the capillary tube under defined physical conditions. The test medium may either contain chemical attractant (chemotaxis medium) for the measurement of chemotaxis, or it may be used simply to measure random motility (motility medium is similar to suspension medium but with no chemical attractant). The number of bacteria that have migrated into the capillary tube over a specified period of

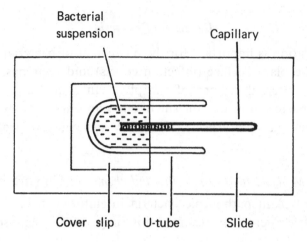

Fig. 5.1 Apparatus used in the assay of bacterial chemotaxis. (Redrawn from Adler, 1973.)

time is determined by making a viable population count from the test medium using standard dilution plating techniques.

The ability to measure bacterial motility and chemotaxis as distinct factors permits a separate assessment of their *in vitro* characteristics and their potential roles in the infection process.

Role of motility in plant infection

Although little detailed information is available on the factors that affect the motility of plant pathogenic bacteria, it is clear that the chemical and physical environment will be of considerable importance.

This has been shown by Hatterman and Ries (1989) in their *in vitro* studies on *Pseudomonas syringae* pv. *glycinea*, where levels of temperature, chelating agent, pH and exogenous energy sources have all been shown to affect motility *in vitro*. The effect of temperature is particularly interesting, since bacterial growth and motility have quite different optimal conditions (Fig. 5.2). The high level of *in vitro* motility at temperatures around 15°C is in line with the known *in vivo* characteristics of this pathogen, since it causes disease and spreads best under cool, wet conditions. Exogenous energy sources which increased motility in the presence of oxygen included asparagine and citrate (both chemo-attractants) and glycerol (a non-attractant).

The role of pathogen motility during leaf infection has been investigated by monitoring disease development in leaves of soybean following separate immersion in suspensions of motile and non-motile strains of *Pseudomonas syringae* pv. *glycinea*. Immersion over a time period of 0–5 minutes resulted in

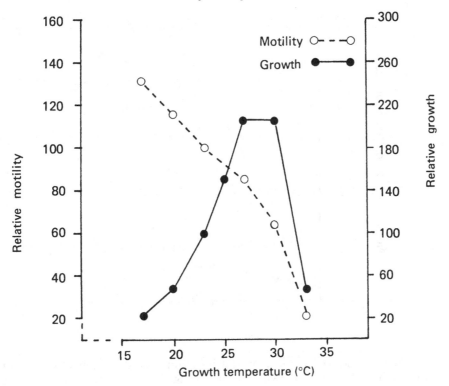

Fig. 5.2 Effect of incubation temperature on motility and growth of *Pseudomonas syringae* pv. *glycinea* cultured *in vitro*. Values for motility and growth are adjusted to an arbitrary value of 100% at 23°C. (Adapted from Hatterman and Ries, 1989.)

an increasing level of lesion development, while immersion in non-motile cells produced a constant low level of disease (Fig. 5.3). The importance of motility for leaf infection has also been shown for infection of bean by *P.s.* pv. *phaseolicola* (Panopoulos & Schroth, 1974), who reported that motile strains of the pathogen caused up to 12 times as many lesions as the non-motile strains.

The role of motility in floral infection has been demonstrated by Bayot & Ries (1986), who showed that greater infection of apple blossoms occurred after inoculation with a motile strain of *Erwinia amylovora* compared with a non-motile strain.

Role of chemotaxis in plant infection

Evidence for a possible involvement of chemotaxis in plant infection is largely based on laboratory experiments which show that a number of phytopathogenic bacteria exhibit a positive *in vitro* response to plant extracts, exudates or organic chemicals, as summarised in Table 5.1. The extent to which these *in*

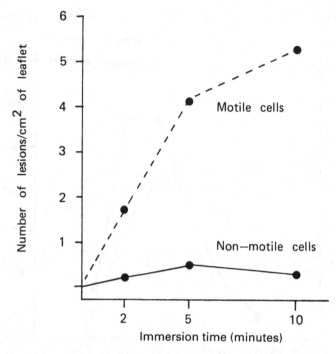

Fig. 5.3 Effect of bacterial motility on leaf infection. The development of necrotic lesions in soybean leaflets is shown with increasing time of immersion in bacterial suspension, for motile and non-motile strains of *Pseudomonas syringae* pv. *glycinea*. (Adapted from Hatterman and Ries, 1989.)

vitro responses operate at the plant surface or within plant tissue has yet to be determined, but the studies show that many plant pathogenic bacteria have the potential for *in vivo* chemotaxis during infection. This may be important for entry of phytopathogenic bacteria both at the root surface (e.g. *Agrobacterium tumefaciens*) or on aerial surfaces (e.g. *Erwinia amylovora*).

Agrobacterium tumefaciens: *chemotaxis and infection at the root surface* The chemotactic activity of this soil-inhabiting bacterium is an important aspect of rhizosphere colonisation, wound infection and host plant transformation (see Chapter 8, Fig. 8.14). On the basis of *in vitro* chemotactic responses (Shaw, 1991), root exudates can be divided into three main categories: strong attractants (requiring the presence of the Ti plasmid for chemotaxis to occur), weak attractants (acting on bacteria with and without the Ti plasmid) and non-attractants (inactive with any strains of the pathogen).

Strong attractants, which are also important in genetic induction of the *vir* locus, include the phenolic compounds acetosyringone, sinapinic acid and syringic acid. The presence of the Ti plasmid is required because chemotaxis is

Table 5.1. In vitro *chemotactic responses of plant pathogenic bacteria*

Pathogen	Chemical attractant	Reference
Pseudomonas syringae pv. *lachrymans*	Extracts from cucumber plants	Chet *et al.*, 1973
Pseudomonas syringae pv. *tomato*		Cuppels & Smith, 1984
Pseudomonas syringae pv. *phaseolicola*	Wounded bean leaves	Mulrean & Schroth, 1979
Xanthomonas campestris pv. *oryzae*	Exudates of susceptible rice	Feng & Kuo, 1975
Erwinia carotovora *Erwinia chrysanthemi*	Variety of sugars, amino acids & organic acids	Hsu & Huang, 1987
Erwinia amylovora	Apple nectar	Raymundo & Ries, 1980
Erwinia amylovora *Erwinia herbicola*	Variety of chemicals (see Table 5.2)	Klopmeyer & Ries, 1987
Agrobacterium tumefaciens	Variety of mono- and oligosaccharides. Phenolic compounds	Shaw, 1991

mediated by the bacterial proteins VirA and VirB, encoded by the *vir*A and *vir*G genes within the *vir* locus. These proteins are involved in reception and transduction of the chemical stimulus, respectively. VirA is an inner membrane chemoreceptor with an acetosyringone binding region in a transmembrane-spanning domain. Signal transduction subsequently involves phosphorylation of VirG as part of an extensive phosphorylation cascade, resulting in either chemotaxis or induction of the *vir* locus. The response of the bacterial cell to acetosyringone depends on the concentration of the chemo-attractant. At low levels it triggers chemotaxis, while at higher levels chemotaxis and motility are suppressed and induction of the *vir* locus occurs. Bacterial sensitivity to acetosyringone in terms of *vir* gene induction appears to be markedly enhanced by the presence of simple pyranose sugars in the root exudate (Mo & Gross, 1991).

Within the soil environment, the role of chemotaxis in root infection by *Agrobacterium tumefaciens* appears to involve a three stage process:

1. Root exudates attract a range of agrobacteria into the rhizosphere and are important in general root surface colonisation (Chapter 4).
2. Wound-induced phenolics attract Ti plasmid-carrying *Agrobacterium tumefaciens* around the wound locality.
3. At the wound site, high levels of phenolics block motility and chemotaxis (preventing exit from the wound surface) and promote induction of the *vir* locus (leading to Ti DNA transfer and plant cell transformation).

Table 5.2. *Chemotactic response of* Erwinia amylovora *and* Erwinia herbicola

Purified chemical or extract	Chemicals which produce a chemotactic response	
	Erwinia amylovora	*Erwinia herbicola*
Nectar extract	Organic acid fraction only	All components
Sugars	None	Many: fructose, glucose, sucrose
Aminoacids	Aspartate only	Many
Organic acids	Malate (& tartrate)	Malate (& fumarate)
Species of surface chemoreceptor	Few, possibly one	Many

Data from Klopmeyer & Ries (1987).

Erwinia amylovora: *chemotactic comparison with other phyllosphere bacteria in relation to aerial infection* Recent studies by Klopmeyer and Ries (1987) have shown that bacteria occupying the same microenvironment may differ very markedly in their response to chemo-attractants, and that this may relate to their mode of nutrition and plant association. The results, obtained for the phytopathogen *Erwinia amylovora* and the saprophyte *Erwinia herbicola*, are summarised in Table 5.2. These *in vitro* experiments show major differences between the two bacteria, with *Erwinia herbicola* having very broad chemotactic responses and *Erwinia amylovora* being highly specific. This difference in chemotactic behaviour probably relates to the variety of chemoreceptors at the cell surface, and may reflect a fundamental difference between pathogenic and saprophytic bacteria. In the case of the pathogen, there is a need only to locate a limited number of specific compounds emitted by the compatible host plant to effect entry. In contrast to this, the broad attractant range of the saprophyte reflects the need to locate a variety of non-specific food sources on a range of plant surfaces.

Bacterial entry into plant tissue

Plant pathogenic bacteria enter host tissue in two main ways: via wounds, or through natural openings, depending on the particular pathogen concerned and the nature of the disease (Table 5.3).

Table 5.3. *Typical mode of entry of phytopathogenic bacteria*

Bacterium	Wounds[a]	Natural openings[b]
Leaf spot diseases		
Pseudomonas syringae pv.		
phaseolicola	−	+s
tabaci	−	+s
pisi	+i	+s
Xanthomonas campestris pv.		
malvacearum	−	+s
oryzae	−	+s.h
translucens	+c	+s
vesicatoria	−	+s
Cankers and die-back		
Pseudomonas syringae pv.		
syringae	+c	+s
Erwinia amylovora	+	+n
Wilt diseases		
Pseudomonas solanacearum	+	−
Xanthomonas campestris pv.		
campestris	+	+h.s
Clavibacter michiganensis subsp.		
michiganensis	+	−
sepedonicum	+	−
Curtobacterium		
flaccumfaciens	+	−
Xylem- and phloem- limited prokaryotes	+i	
Soft rot diseases		
Erwinia carotovora "group"	+	+l
Pseudomonas marginalis	+	−
Tumour diseases		
Agrobacterium		
tumefaciens	+	−
rhizogenes	+	−
Rhodococcus fascians	+	−

[a] i, insects feeding; c, ice damage.
[b] h, hydathodes; n, nectarthodes; s, stomata; l, lenticels.

Adapted from Klement (1990).

Wound entry

Wounds to aerial or subterranean parts of plants are generally caused either by animals, physical environmental factors or cultural practices. In the first situation, the animal may act as vector for the pathogen, with wound entry as the final part of the transmission process. The spread of xylem-limited and phloem-limited fastidious prokaryotes (Mollicutes and phloem-limited bacteria) by leafhoppers provides a good example of this, with direct transmission of the pathogen to internal (vascular) sites of multiplication.

A range of environmental factors may be important in causing wound entry of plant pathogenic bacteria, including wind, hailstorm and frost damage. The first two factors have been cited as important in the spread of a number of pathogens, including *Erwinia amylovora* (Van der Zwet & Keil, 1979).

Frost damage

Frost damage is a major factor in bacterial infection in temperate zones, acting both within tissues and at the plant surface.

The potential importance of internal freezing is illustrated by its effect on the infection of grapevine by *Agrobacterium tumefaciens*. This is particularly serious in the case of spring frosts, when freezing of water in xylem vessels causes rupture of the vascular bundle and injury to large numbers of surrounding parenchyma cells. Bacterial cells are subsequently carried in large numbers within the xylem to the sites of injury (and potential infection), due to the strong root pressure that occurs at that time of year.

Frost injury is also important in the entry of bacteria at the plant surface, particularly where the bacterial pathogens themselves are acting as sites of ice nucleation (INA$^+$ bacteria). This is indicated by a higher incidence of disease (or increased disease severity) under freezing conditions when epiphytic INA$^+$ bacteria are present, and has been shown to occur with a number of foliar diseases, including black chaff of wheat, caused by *Xanthomonas campestris* pv. *translucens* (Azad and Schaad, 1988). Strains of this bacterium isolated from wheat leaf surfaces showed positive ice nucleation activity, and wheat plants that had been sprayed with INA$^+$ bacteria sustained greater frost damage than control plants. Field observations also showed that disease lesions developed more rapidly in plants exposed to frost than unexposed plants, and that frost damage and severity of disease were both directly related to the number of pathogenic INA$^+$ bacteria present on the leaf surfaces during the frost period. Although frost damage is not essential for the development of black chaff disease, the above experimental and field observations would suggest that it is an important contributory factor. This is further borne out by the increased incidence of disease in spring wheat and in crops grown at high elevation.

Ice formation is also important in certain non-foliar diseases such as blossom blast of pear, caused by *Pseudomonas syringae* pv. *syringae*. Experimental studies by Whitesides and Spotts (1991) using INA$^+$ bacteria have shown that reducing the ambient temperature at which inoculation takes place leads to a marked increase in infection at the time of freezing. This was indicated by an exothermal generation of heat and a small rise in temperature at the transition point of ice formation.

In view of the direct role of INA$^+$ pathogenic bacteria in promoting frost injury and disease occurrence, the use of ice nucleation tests to monitor bacterial populations for disease prediction would seem particularly appropriate. In line with this, Hirano *et al.* (1987) have successfully used the tube nucleation test to determine populations of INA$^+$ *Pseudomonas syringae* pv. *syringae* on bean leaves for disease prediction, and have also used the test to predict the incidence of halo blight on oats caused by *Pseudomonas syringae* pv. *coronafaciens*.

Entry through natural openings

Entry of pathogenic bacteria through natural openings such as stomata, hydathodes, lenticels, nectarthodes and the dehiscence zone of anthers is important in the initiation of many plant diseases, since bacteria are not able to pass directly across the epidermis (with its outer layer of cuticle) under normal circumstances. In some cases (e.g. stomata, lenticels), the route of entry is widely used by many different types of pathogen, while in others (e.g. infection of hydathodes, floral surfaces) the route of entry is more specialised and limited to particular bacteria.

Stomata

Stomata are generally considered to be of particular importance in the induction of foliar diseases where pathogens are normal residents of the leaf surface and probably enter via rain-splash or by their own motility (see previously).

The infection of tomato by *Xanthomonas campestris* pv. *vesicatoria*, causing leaf spot, is a good example of where this may occur. The potential importance of stomatal entry may be investigated by examining the relationship between disease occurrence and a range of stomatal factors such as stomatal opening, frequency and arrangement of stomata on the leaf surface. Although some of these factors are difficult to determine under natural conditions, they may be investigated experimentally by spray inoculation in the laboratory under controlled conditions. Using this approach, Ramos and Volin (1987) were able to show that pathogen entry and disease occurrence in tomato was directly related to:

1. Stomatal opening. Stomatal closure was induced either physiologically (by placing plants in the dark four hours prior to inoculation) or chemically, resulting in reduced infection.
2. Stomatal frequency. Different *Lycopersicon* species, hybrids and breeding lines with a wide variation in stomatal frequency were shown to vary proportionally in their resistance to spray infection.

Lenticels

Lenticels are derived from stomata during the development of periderm, and represent an important site of bacterial entry in woody stems and some subterranean organs. In potato, for example, they are a major site for the occurrence of *Erwinia carotovora* and infection by the pathogen under wet, anaerobic conditions (Perombelon & Lowe, 1975). In these circumstances, water is absorbed causing the lenticels to open, and oxygen lack brings about the release of cellular fluids from the parenchyma cells leading to a continuous aqueous phase from the lenticels into the cortex. Bacteria present in lenticels at the tuber surface are then able to migrate into the potato, with the development of soft rot.

In the soil environment generally, natural openings such as lenticels and growth cracks (caused by the emergence of lateral roots) are major sites of infection for rhizosphere pathogens such as *Xanthomonas campestris* pv. *oryzeae* (causing bacterial blight of rice).

Hydathodes

Invasion of the host plant via hydathodes leads directly to infection of the vascular bundles, and is particularly important where the pathogen is able to migrate in the xylem. This is one route for the entry and systemic spread of *Erwinia amylovora*, and is also important in the infection of rice by *Xanthomonas campestris* pv. *oryzeae* (Guo & Leach, 1989), which is the causal agent of bacterial blight, and is primarily a vascular pathogen. In rice, 10–15 hydathodes are present near to the edge of each leaf, and each hydathode has 10–20 water pores. These water pores are 2–4 times larger than leaf stomata, from which they are phylogenetically derived. Once bacteria have entered a water pore, they multiply in the parenchyma tissue (epithem) and subsequently invade the xylem vessels through the vascular pass.

Flower infection: specialised routes of bacterial entry

Flowers present a short-term site for epiphytic growth and infection in certain specific diseases, such as pear blast (caused by *Pseudomonas syringae* pv. *syringae*) and fireblight of rosaceae (*Erwinia amylovora*).

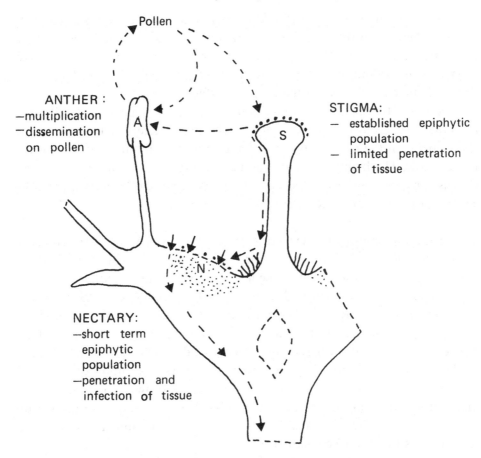

Fig. 5.4 Infection of hawthorn flower by *Erwinia amylovora*. The diagram shows the flower in half-section, and illustrates potential sites of bacterial multiplication, dissemination and tissue penetration. (See also Figs. 4.1 and 4.2.)

The epiphytic build-up and entry of *Erwinia amylovora* during blossom infection has been investigated particularly in flowers of apple, hawthorn and pear and involves three major sites: the stigma, anthers and nectary. Although the detailed course of infection depends on the type of flower and the environmental conditions, each of these sites has a particular role in the infection process, as summarised in Fig. 5.4.

Stigmatic surface The stigmatic surface is important as a site of major epiphytic inoculum, but is generally a region of limited tissue penetration.

In arid environments, the permanently moist stigmatic surface is the only part of the flower supporting an epiphytic population of the pathogen, which may reach levels of 10^7 CFU. Thomson (1986) observed high stigmatic

populations in apparently healthy blossoms under such conditions, suggesting that there is no resulting infection at low humidity. In contrast to this,
epiphytic build-up on the stigmatic surface in wet environments almost
always leads to floral infection, since the bacteria are washed off by rain or
dew and enter the flower via the nectary surface.

Infection of the stigma has been examined in detail in hawthorn (Wilson *et
al.*, 1989a), where penetration involves passage of bacteria into intercellular
spaces between the outer papillary cells, followed by limited invasion of the
tissue below. The highly unusual passage of bacteria through the outer layer
of cells can occur in this case because of secretory activity which ruptures the
covering of cuticle and allows entry of the pathogen.

The nectary In contrast to the stigma, the nectary appears to be a major site
of bacterial entry, but with limited epiphytic populations. This is particularly
the case in arid conditions, where the nectary surface may be either too dry or
be producing nectar of too high a concentration to support growth of the
pathogen (Ivanoff & Keit, 1941). This part of the flower probably supports
epiphytic populations of the bacterium under limited conditions of high
humidity or free water, when there is plenty of dilute nectar on the surface.

The entry of bacteria into the nectary occurs via modified stomatal
openings or nectarthodes (Wilson *et al.*, 1989b), as shown in Fig. 4.2*b*. High
humidity is important for infection not only in terms of population build-up
on the nectary surface, but also for bacterial migration and chemotaxis in a
surface moisture film (see Chapter 4).

The anther Under experimental conditions, inoculation of the anther surface
leads to bacterial entry and multiplication in the locule, with the production
of contaminated pollen (see Fig. 4.2*a*). Infection of this part of the flower
occurs mainly via the dehiscence zone, but some entry may also occur via
stomata, which in hawthorn are limited to a region near the filament insertion
(Wilson *et al.*, 1989c).

In hawthorn, anther infection does not appear to lead to direct infection of
the whole flower (via the filament). Its importance under natural conditions is
not clear, though it may have an indirect and short-term role in the rapid
boosting of inoculum levels coupled with vector dispersal of contaminated
pollen (Fig. 5.4).

Movement of bacteria within the plant

Movement of pathogenic bacteria within the plant is an important aspect of the systemic development of many plant diseases, and is particularly relevant to the spread of infection from primary sites of entry into the plant and from primary sites of population increase. Such movement of bacteria does not always occur, however, and in some cases – particularly where the pathogen is relatively non-aggressive – there may be little migration within infected tissue away from the initial site of infection. This has been reported recently, for example, in the infection of *Syngonium* by *Xanthomonas campestris* pv. *syngoni* (Dickey & Zumoff, 1987), where spread of the bacterial blight pathogen from one leaf to the next was almost entirely by external means.

In the majority of diseases, where migration of bacteria within the plant does occur, the process of pathogen movement has been little investigated. In general, there are three main mechanisms: bacterial motility, bulk flow of bacterial mass, and passive transport of bacteria through vascular tissue.

Bacterial motility

Bacterial motility may be important in the movement of bacteria (as individual cells) through intercellular spaces, possibly on the surface film of parenchyma cells.

Recent studies by Sigee and El-Masry (1989) on wildfire disease of tobacco have shown that bacterial cells directly isolated from infected leaf tissue are flagellate, suggesting that the pathogen is motile *in planta*. The degree of flagellation varies with the growth phase of the pathogen, showing a marked increase during the exponential growth phase in the plant tissue (Fig. 5.5). If the spread of bacteria in the leaf mesophyll does relate to the presence of flagella and cell motility, then the results suggest that migration from the region of infiltration will be greatest during the phase of maximal population growth.

Just how important motility is for bacterial spread within infected tissue is not clear, and may vary with the physiological state of the plant, water availability and the particular host–pathogen combination. Recent studies by Hatterman and Ries (1989), comparing motile and non-motile strains of *Pseudomonas syringae* pv. *glycinea*, showed that although motility was important for bacterial entry into soybean leaf, it made no difference to disease development once bacteria were inside. Stab-inoculated motile and non-motile strains were equally pathogenic and produced identical symp-

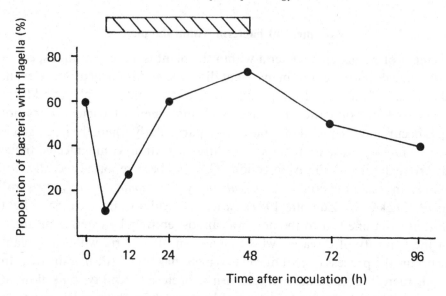

Fig. 5.5 Changes in bacterial flagellation during disease development. Changes in the percentage occurrence of bacterial cells with flagella are shown after inoculation of tobacco leaves with *Pseudomonas syringae* pv. *tabaci*. Cell counts were made by transmission electron microscope examination of filtered leaf macerates. The major phase of logarithmic growth within the inoculated tissue is shown by the bar. Counts of bacterial cells with just one, two or three flagella showed a similar trend to the total flagellar count. (Adapted from Sigee and El-Masry, 1989.)

toms, suggesting that bacterial motility inside the leaf was not important in this particular situation.

Bulk flow of bacterial mass

In some cases, pathogenic bacteria are present within intercellular spaces as a mass of cells embedded in a matrix of extracellular material. In this situation, the spread of the pathogen within the tissue appears to result from an increase in cell population and extracellular material, which causes a positive pressure flow of the bacterial mass throughout the intercellular spaces. The spread of *Erwinia amylovora* through the receptacle of infected blossom appears to occur in this manner, with the development of lysigenous cavities due to the pressure exerted by the increasing mass of bacteria.

Passive transport of bacteria through vascular tissue

A number of plant pathogenic prokaryotes are thought to move passively within the phloem or xylem of the vascular tissue, including both fastidious prokaryotes (mycoplasma-like organisms, spiroplasmas, xylem-limited bacteria) and non-fastidious prokaryotes such as *Agrobacterium tumefaciens* and *Erwinia amylovora*.

In the case of fastidious prokaryotes, where the organism is entirely adapted to a vascular existence *in planta*, the presence of flagella may be of limited use under conditions of passive flow of the surrounding medium. In this context, it is interesting to note that the xylem-limited bacterial pathogen causing Sumatra disease of cloves appears to be entirely non-flagellate (Bennett *et al.*, 1987).

The passive transport of *Agrobacterium tumefaciens* in the transpiration stream of vines has been implicated in the transfer of bacteria from subterranean galls to aerial parts of the plant (Tarbah & Goodman, 1987), with the development of secondary tumours at some distance from the primary root infection. Evidence for such rapid, long-distance transfer has been obtained by root inoculation of antibiotic-resistant labelled bacteria, followed by recovery in aerial regions within a short space of time. The use of light and electron microscopy of stem tissue has demonstrated the exclusive presence of the pathogen in the lumen of xylem vessels.

In contrast, *Erwinia amylovora* is transported in both phloem sieve tubes and xylem vessels (Van der Zwet & Keil, 1979). Evidence for this comes from the demonstration of rapid movement of pathogen cells over long distances in symptomless stems and petioles, together with direct electron microscope examination of phloem and xylem tissue. The long-distance transport of these pathogen cells in xylem is prevented when the lumen of the vessels is blocked, as occurs when bacterial aggregates are formed during the hypersensitive reaction.

References

Adler J. (1973). A method for measuring chemotaxis and use of the method to determine optimum conditions for chemotaxis in *Escherichia coli*. *J. Gen. Microbiol.* **74**, 77–91.

Azad H. & Schaad N. W. (1988). The relationship of *Xanthomonas campestris* pv. *translucens* to frost and the effect of frost on black chaff development in wheat. *Phytopathology* **78**, 95–100.

Bayot R. G., & Ries S. M. (1986). Role of motility in apple blossom infection by *Erwinia amylovora* and studies of fireblight control with attractant and repellant compounds. *Phytopathology* **76**, 441–5.

Bennett C. P., Jones P. & Hunt P. (1987). Isolation, culture and ultrastructure of a xylem-limited bacterium associated with Sumatra disease of cloves. *Plant Pathol* **36**, 45–52.

Chet I., Zilberstein Y. & Henis Y. (1973). Chemotaxis of *Pseudomonas lacrymans* to plant extracts and to water droplets collected from the leaf surfaces of resistant and susceptible plants. *Physiol. Plant Pathol.* **3**, 473–9.

Crosse J. E. (1957). Bacterial canker of stone fruits. III. Inoculum concentration and time of inoculation in relation to leaf scar infection of cherry. *Ann. Appl. Biol.* **45**, 19–35.

Cuppels D. A. & Smith W. (1984). Chemotaxis by *Pseudomonas syringae* pv. *tomato. Phytopathology* **74**, 798 (abstract).

Dickey R. S. & Zumoff C. H. (1987). Bacterial leaf blight of *Syngonium* caused by a pathovar of *Xanthomonas campestris. Phytopathology* **77**, 1257–62.

Feng T. Y. & Kuo T. T. (1975). Bacterial leaf blight of rice plants. VI. Chemotactic responses of *Xanthomonas oryzeae* to water droplets exudated from water pores on the leaf of rice plants. *Biol. Abstr.* **61**, 4739.

Guo A., & Leach J. E. (1989). Examination of rice hydathode water pores exposed to *Xanthomonas campestris* pv. *oryzeae. Phytopathology* **79**, 433–6.

Hatterman D. R. & Ries S. M. (1989). Motility of *Pseudomonas syringae* pv. *glycinea* and its role in infection. *Phytopathology* **79**, 284–9.

Hirano S. S., Rouse D. I. & Upper C. D. (1987). Bacterial ice nucleation as a predictor of bacterial brown spot disease on snap beans. *Phytopathology* **77**, 1078–84.

Hsu S. T. & Huang H. (1987). Chemotaxes of *Erwinia carotovora* subsp. *carotovora* and *Erwinia chrysanthemi*. In *Proceedings of the 6th International Conference on Plant Pathogenic Bacteria*, ed. E. L. Civerolo, A. Collmer, R. E. Davis & A. G. Gillaspie. Dordrecht: Martinus Nijhoff.

Ivanoff S. S. & Keit G. W. (1941). Relation of nectar concentration to growth of *Erwinia amylovora* and fireblight infection of apple and pear blossoms. *J. Agric. Res.* **62**, 732–43.

Klement Z. (1990). Inoculation of plant tissues. In *Methods in Phytobacteriology*, ed. Z. Klement, K. Rudolph & D. C. Sands. Budapest, Akademiai Kiado, pp. 96–102.

Klopmeyer M. J. & Ries S. M. (1987). Motility and chemotaxis of *Erwinia herbicola* and its effect on *Erwinia amylovora. Phytopathology* **77**, 909–14.

Legard D. E. & Schwartz H. F. (1987). Sources and management of *Pseudomonas syringae* pv. *phaseolicola* and *Pseudomonas syringae* pv. *syringae* epiphytes on dry beans in Colorado. *Phytopathology* **77**, 1503–9.

Lindeman J., Arny D. C. & Upper C. D. (1984). Use of an apparent threshold population of *Pseudomonas syringae* to predict incidence and severity of brown spot of bean. *Phytopathology* **74**, 1334–9.

Malvick D. K. & Moore L. W. (1988). Population dynamics and diversity of *Pseudomonas syringae* on maple and pear trees and associated grasses. *Phytopathology* **78**, 1366–70.

Miller T. D. & Schroth M. N. (1972). Monitoring of the epiphytic population of *Erwinia amylovora* on pear with a selective medium. *Phytopathology* **62**, 1175–82.

Mo Y. & Gross D. C. (1991). Plant signal molecules activate the *syr*B gene, which is required for syringomycin production by *Pseudomonas syringae* pv. *syringae. J. Bacteriol.* **173**, 5784–92.

Mulrean E. M. & Schroth M. N. (1979). *In vitro* and *in vivo* chemotaxis by *Pseudomonas phaseolicola. Phytopathology* **69**, 1039.

Norelli J. R. & Beer S. V. (1984). Factors affecting the development of fireblight blossom infection. *Acta Hort.* **151**, 37–39.

Panopoulos N. J. & Schroth M. N. (1974). Role of flagellar motility in the invasion of bean leaves by *Pseudomonas phaseolicola*. *Phytopathology* **64**, 1389–97.

Perombelon M. C. M. & Lowe R. (1975). Studies on the initiation of bacterial soft rot in potato tubers. *Potato Res.* **18**, 64–82.

Ramos L. J. & Volin R. B. (1987). Role of stomatal opening and frequency on infection of *Lycopersicon* spp. by *Xanthomonas campestris* pv. *vesicatoria*. *Phytopathology* **77**, 1311–17.

Raymundo A. K. & Ries S. M. (1980). Chemotaxis of *Erwinia amylovora*. *Phytopathology* **70**, 1066–9.

Shaw C. H. (1991). Swimming against the tide: Chemotaxis in *Agrobacterium*. *BioEssays* **13**, 25–9.

Sigee D. C. & El-Masry M. H. (1989). Changes in cell size and flagellation in the phytopathogen *Pseudomonas syringae* pv. *tabaci* cultured *in vitro* and *in planta*: A comparative electron microscope study. *Phytopathol. Z.* **125**, 217–30.

Tarbah F. & Goodman R. N. (1987). Systemic spread of *Agrobacterium tumefaciens* Biovar 3 in the vascular system of grapes. *Phytopathology* **77**, 915–20.

Thomson S. V. (1986). The role of the stigma in fireblight infection. *Phytopathology* **76**, 476–82.

Van der Zwet & Keil H. K. (1979). Fireblight: A bacterial disease of rosaceous plants. Agriculture handbook 510. Washington, D.C.: U.S.D.A. Science and Education Administration.

Whitesides S. K. & Spotts R. A. (1991). Induction of pear blossom blast caused by *Pseudomonas syringae* pv. *syringae*. *Plant Pathol* **40**, 118–27.

Wilson M., Epton H. A. S. & Sigee D. C. (1989a). *Erwinia amylovora* infection of hawthorn blossom. 2. The stigma. *Phytopathol. Z.* **127**, 15–28.

Wilson M., Sigee D. C. & Epton H. A. S. (1989b). *Erwinia amylovora* infection of hawthorn blossom. 3. The nectary. *Phytopathol. Z.* **128**, 62–74.

Wilson M., Sigee D. C. & Epton H. A. S. (1989c). *Erwinia amylovora* infection of hawthorn blossom. 1. The anther. *Phytopathol. Z.* **127**, 1–14.

6

Compatible and incompatible interactions: the hypersensitive response

The entry of bacteria into the plant during the infection process leads to various types of interaction, observable at the level of the whole plant, constituent tissues or individual cells. These interactions have been investigated experimentally by artificial infiltration of intact plants ('Inoculation of intact plants', this page) or by the use of *in vitro* systems (including micropropagates, excised organs and cell suspensions; see 'Use of *in vitro* systems', p. 132).

Inoculation of intact plants

The effect of different bacteria in determining the nature of the plant response was initially demonstrated by Klement *et al.* (1964), who artificially infiltrated leaves of tobacco with a range of bacterial species (Fig. 6.1) and observed three main types of result:

1. Hypersensitive reaction (HR): where there is typically a rapid death of the plant cells, with no spread of bacteria to surrounding tissues. This reaction was induced by a range of bacteria comprising various pathovars of *Pseudomonas syringae*.
2. Disease reaction: involving a delayed host cell response, with spread of bacteria to other parts of the plant. This reaction was induced by *Pseudomonas syringae* pv. *tabaci* and resulted in wildfire disease.
3. No observable reaction, after infiltration of the saprophytic bacterium *Pseudomonas fluorescens*.

The results obtained by Klement *et al.* with tobacco have been corroborated by subsequent work on other plants, each with their own particular pathogens. The major implications of these general results are:
   1. The interaction that occurs between bacteria and plant tissue depends

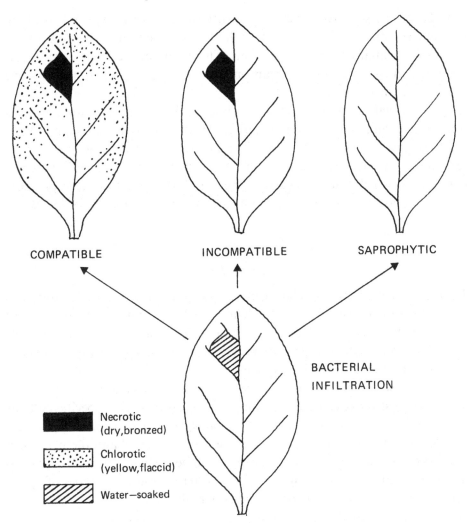

Fig. 6.1 Leaf response to bacterial infiltration. Interpretation of experiment by Klement *et al.* (1964). Local infiltration of leaf tissue leads to compatible, incompatible or saprophytic reactions (shown about 5 days after inoculation), depending on bacterial species or pathovar. (Adapted from Sigee, 1984.)

primarily on the type of bacterium and plant, rather than on other characteristics such as environmental, developmental or physiological features.

2. Ability to cause disease (pathogenicity) can be characterised either by the direct induction of disease (after infiltration into a host plant) or by the induction of a hypersensitive reaction (after infiltration into a non-host plant). The induction of a hypersensitive reaction does not occur with a non-phytopathogen, and provides a useful test (pathogenicity test) to determine whether a particular bacterial isolate has pathogenic potential.

3. The hypersensitive reaction constitutes a general type of localised resistant response which enables the plant to overcome the entry of a wide range of potentially harmful organisms. The ability of the plant to recognise and respond to these foreign organisms shows some analogy to the immune system of higher animals.

4. The specific ability of a plant pathogenic bacterium to induce either a disease reaction or a resistant reaction in plant tissue implies a common basis for these two types of response. Recent studies have shown that these reactions are controlled by a common gene system (*hrp* genes, see Chapter 9), mutation of which leads to a loss in the ability of the pathogen to promote either a disease or a resistant reaction (e.g. Atkinson & Baker, 1987a; Baker *et al.*, 1987).

Terminology

The fundamental distinction between hypersensitive and disease reactions has given rise to a range of terms to define the bacterium, plant and type of interaction that occurs in these two situations. These are summarised in Table 6.1, which relates to the general interactions occurring between bacterial species or pathovars and plant species.

Where a strain or race of a bacterium that is normally virulent on a particular host plant gives an incompatible response with a specific cultivar (e.g. *Pseudomonas syringae* pv. *phaseolicola* race 1 on *Phaseolus vulgaris* cv. Red Mexican) it is referred to as 'race-specific resistance' and the plant as a 'resistant host'. Although there is a distinction between race-specific and non-race-specific incompatibility in experimental terms, these two types of resistant reaction are regarded as being fundamentally similar.

Changes in bacterial population

Changes in bacterial population following infiltration into leaf tissue are normally determined as changes in viable count, which involves tissue maceration at defined time intervals followed by plating and colony counting at serial dilution. As an alternative approach, changes in total population may be determined by counting individual bacterial cells in 2 μm-thick sections of leaf tissue under phase-contrast microscopy (Sigee & Al-Issa, 1982). Whichever parameter is used, the multiplication of infiltrated bacteria shows a marked difference between compatible and incompatible interactions (Allington & Chamberlain, 1949; Diachun & Troutman, 1954) and is correlated with the onset and progression of mesophyll cell death.

Table 6.1 *General distinctions between disease and resistant reactions*

	Disease reaction	Resistant Reaction
Terminology: type of bacterial/ plant association		
Plant (A)	Host to bacterium (B)	Non-host to bacterium (D)[a]
Bacterium (B)	Pathogen to plant (A)	Non-pathogen to plant (C)[a]
Type of interaction		Hypersensitive reaction
	Compatible	Incompatible
	Homologous	Heterologous
Examples		
Tobacco	*Pseudomonas syringae* pv. *tabaci*	*Pseudomonas syringae* pv. *phaseolicola, Erwinia amylovora*, etc.
Bean	*Pseudomonas syringae* pv. *phaseolicola*	*Pseudomonas syringae* pv. *tabaci, Erwinia amylovora*, etc.
Characteristics		
Specifity	Highly specific (in case of leaf pathogens)	Typically non-specific (induced by wide range of phytopathogenic bacteria)
Plant reaction[a]	Delayed cell death	Rapid cell death
Bacterial multiplication[b]	Considerable increase over long time	Moderate rise of short duration
	Bacteria spread to other parts of plant	Limited to region of infiltration

[a] Bacteria B and D are both phytopathogens.
[b] Under experimental conditions, with high levels of bacterial inoculum.

The distinction between compatible and incompatible reactions is shown in Figs. 6.2 and 6.3. Infiltration of compatible *Pseudomonas syringae* pv. *tabaci* into tobacco leaf tissue leads to a prolonged multiplication (over 48 hours) of the pathogen, with an increase from 10^5 to 10^9 cells/ml (Fig. 6.2). In contrast, infiltration of incompatible *Pseudomonas syringae* pv. *pisi* results in a limited population increase, from 10^5 to 10^6 cells/ml, reaching a maximum at about 12 hours after infiltration (Fig. 6.3). In both cases, visible symptoms of necrosis appear at about the time of maximal population level.

Variation in the expression of incompatibility

The distinction that Klement *et al.* (1964) observed between compatible and incompatible bacteria is not always as clear-cut as their initial results would suggest. This is particularly the case in race-specific resistance, where the plant–pathogen combination is obviously very close to the susceptible inter-

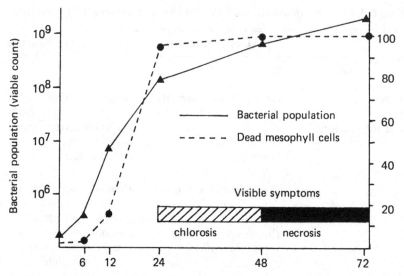

Fig. 6.2 Changes in bacterial population during the compatible reaction. Viable counts of *Pseudomonas syringae* pv. *tabaci* in tobacco leaf tissue are shown over a 72-hour period, after inoculation of bacterial suspension at 10^8 cells/ml. The percentage of dead mesophyll cells (determined under phase-contrast light microscopy) is also shown, together with the timing of visible symptoms. (Adapted from Sigee, 1984.)

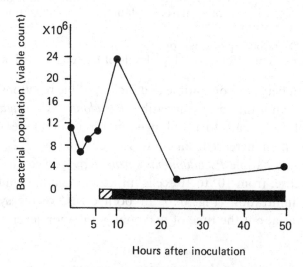

Fig. 6.3 Changes in bacterial population during the incompatible reaction. Viable counts of bacteria are shown after inoculation of tobacco leaf tissue with a suspension of *Pseudomonas syringae* pv. *pisi* at 10^8 cells/ml. The timing of visible symptoms is indicated as in Fig. 6.2. (Adapted from Sigee & Al-Issa, 1982.)

action. In this situation, the resistant response often appears intermediate in its characteristics, though the development of disease is genuinely prevented. For example, infiltration of *Pseudomonas syringae* pv. *tabaci* into leaves of *Nicotiana longiflora* (closely related to the bacterial host *Nicotiana tabacum*) results in a resistant response which is characterised by a relatively late death of the mesophyll cells and a build-up of bacterial population to quite high levels (Sigee, 1984). Similarly, Lyon & Wood (1976) have shown that infiltration of *P.s.* pv. *phaseolicola* race 1 into resistant bean leaves leads to a delay in electrolyte leakage and a greater bacterial multiplication compared with *P.s.* pv. *mors-prunorum*.

HR test for pathogenicity

Applicability

The ability of plant pathogenic bacteria to induce an HR in a non-host plant provides a useful test for pathogenicity. The test can be used for most of the major types of plant pathogenic bacteria, including bacteria that cause necrosis and vascular wilt diseases and some bacteria that cause soft rots (see Table 7.1).

Although the HR test is broadly applicable over a wide spectrum of bacteria, it does not work with opportunistic phytopathogens. In the grouping of green-fluorescent pseudomonads according to the LOPAT scheme (Lelliott *et al.*, 1966), for example, groups I, II and III – which are strongly pathogenic – give a clear HR reaction, while weakly pathogenic groups IV and V do not. Thus among the soft rot pseudomonads, *Pseudomonas viridiflava* (LOPAT group II) gives a positive HR but not *Pseudomonas marginalis* (LOPAT group IV). The latter is a pectolytic form of the saprophyte *Pseudomonas fluorescens*, and qualifies as an opportunistic pathogen. Some other phytopathogens also normally fail to induce an HR in incompatible plants, including soft rot erwinias plus tumour-inducing agrobacteria and *Pseudomonas savastanoi* subsp. *savastanoi*. The relationship between type of disease and HR induction is considered further in Chapter 7 (Table 7.1).

Experimental procedures

The HR test normally involves the infiltration of high levels (10^8–10^9 cells/ml) of actively-growing (log phase) inoculum into the intercellular spaces of a leaf, using a hypodermic syringe or spray infiltration. Tobacco plants are normally employed for this due to the ease of infiltrating the fleshy leaves, and the convenience of growing and maintaining the plants under laboratory conditions. Although tobacco gives a clear HR with most incompatible phyto-pathogens, it does not give a good reaction with xanthomonads, where other plants such as tomato and pepper are used in preference. In contrast to tobacco and other dicotyledons, monocots may give only a weak HR. Infiltration of temperate cereals with *Pseudomonas syringae* pv. *phaseolicola* or *P.s.* pv. *tabaci*, for example, leads to only a faint chlorosis.

Use of *in vitro* systems

Although interactions involving plant pathogenic bacteria have in the past frequently been tested in the glasshouse using intact plants, an increasing amount of research now involves the use of *in vitro* cultured material, including aseptically grown seedlings, micropropagated plants, excised organs such as leaves and fruits, excised discs or segments and cell suspensions.

Applications

Aseptically grown seedlings have been frequently used for pathogenicity testing. Daniels *et al.* (1987), for example, tested the pathogenicity of *Xanthomonas campestris* pv. *campestris* isolates by stab-inoculation of *Brassica campestris* seedlings and were able to screen hundreds of bacterial clones per day.

An example of a micropropagated plant is shown in Fig. 6.4. This type of material has proved useful not only for pathogenicity testing, but also for selection of resistant plants (e.g. to fireblight, Norelli *et al.*, 1988) and testing of biocontrol agents (Chapter 9).

Cell suspensions are being employed increasingly in studies involving plant pathogenic bacteria, and have been used by Baker *et al.* (1987), for example, to screen mutants of *Pseudomonas syringae* pv. *syringae* for their ability to generate a hypersensitive reaction with cultured tobacco cells, using a microtiter plate. Where bacteria have induced an HR in individual microtiter wells, the uptake of H^+ ions by the cultured cells results in a rise in pH and a colour change from yellow to purple. Using this technique, Baker *et al.* (1987) were able to screen 1600 mutants and isolate 9 which either failed or were only partly able to induce an HR.

Suitability of in vitro *systems*

The use of such *in vitro* systems has a number of general advantages over the use of whole plants, including the ability:

Fig. 6.4 Micropropagated plant of hawthorn (*Crataegus monogyna*), cultured under sterile conditions on an agar medium. Large numbers of these plants can be used for *in vitro* laboratory testing of bacterial pathogenicity.

1. To carry out a large number of tests (e.g. pathogenicity tests) under laboratory conditions. This can be particularly useful where a large number of bacterial mutants are being investigated.
2. To carry out tests on genetically uniform higher plant material.
3. To carry out experiments under precisely controlled and monitored chemical and physical microenvironmental conditions. Changes in extracellular pH during the hypersensitive reaction between bacteria and tobacco mesophyll cells, for example, can be determined much more readily in suspension cultures than in intact leaves.

One major disadvantage of using *in vitro* systems is that the plant material being used may be in a highly abnormal condition and situation. This is particularly true for cells cultured in suspension, which have little close resemblance to cells in intact tissue and which are being challenged with bacteria in an environment that is quite different from that inside the plant.

Induction of the hypersensitive reaction

Although most studies carried out on the hypersensitive reaction have involved leaf infiltration, resistance to pathogenic bacteria has also been investigated in other plant parts, including bean pods (Harper *et al.*, 1987) and potato tubers (Lyon, 1989). The appearance of the localised necrosis that

Fig. 6.5 Hypersensitive reaction in tobacco leaf inoculated with *Pseudomonas syringae* pv. *pisi* (10^8 cells/ml). The localised region of necrosis (arrow) appears pale and bronzed, and is shown 24 hours after bacterial infiltration.

occurs after inoculation of leaf tissue with incompatible bacteria is shown in Fig. 6.5. The changes that occur within the mesophyll tissue follow a well-defined time course and are influenced by a number of factors.

Time course of the hypersensitive reaction: induction and latent periods

A typical time course for the hypersensitive reaction (as occurring in tobaco leaf tissue) is shown in Fig. 6.6. Within this, the time between bacterial infiltration and symptom appearance (mesophyll cell death) can be divided into two distinct periods: the induction period (during which the effect of bacteria can be reversed by infiltration of antibiotics), and the latent period (during which antibiotics can inhibit bacterial growth but not the onset of

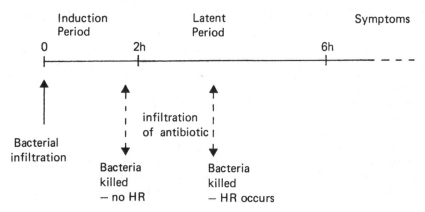

Fig. 6.6 Major phases of the hypersensitive reaction. The times given are appropriate for the rapid reaction that occurs following infiltration of tobacco leaves with *Pseudomonas syringae* pv. *pisi*.

HR). As far as mesophyll death is concerned, the induction period contains the sequence of events leading to irreversible cell change, while the latent period is the time during which irreversible change leads to the appearance of visible symptoms.

The duration of the induction period varies with the experimental situation and the particular bacterium–plant combination. For example, infiltration of tobacco leaves with incompatible bacteria leads to reported induction periods which vary from about 2 hours (Al-Issa & Sigee, 1982b, for *Pseudomonas syringae* pv. *pisi*) to 5 hours (Klement, 1982, for *Xanthomonas campestris* pv. *vesicatoria*). Studies by Harper *et al.* (1987) on induction of HR in bean leaves have shown that measurement of the induction time varies slightly with the antibiotic used, and that the induction period of specific race–cultivar combinations is longer than other incompatible situations. The time taken for HR necrosis to develop also varies with the particular incompatible combination and may be particularly long in the case of race-specific resistance (e.g. 14–20 hours for *P.s.* pv. *phaseolicola*-incompatible races in French bean).

Factors affecting induction and expression of HR

Infiltration of incompatible bacteria into plant tissue does not always lead to a visible hypersensitive response. Variations in relation to the particular plant–bacterial combination have already been noted ('Inoculation of intact plants', above), and other factors are also important, including environmental factors, age of the inoculated leaf, bacterial metabolic state, inoculum level and bacteria–mesophyll cell contact.

Table 6.2 *Antibiotic inhibition of HR-induction by* Pseudomonas syringae *pv.* pisi *in tobacco leaves*

Metabolic effect	Inhibition	Effective antibiotics
Cell wall synthesis[a]	−	
Membrane integrity	+	Cetrimide, polymixin B
DNA replication[a]	−	
Uptake of nucleotide analogue	+	Tubericidin
DNA degradation	+	Mitomycin-c
DNA template intercalcation	+	Ethidium bromide, proflavine, acriflavine
Inhibit RNA polymerase	+	Rifampycin streptovaricin
Protein synthesis	+	Chloramphenicol, kanamycin, tetracycline

[a] Antibiotics that were tested and failed to inhibit the induction of HR included ampicillin and bacitracin (inhibition of cell wall synthesis) and nalidixic acid and phenethanol (inhibition of DNA replication).
(Information from Sasser, 1982.)

Environmental factors

Temperature, light and humidity all influence the expression of HR in tobacco leaves. As the ambient temperature is raised from 16 °C to 28 °C the induction period becomes progressively shorter. Exposures to high temperature (37 °C) during the induction period lead to complete suppression of HR, possibly by inhibiting bacterial division (Sequeira, 1984). The effect of light is discussed below (p. 150).

Chemical aspects of the cellular environment may also be important. The presence of divalent cations such as Ca^{2+} in the inoculation medium appears to either increase or decrease the loss of electrolytes from the plant cells, depending on the particular plant–bacterium combination (Sequeira, 1984).

Bacterial metabolic state

The hypersensitive reaction is only induced where the infiltrated bacteria are viable and metabolically active (Klement & Goodman, 1967). Those aspects of bacterial metabolism which are important in the induction of the hypersensitive reaction have been investigated (Sasser, 1982) by applying metabolically specific antibiotics with the bacterial inoculum and observing the presence or absence of rapid leaf cell necrosis. The results obtained with a broad range of differential antibiotics (active against bacteria but not plant cells) are shown in Table 6.2. The differential nature of the compounds used was shown by the fact that bacterial mutants resistant to the specific antibiotic induced a hypersensitive reaction in the presence of the chemical, demonstrating that a direct effect on the plant cells was not involved.

The results show that certain aspects of bacterial metabolism such as cell wall synthesis and DNA replication are not directly involved in the induction of the hypersensitive reaction. Other aspects, however, are important, including the maintenance of cell membrane integrity, synthesis of RNA (blocked by DNA degradation, DNA template intercalation and inhibition of DNA-dependent RNA polymerase) and synthesis of proteins.

Recent studies by Somlyai *et al.* (1988) have shown that the kinetics of HR are directly correlated with the *in vitro* ability of the bacteria to grow and synthesise proteins. Leaf-infiltration of bacteria that had been treated with chloramphenicol or starved of amino acids leads to a delayed hypersensitive reaction, where the length of the delay was equivalent to the length of the bacterial lag phase *in vitro*.

Bacterial inoculum level

The quantitative relationship between inoculum level (number of viable cells/ml) and the induction of HR may be considered either at the level of symptom development (tissue necrosis) or at the cellular level (single cell death).

Studies by Ercolani (1973) have shown that compatible and incompatible bacteria show quite different dose–response curves in terms of symptom development, resulting in the conclusion that compatible bacteria have independent action (infiltration of a single bacterium will theoretically lead to symptom development) while incompatible bacteria have co-operative action. In the latter case, a minimal level was required for visible tissue necrosis, below which no symptom response occurred. In differentiating between compatible and incompatible reactions, the use of dose–response analysis (also referred to as infectivity titration) provides a useful criterion for assessing the presence of resistant genes in the testing of new cultivars (Knoche *et al.*, 1987).

The above distinction between compatible and incompatible bacteria at the tissue level relies on the fact that death of all the cells (confluent necrosis) is required for visible symptoms. If the death of individual mesophyll cells is considered, then incompatible bacteria do not have co-operative action, and a single bacterial cell can theoretically lead to the death of a single plant cell. The level of bacterial inoculum required to induce confluent necrosis appears to vary with the particular plant–bacterial combination, ranging in tobacco tissue from critical levels of about 10^4 cells/ml (infiltration of *Pseudomonas syringae* pv. *pisi*, Turner & Novacky, 1974) to 10^6 cells/ml (infiltration of *P.s.* pv. *syringae*; Klement & Goodman, 1967).

Turner & Novacky (1974) investigated the ability of incompatible bacteria to induce death of individual mesophyll cells using the dye Evans blue, which is excluded by viable but not dead mesophyll cells. Light microscope

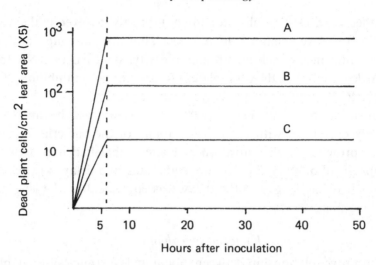

Fig. 6.7 Relationship between mesophyll cell death and bacterial inoculum level. Numbers of dead mesophyll cells per unit leaf area have been determined from sections of tobacco leaves that were infiltrated with *Pseudomonas syringae* pv. *pisi* at levels of 5×10^3 cells (A); 5×10^2 cells (B); and 5×10^1 (C) cells/cm² leaf area. (Adapted from Turner & Novacky, 1974.)

examination of tobacco leaf tissue infiltrated with levels of bacteria below those required to induce confluent necrosis showed subconfluent cell death, in which the number of dead mesophyll cells (but not the time of cell death) was directly related to inoculum level (Fig. 6.7) at a bacteria–plant cell ratio of 1 : 1.

Cell contact

Infiltration of a suspension of bacteria into leaf tissue leads to their eventual deposition onto mesophyll cell surfaces as the water becomes dispersed. Experiments by Cook & Stall (1977) and Stall & Cook (1979) have suggested that this contact between bacteria and mesophyll cells is essential for induction of the incompatible but not the compatible reaction. When incompatible bacteria were prevented from deposition by maintenance of water-soaking or suspension in agar they did not induce the hypersensitive reaction but did induce disease symptoms.

The differential importance of cell contact and attachment in the incompatible but not the compatible reaction has also been shown by quantitative studies on the association of *Pseudomonas solanacearum* with tobacco suspension culture cells (Sequeira, 1984). C[14]-labelled incompatible (strain B1) and compatible (strain K60) bacterial cells were incubated with cultured plant cells, and the number of bacterial cells sticking to the tobacco cells subsequently monitored by radioassay. B1 cells attached rapidly to the plant cells, reaching saturation at about 90 minutes after addition, while K60 cells showed no significant attachment. The initial (first 15 minutes) docking of B1 cells was probably an ionic interaction, since cells were easily washed off by phosphate buffer, but irreversible binding followed within about 30 minutes.

Although cell contact and attachment do not appear to be essential in generating the compatible reaction, it should be noted that:

1. Attachment of virulent bacteria to plant cells does occur in the intact plant under experimental conditions, as shown by work on bacterial cell recovery and ultrastructural observation. Infiltration of pseudomonads into tobacco leaves followed by attempted recovery by centrifugation showed that a high proportion of virulent, avirulent and also saprophytic bacteria were retained within the tissue (Atkinson *et al.*, 1981). Recent ultrastructural studies by Brown & Mansfield (1988) on the frequency of mesophyll cell contact between different races of *Pseudomonas syringae* pv. *phaseolicola* infiltrated into bean leaves, did not reveal any difference in the frequency of cell contacts between the compatible and incompatible situations.

2. Bacterial contact and recognition are essential in compatible interactions involving *Agrobacterium* and *Rhizobium* (see Chapter 8).

In the hypersensitive reaction, cell contact appears to be of potential importance in three specific ways: recognition of the incompatible bacterium by plant cells, biochemical induction of the hypersensitive reaction and immobilisation of incompatible bacteria on the mesophyll cell surface.

Bacterial recognition and activation of the hypersensitive reaction

The molecular events that lead to activation of the hypersensitive reaction are not clear, but appear to involve some type of recognition between the incompatible pathogen and the plant cell. The process of recognition, and its involvement in the induction of the HR, may operate in various ways.

Ercolani (1970) initially proposed that incompatible bacteria become specifically attached to plant cells (recognition) and are thereby activated to release a biochemical inducing factor (elicitor) which subsequently triggers the plant cell response. This process of recognition might involve complementary surface molecules (agglutinins) on the two types of cell, for example, between bacterial surface sugars and plant cell wall lectins (Sequeira, 1978), leading to a specific adhesion in the case of incompatible (but not compatible) bacteria.

An alternative proposal (the elicitor–receptor model) suggests that recognition occurs directly between the released elicitor itself and a surface receptor

molecule on the plant cell, specific adhesion of which leads to the hypersensitive reaction. This simple model has provided a useful basis for studying the
molecular genetics of HR induction (see Chapter 8).

Whatever the process of recognition involves, there is now a wide range of
evidence to support its role in the hypersensitive reaction (see below). The
roles of diffusible substances (bacterial elicitors and plant phytoalexins)
involved in the HR, and resulting changes in plant and bacterial cells, are
discussed later in this chapter.

Evidence for a specific recognition process

The occurrence of some sort of specific recognition process between incompatible bacteria and plant cells is suggested by a number of factors.

Role of cell contact As previously noted, apart from *Agrobacterium* and
Rhizobium, direct cell contact is essential for the incompatible but not the
compatible reaction.

Genetic determination of incompatibility Interactions between bacterial races
and plant cultivars involve gene-for-gene systems, in which resistance is
determined by the joint presence of bacterial genes for avirulence and plant
genes for resistance (see Chapter 8). It seems likely that these genes directly or
indirectly specify complementary cell surface molecules that are involved in
the recognition process.

Isolation of plant recognition factors The possibility that plant cell surface
molecules may be involved in the recognition and specific adhesion of
incompatible bacteria has been investigated by testing the ability of plant cell
wall components (e.g. lectins) and tissue extracts to agglutinate incompatible/
compatible bacteria *in vitro*.

The presence of bacterial surface sugars which may act as potential
receptors for plant recognition molecules has been examined by Anderson
and Jasalavich (1979). Using cultures of various *Pseudomonas* species, these
workers demonstrated clear agglutination of bacterial cells by a range of
lectins which are able to bind to specific surface sugars (Table 6.3). Of the
lectins tested, only ricin failed to agglutinate the bacterial cells, indicating that
a range of surface sugars (with the exception of galactose) was available for a
possible recognition process.

Studies involving tissue extracts have given variable results. In some cases,
the recognition hypothesis appears to be supported by the clear agglutination

Table 6.3 *Agglutination of* Pseudomonas *species by lectins*

Lectin	Surface sugar receptors	Agglutination
Concanavalin A	Alpha-mannosyl and alpha-glycosyl units	+
Wheat germ agglutinin	*N*-acetylglucosamine	+
Phytohaemagglutinin	Galactosamine	+
Ricin	Galactose	−

Data from Anderson & Jasalavich (1979). Bacteria tested were *Pseudomonas fluorescens. P. putida. P. syringae* pv. *phaseolicola* and *P.s.* pv. *tabaci*

of incompatible (but not compatible) bacteria (Sequeira & Graham, 1977); while in others (Anderson & Jasalavich, 1979; Fett & Sequeira, 1980; Slusarenko & Wood, 1983) a differential agglutination was not demonstrated.

Sequeira and Graham (1977) obtained agglutination of avirulent but not virulent strains of *Pseudomonas solanacearum* by an agglutinin. This has been identified as glycoprotein (Sequiera, 1984) and is now thought to mediate a relatively weak, possibly preliminary binding of bacterial cells.

Anderson & Jasalavich (1979) demonstrated agglutination of saprophytic (but not pathogenic) bacteria by crude bean leaf homogenates. Extraction of purified cell wall polysaccharides showed that pectin and galacturonic acid agglutinated both virulent and avirulent phytopathogens, but other polysaccharides had no effect. The absence of a positive result in this type of experiment does not disprove the recognition hypothesis, since plant surface receptors might be at low level or absent from the extract, masked by other molecules or be monovalent (i.e. only one reactive group), in which case they would not bridge and agglutinate bacterial cells.

Identification of HR-inducing bacterial surface molecules The advent of molecular biology, with the capability to specifically isolate or mutate avirulence genes, has considerable potential for the identification of any genetically determined cell surface receptors that may occur. Details of this are discussed in Chapter 8, but so far, sequencing of cloned avirulence genes has led to predicted amino acid sequences which are not typical of membrane proteins. This would either suggest that the primary gene product is not a receptor, or that a genetically determined receptor is not part of the surface membrane.

An alternative approach is to make a direct and comparative analysis of cell surface components in bacteria with and without avirulence genes to determine specific HR-inducing factors.

This has been carried out by Neipold & Huber (1988), who used immunological techniques to compare surface receptors (antigens) on a pathogenic

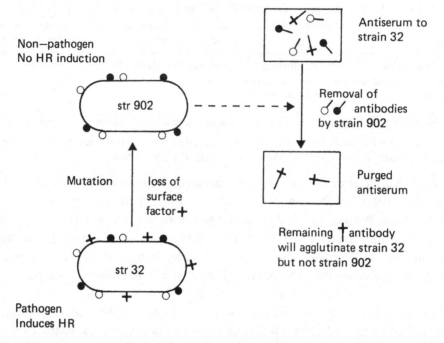

Fig. 6.8 Demonstration of a surface pathogenicity factor in *Pseudomonas syringae* pv. *syringae*. Pathogenic strain 32 of the bacterium has a range of cell surface antigens (●○+), one of which (+) is lost during mutation to the non-pathogenic state (strain 902). This is demonstrated by treating strain 902 with antiserum containing antibodies (●○+) to strain 32, then testing the purged antiserum for agglutination of strains 32 and 902. (Summary of experiment by Niepold & Huber, 1988.)

strain (HR-inducing) and a non-pathogenic (non-HR-inducing) transposon mutant of *Pseudomonas syringae* pv. *syringae*, as shown in Fig. 6.8. The study shows that deletion of an 8.5 kb DNA fragment in the mutant, which leads to the loss of pathogenicity, results in the loss of a surface molecule, as identified by a specific antibody. Restoration of HR gene function by complementation with the wildtype DNA fragment restored both pathogenicity and the production of the surface antigen, which could now be identified by cell agglutination with the specific antibody. The results suggest that ability to induce a hypersensitive reaction is correlated with a specific surface component, but whether this is directly involved in the induction process is not clear.

Biochemical interactions during the HR: elicitors and phytoalexins

During the hypersensitive reaction, bacterial cells are separated from the plant protoplast by a cellulose cell wall, so that interaction between the two types of cell must be by remote biochemical means involving diffusible

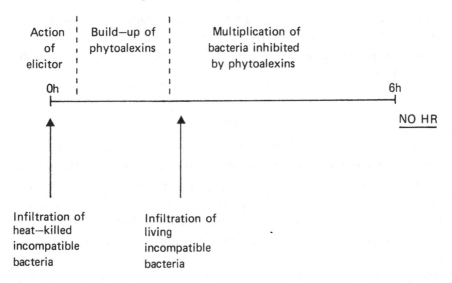

Fig. 6.9 Modification of the visible hypersensitive reaction in leaf tissue by pretreatment with heat-killed bacteria. (Summary of experiment by Sequeira, 1976.)

substances. These are either elicitors (released by bacterial cells) or phytoalexins (produced by the plant cells).

Evidence for the involvement of bacterial and plant diffusible compounds in the hypersensitive reaction initially came from studies by Sequeira (1976) on the suppression of induction of this interaction, using dead bacterial cells. Sequeira showed that the induction of HR in tobacco leaves could be prevented by pre-infiltration with heat-killed bacteria (Fig. 6.9). This pretreatment led to the release and build-up of phytoalexins, which inhibited the growth of living bacteria and the subsequent development of HR. The production of phytoalexins in this situation apparently required ATP from photophosphorylation (no inhibition of HR induction in the dark) and involved active protein synthesis (inhibited by actinomycin D). In these experiments, the suppression of HR by the heat-killed bacterial cells was mediated by specific components of the bacterial cell wall (elicitors or 'protection inducers') which were identified as glycoproteins from the periplasmic space. Removal and analysis of the intercellular fluid from leaves infiltrated with dead bacteria led to the identification of a terpenoid phytoalexin (referred to as a 'protection factor') which inhibited bacterial growth *in vitro*. No such inhibition occurred with intercellular fluid isolated from leaves infiltrated with distilled water.

In these experiments, the use of heat-killed bacteria permits phytoalexin production without any possible build-up in bacterial population. Under the experimental conditions usually employed to investigate HR induction, the very high levels of inoculum used (required for confluent necrosis) completely overwhelm the phytoalexin response and the local population of bacteria rises to quite high levels.

Elicitors

The elicitation of the plant resistance response has been intensively studied for a range of fungal and bacterial phytopathogens (Darvill & Albersheim, 1984). These studies have shown in general that there is no universal inducing chemical that will elicit a response in all plants and that the elicitor acting in a particular incompatible situation may be derived either directly from the pathogen itself or secondarily (as a result of pathogen activity) from the plant cell wall.

Although a substantial amount of work has been carried out on fungi, where a number of elicitors have been shown to be derived from the pathogen cell wall (including glucans, lipids such as eicosapentaenoic and arachidonic acids and surface glycoproteins) relatively few studies have been carried out on phytopathogenic bacteria.

The ability of particular bacterial cell or plant tissue extracts to elicit a hypersensitive response can be monitored experimentally in various ways, including suppression of a future hypersensitive response by pre-infiltration of tissue, phytoalexin production and production of specific mRNA transcripts.

Pre-infiltration of tissue The identification of potential chemical elicitors by prevention of subsequent bacterially induced tissue necrosis follows on from the work of Sequiera (1976) using whole bacterial cells.

Recent studies by Baker *et al.* (1990) have shown that infiltration of tobacco leaf discs with pectate lyase or pectate lyase digests of polygalacturonic acid prevents the subsequent induction of necrosis by live bacterial cells. The results suggest that pectate lyase releases a heat stable oligomer from cell surface pectic material that can act as an elicitor, and is in line with other studies showing that release of oligogalacturonides from plant cells by pectate enzymes promotes a hypersensitive response.

Phytoalexin production Bruegger & Keen (1979) have examined the production of phytoalexins in the *Pseudomonas syringae* pv. *glycinea* – soybean system, where infiltration of living bacteria leads to the race-specific elicitation of glyceollin in cotyledon tissue. Elicitor activity in these bacteria appears to reside at the cell surface, since solubilised cell envelope preparations had a broadly similar effect to living cell suspensions. Subsequent work has failed to identify race-specific elicitors in purified cell envelope preparations, though Barton-Willis *et al.* (1984) have shown that lipopolysaccharides act as non-specific elicitors at high concentrations.

Induction of mRNA transcripts The ability of elicitors to induce the formation of specific mRNA transcripts (derived by activation of disease resistance response genes) provides a very useful biochemical approach to determining potential elicitor activity. Using this technique, Daniels *et al.* (1988) showed that complete bacterial envelope preparations of *Pseudomonas syringae* pv. *pisi* had elicitor activity in incompatible pea tissue, but were not able to demonstrate elicitation by purified preparations.

Phytoalexins

The widespread ability of higher plants to produce anti-microbial compounds (particularly phytoalexins) in response to organisms which are recognised as being foreign (i.e. are incompatible) is analogous to the immune system of vertebrates, and is probably of equivalent importance in combating disease. However, the production of phytoalexins differs from the production of antibodies in a number of fundamental ways:

1. Phytoalexin production is almost always a localised response, and remains localised.
2. Phytoalexins are low molecular weight compounds.
3. The chemical nature of the phytoalexin is determined entirely by the plant and bears no relationship to the chemical nature of the pathogen.

Phytoalexins are normally defined as 'low molecular weight anti-microbial compounds that are produced by, and accumulate in, plants which have been exposed to micro-organisms'. This definition excludes certain high molecular weight anti-microbial compounds, such as phenolics and proteins. The latter include hydroxyproline-rich glycoproteins (HRGP, contained in plant primary cell walls), lectins (widely occurring in plants), purothionins (produced by cereals), hydrolases, and inhibitors of wall-degrading enzymes and proteinases (Lyon, 1989).

Phytoalexin diversity

Although the production of phytoalexins is a widespread (if not universal) feature of higher plants, there is considerable diversity in their chemical

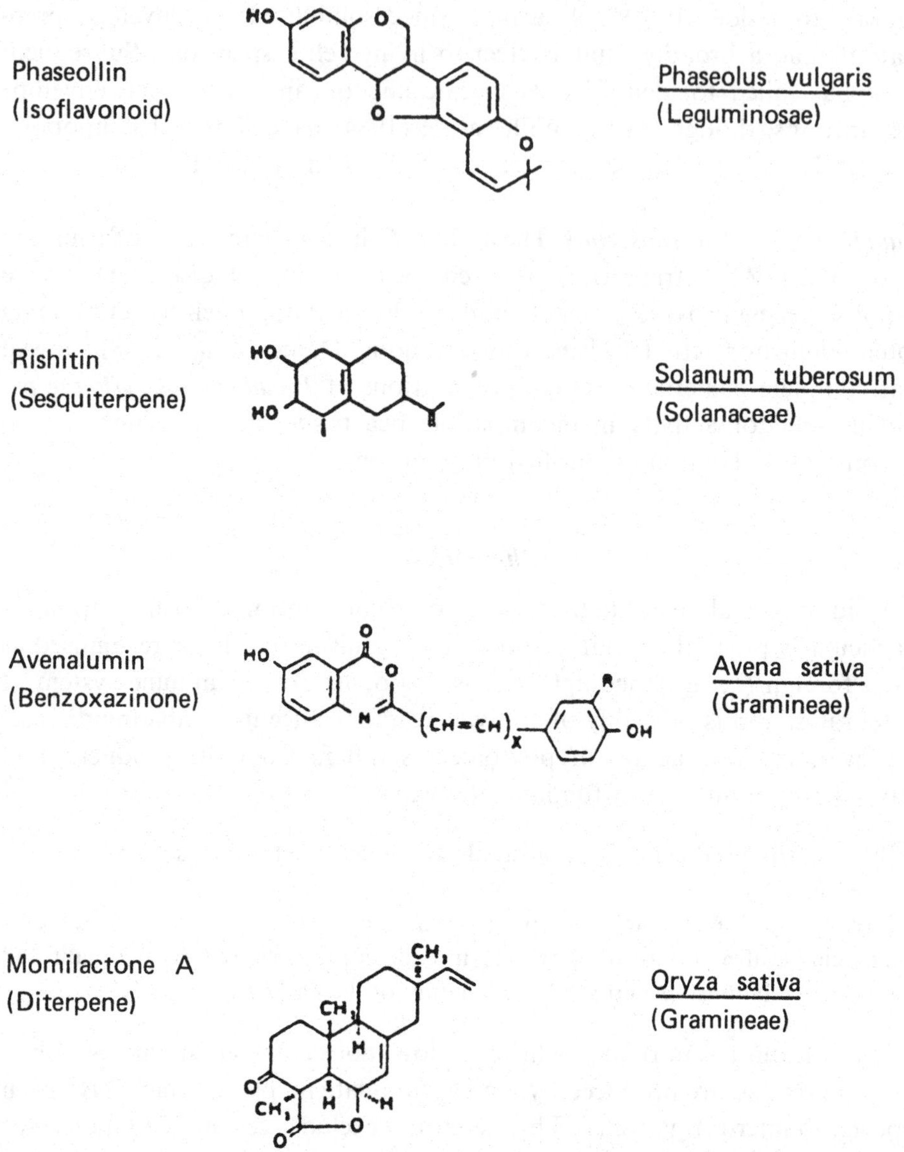

Phaseollin
(Isoflavonoid)

Phaseolus vulgaris
(Leguminosae)

Rishitin
(Sesquiterpene)

Solanum tuberosum
(Solanaceae)

Avenalumin
(Benzoxazinone)

Avena sativa
(Gramineae)

Momilactone A
(Diterpene)

Oryza sativa
(Gramineae)

Fig. 6.10 Diversity of phytoalexin origin and structure. Variations in avenalumin structure: (*a*), $\times = 1$, R = H; (*b*), $\times = 1$, R = OMe; (*c*) $\times = 2$, R = H. (Adapted from Mansfield, 1986.)

nature. Some of this diversity is shown in Fig. 6.10, which gives representative examples of these compounds and emphasises the point that particular plant families produce particular types of phytoalexin. So far, phytoalexins have been chemically identified in at least 17 families of higher plants, with uncharacterised phytoalexins being demonstrated in a similar number (Mansfield, 1986).

Detection and quantitation of phytoalexins

Although the majority of studies on phytoalexins have been with fungal pathogens, similar principles apply to their quantitation and detection in bacteria-infiltrated tissues. In general, phytoalexins may be detected either chemically (following bulk extraction from infected tissue) or microscopically (in tissue sections).

Chemical analysis involves three main stages:

1. Homogenisation of tissue, extraction in methanol then partition with chloroform.
2. Separation and identification of phytoalexin fractions by thin layer or HPLC chromatography.
3. Elution of HPLC fractions, with quantitative analysis by spectrophotometry.

Application of these techniques to infiltrated bean pod or leaf tissue, for example, reveals the production of three major phytoalexins: phaseollin, phaseollin-isoflavin and phaseollidin (Harper *et al.*, 1987). These compounds were detected as three separate bands on thin layer chromatograms, where they were identified by comparison with authentic markers and by differential staining with diazotised *p*-nitroaniline (DPN).

Phytoalexin detection by microscopy usually involves the use of fluoresence techniques and is based on the spectral similarity between individual cells or groups of cells and the fluorescence of purified phytoalexin compounds. This approach has been used, for example, to localise glyceollin production in soybean leaves (Holliday *et al.*, 1981) and sesquiterpenoid phytoalexins in cotton (Pierce & Essenberg, 1987; see Fig. 6.11). Other techniques that have been used to detect phytoalexins in tissue section include radioimmunoassay (Moesta *et al.*, 1983) and laser microprobe analysis (LAMMA) (Moesta *et al.*, 1982).

Critical evidence for the localised production of phytoalexins by plant cells in the immediate vicinity of incompatible bacteria has recently been provided by Pierce & Essenberg (1987) using combined microscopical and spectroscopical techniques (Fig. 6.11). Sections of cotton leaves infiltrated with *Xanthomonas campestris* pv. *malvacearum* revealed localised groups of bacteria surrounded by necrotic cells which showed specific autofluorescence. Separation of fluorescing and non-fluorescing mesophyll cells in leaf homogenates by activated sorting, followed by chemical analysis, demonstrated a direct correlation between autofluorescence and phytoalexin level, indicating a restricted production of phytoalexins by cells adjacent to bacteria.

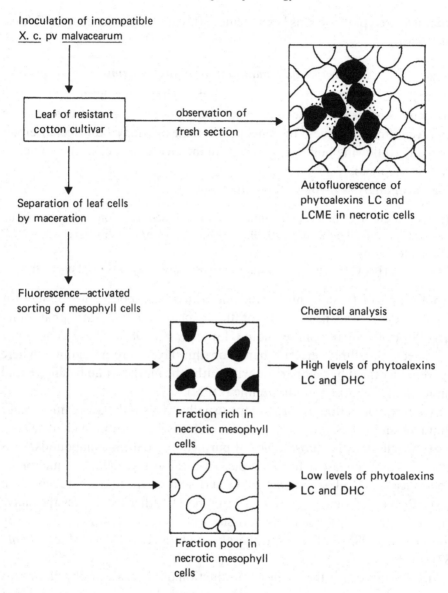

Fig. 6.11 Demonstration of localised phytoalexin production by leaf mesophyll cells. The use of fluorescence-activated sorting technology permits the combined use of microscopy and chemical analysis, following inoculation of resistant leaves of cotton with incompatible *Xanthomonas campestris* pv. *malvacearum*. Phytoalexins being assayed are: lacinilene C (LC), lacinilene C 7-methyl ester (LCME) and 2,7-dihydroxycadalene (DHC). (Summary of experiment by Pierce & Essenberg, 1987.)

Variation in phytoalexin production

The amount of phytoalexin synthesised by plant cells during the hypersensitive reaction varies considerably and depends on a number of factors, including the plant species or cultivar, type of tissue and type of pathogen. This is illustrated in studies by Harper *et al.* (1987) on the induction of phaseollin by races of *Pseudomonas syringae* pv. *phaseolicola* and by *P. coronafaciens* in cultivars of bean (*Phaseolus vulgaris*). These studies indicated that the production of phaseollin was higher in cultivar Tendergreen compared with Canadian Wonder and Red Mexican, and that infiltration of *P. coronafaciens* generated a more rapid initial production of phytoalexin compared with incompatible races of *P.s.* pv. *phaseolicola*. The production of phaseolin was much greater in pod tissue (reaching levels of approximately 1000 µg g^{-1} fresh weight) compared to leaf tissue (maximum level about 150 µg g^{-1}). Although the large-scale production of phytoalexins was restricted to incompatible race–cultivar combinations, small levels of phytoalexin were recorded with some compatible bacteria, possibly arising as a non-specific effect of the experimental procedure.

Metabolic control of phytoalexin production

The triggering of phytoalexin production by natural or artificial elicitors involves the induction of *de novo* mRNA transcription and the translation of enzymes involved in phytoalexin synthesis. At the present time, the most detailed information on the biosynthetic pathway of these molecules relates to the production of isoflavonoid phytoalexins such as phaseollin, and is summarised in Fig. 6.12.

The isoflavonoid phytoalexin pathway involves the synthesis of three key enzymes: phenylalanine ammonia lyase (PAL), chalcone synthase (CHS) and chalcone isomerase (CHI). Recent studies by Cramer *et al.* (1985) on bean cell suspension cultures, have shown that addition of fungal wall elicitors results in a rapid promotion of phytoalexin synthesis:

1. Use of specific antisera to PAL, CHS and CHI showed that these enzymes were synthesised prior to phytoalexin formation, and were detected at low levels within 20 minutes of elicitor treatment; maximum levels being reached at three to four hours after elicitation.

2. *In vitro* translation of isolated RNAs showed that induction of phytoalexin synthesis was promoted by an increase in the level of mRNAs specifically encoding the three enzymes. Maximum rates of phytoalexin-mRNA synthesis occurred 2.5–3.5 hours after addition of elicitor, and came at a time of rapid and extensive change in general mRNA synthesis.

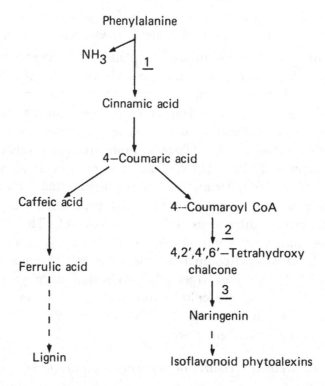

Fig. 6.12 Pathways of isoflavonoid phytoalexin production and lignin formation. Enzymes involved in the phytoalexin pathway include phenylalanine ammonia lyase (1); chalcone synthase (2); and chalcone isomerase (3). (Adapted from Mansfield, 1986.)

3. Specific mRNAs were further analysed by blot hybridisation using 32P-labelled cloned enzyme DNA sequences as probes, indicating that the pathogen elicitor causes rapid transient activation of genes involved in phytoalexin synthesis.

Role of environmental conditions in phytoalexin synthesis and the resistant response

Various environmental factors have been shown to affect the synthesis of particular phytoalexins – including light and oxygen tension.

The effect of light on phytoalexin production The importance of light for the expression of HR in aerial parts of the plant appears to vary according to the experimental situation, being essential in some cases – but inhibitory or neutral in others (Sequeira, 1984). In those cases where light is required, it appears to be specifically essential for phytoalexin production. The failure of

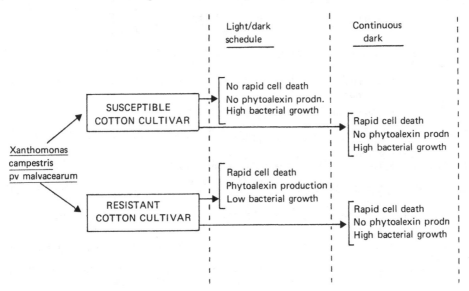

Fig. 6.13 Importance of light in the response of leaf cells to compatible and incompatible bacteria. (Summary of experiment by Morgham *et al.*, 1988.)

a plant to limit the growth of incompatible bacteria in continuous dark has been shown for a number of species, including tobacco (Lozano & Sequeira, 1970), pepper (Sasser *et al.*, 1974) and cotton (Morgham *et al.*, 1988).

Although light may be essential for phytoalexin production, it is not involved in other aspects of the hypersensitive response such as cell death. This differential effect permits assessment of the relative importance of phytoalexin production and cell death in limiting bacterial population, as shown by Morgham *et al.* (1988). These authors infiltrated susceptible and resistant cultivars of cotton with *Xanthomonas campestris* pv. *malvacearum*, and observed the development of cell death, sesquiterpenoid phytoalexin production and bacterial growth under respective conditions of intermittent light and continuous dark (Fig. 6.13). The results show that synthesis of phytoalexins only occurs in resistant plants in the light, and that limitation in the growth of bacteria relates to phytoalexin production and not cell death.

The importance of oxygen levels in phytoalexin production Studies by Lyon (1989) have shown that production of the terpenoid phytoalexins rishitin (see Fig. 6.10) and phytuberin by potato tubers only occurs under aerobic conditions. These anti-microbial compounds are produced as part of an active resistance response to both fungal (e.g. *Phytophthora infestans*) and bacterial (e.g. *Erwinia carotovora* subsp. *atroseptica*) pathogens, and are not detectable in non-inoculated tubers.

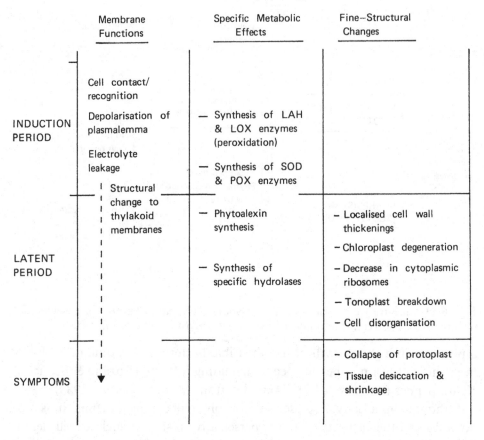

	Membrane Functions	Specific Metabolic Effects	Fine–Structural Changes
INDUCTION PERIOD	Cell contact/ recognition Depolarisation of plasmalemma Electrolyte leakage	— Synthesis of LAH & LOX enzymes (peroxidation) — Synthesis of SOD & POX enzymes	
LATENT PERIOD	Structural change to thylakoid membranes	— Phytoalexin synthesis — Synthesis of specific hydrolases	— Localised cell wall thickenings — Chloroplast degeneration — Decrease in cytoplasmic ribosomes — Tonoplast breakdown — Cell disorganisation
SYMPTOMS			— Collapse of protoplast — Tissue desiccation & shrinkage

Fig. 6.14 Mesophyll cell changes during the hypersensitive reaction. Specific enzymes quoted are lipolytic acyl hydrolase (LAH); lipoxygenase (LOX), superoxide dismutase (SOD) and peroxidase (POX).

In addition to their role in resistance to plant pathogens, rishitin and other terpenoids are also produced at low level as a response to wounding of plant tissue. They are toxic to plant cells as well as bacteria, though the degree to which their phytotoxicity contributes to plant cell death during the resistant response is not known.

Pathogenic changes in plant cells

Contact between plant cells and incompatible bacteria results in a cascade of degenerative changes which lead to cell death. This transition can be considered in three main categories: membrane changes, alterations in cell metabolism and changes in fine structure (Fig. 6.14). These three aspects are interrelated and each follows a well-defined pattern, suggesting that the induction of cell death during the HR is a programmed event that is triggered by recognition of the incompatible pathogen. This type of cell death is quite different from degenerative changes resulting simply from injury, and is similar in type to the programmed cell death that occurs in other situations for both plant cells (leaf senescence, fruit ripening, differentiation of phloem and xylem) and animal cells (for references, see Davies & Sigee, 1984).

Membrane changes

Incompatible reaction

Degeneration of the plant cell appears to be initiated by changes occurring at the plasmalemma. These are first apparent at the end of the induction period and initially involve generation of oxygen radicals and peroxidation of membrane lipids – followed in sequence by a fall in membrane potential, electrolyte leakage then water efflux and cell plasmolysis. These changes are summarised in Fig. 6.15, where they are also compared to cell surface membrane changes during the compatible interaction.

Lipid peroxidation Keppler and Novacky (1986) monitored levels of lipid peroxidation in cucumber cotyledons infiltrated with incompatible (*Pseudomonas syringae* pv. *pisi*) compatible (*P.s.* pv. *lachrymans*) and saprophytic (*P. fluorescens*) bacteria. Lipid peroxidation increased significantly only in the incompatible combination (leading to an HR) when plants were infiltrated with live (not heat-killed) bacteria, and was caused by the *de novo* synthesis of two enzymes (Croft *et al.*, 1990). One of these (lipolytic acyl hydrolase or LAH) actively releases fatty acids from the cell membrane which the other (lipoxygenase or LOX) can then oxidise. The increase in level of these enzymes only occurs in the incompatible situation and is abolished by treatment with cycloheximide, suggesting that it represents a genuine protein synthesis rather than release from a pre-existing pool.

In vivo peroxidation of cell membranes during the incompatible (but not compatible) reaction has also been demonstrated by Slusarenko *et al.* (1989) for race-specific hypersensitivity in leaves of *Phaseolus vulgaris*. Early rises in the level of LAH and LOX (isoenzymes 1 and 2) occurred as previously, plus an increase in the level of ethane. This is produced as an end-product of fatty acid hydroperoxidase breakdown from linolenic acid, and is a specific indicator of membrane-lipid peroxidation.

Peroxidation of the membrane lipid phase results in perturbation of the ion-transport mechanisms of the plasmalemma, leading in turn to changes in

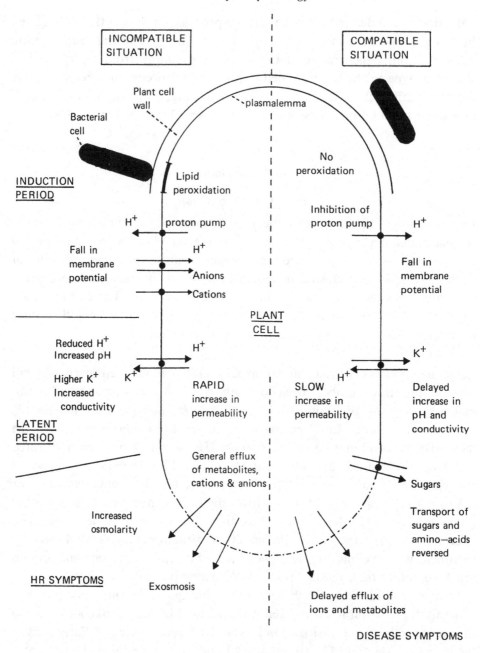

Fig. 6.15 Comparative effects of compatible and incompatible bacteria at the plant cell surface. Showing changes in plasmalemma transport activity after mesophyll infiltration with compatible (right side) and incompatible (left side) bacteria. Changes in membrane function with time are presented in downward progression, leading to disease or HR symptoms respectively.

the distribution of protons and cations across the membrane and to changes in the membrane potential.

Membrane potential At the surface of the plant cell, the membrane potential is maintained to a large extent by an imbalance of protons (H^+) across the plasmalemma which is effected by a proton ATPase extrusion pump. This major electrogenic pump creates a H^+ gradient which is balanced to some extent by passive diffusion of protons and cations back across the plasmalemma.

The transmembrane potential (Em) can be measured electrophysiologically following insertion of microelectrodes into the plant cell and is composed of two separate components: an energy-dependent component (Ep), which is maintained by the electrogenic pump, and a diffusion component (Ed), which depends on the integrity of the plasmalemma, particularly the phospholipid bilayer. The value for Ed may be determined under experimental conditions where the level of adenosine triphosphate (ATP) – required for functioning of the electrogenic pump – is maintained at a low level. Ep can then be calculated by subtraction of Ed from Em.

During the HR the presence of incompatible bacteria leads to a rapid decline (commencing within 2 hours) in the transmembrane potential (Em). This has been shown by Pavlovkin *et al.* (1986) and Keppler and Novacky (1986) following infiltration of cotton (*Gossypium hirsutum*) cotyledons with *Pseudomonas syringae* pv. *tabaci*. Although the decline in transmembrane potential correlates with a fall in the value of both Ep and Ed, the primary effect of incompatible bacteria is on the diffusion component rather than the proton pump. The decline in value of the diffusion component (Ed) continues beyond the induction period up to the time of symptom development, falling to 45% of the level in control tissue.

Electrolyte leakage – membrane permeability As lipid peroxidation proceeds, the continuing alteration of the cell membrane is paralleled by a continued depolarisation and by the occurrence of electrolyte leakage. This was first reported by Cook & Stall (1968) and Goodman (1968), and results from an increase in membrane permeability. Electrolyte leakage initially involves a particularly high efflux of K^+ from the plant cell, and Atkinson & Baker (1987b) have suggested that this involves specific activation of a passive K^+/H^+ exchange mechanism. Further increases in membrane permeability lead to a more general loss of electrolytes from the cell.

Information on changes in membrane permeability comes from several directions:

1. *Plasmolytic activity* A clear indication of early membrane disfunction is given by the failure of cells to plasmolyse properly under appropriate hypertonic conditions (Croft *et al.*, 1990).

2. *Retention of vital dyes* Changes in membrane permeability can be directly monitored by measuring the retention of polar dyes. Keppler *et al.* (1988) used fluorescein diacetate to study changes in membrane permeability during bacterially induced HR in tobacco suspension cells. Fluorescein diacetate is a non-polar molecule which moves across the cell membrane by passive diffusion through the lipid phase. Inside the cell, esterase activity results in the release of fluorescein, which is a polar molecule and accumulates due to its inability to diffuse back across the plasmalemma. The accumulation of fluorescein in tobacco cells challenged with incompatible bacteria is much less than the compatible situation. In a situation where the uptake of dye and the level of esterase is the same, this difference reflects a greater permeability in the incompatible case with resulting loss of the polar molecules.

3. *Analysis of medium surrounding the cells* The leakage of electrolytes is normally determined by measuring changes in the medium surrounding the cells. Where intact plants have been inoculated with bacteria, this usually involves placing segments of infiltrated tissue in de-ionised water and analysing changes in this. Using this approach, the onset of detectable electrolyte leakage appears to vary from about 90 minutes (Goodman, 1968) to 4 hours (Keppler & Novacky, 1986) after bacterial challenge, depending on the experimental system that is operating.

More recently, a number of studies have been carried out with cell suspensions (Baker *et al.*, 1987; Keppler *et al.*, 1988), where electrolyte leakage can easily be monitored by analysis of the suspension fluid.

Whether whole tissues or single cells are being investigated, the K^+/H^+ interchange can be measured either in terms of K^+ efflux or H^+ influx. K^+ efflux may be determined as increase in conductivity (Fig. 6.16) or by direct cation assay using atomic absorption spectrophotometry. H^+ influx can be monitored using a pH meter (Fig. 6.16), pH indicator dye or by acid–base titration.

Water efflux and plasmolysis In the intact tissue, the continued damage to the plasmalemma during HR results in a continued loss of electrolytes into the intercellular space. This leads to an increased osmolarity outside the cells, resulting in a flow of water into the intercellular space by exosmosis. Eventually the cells become plasmolysed, tissue collapse occurs and visible symptoms of HR are apparent.

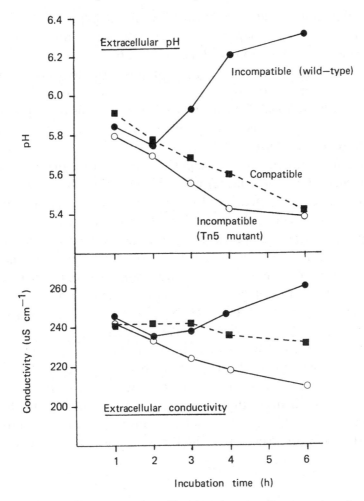

Fig. 6.16 Induction of H$^+$ influx/K$^+$ efflux in tobacco cell suspensions by incompatible bacteria. Increases in extracellular pH (H$^+$ influx) and conductivity (K$^+$ efflux) occur after infiltration of incompatible (HR-inducing) *Pseudomonas syringae* pv. *syringae* (wild-type cells) but not compatible *P.s.* pv. *tabaci*. Infiltration of cells of a non-pathogenic Tn5 mutant derived from the pv. *syringae* has no effect. (Adapted from Atkinson & Baker, 1987b.)

Compatible reaction

In the compatible interaction (Fig. 6.15) there is no significant lipid peroxidation and no effect on the diffusion component of membrane potential (Ed). However, the overall membrane potential (Em) still falls due to an inhibition of the electrogenic pump component (Ep) – as shown by Novacky & Ullrich-Eberius (1982) during infection of cotton leaf tissue with *Xanthomonas campestris* pv. *malvacearum* (susceptible reaction). The general loss of electrolytes and cell fluid is considerably slower than in the hypersensitive reaction, though ultimately tissue collapse does occur, resulting in necrotic lesions.

Recent studies by Atkinson & Baker (1987a,b) have suggested that the maintenance of membrane integrity in the compatible reaction may be important in preventing the release of toxic chemicals from the cells, but permit the specific leakage of cell nutrients. The loss of sucrose appears to be particularly important and is mediated by a specific effect of the bacteria on an H^+ sucrose active transport system.

In a normal plant cell, there is a pH gradient across the plasmalemma (pH 5.5 outside, pH 7.0–7.7 inside) which is maintained by the H^+ ATPase pump. This gradient is thought to drive the active uptake of inorganic ions (K^+, PO_4^{2+}), amino acids and sucrose from the apoplast into mesophyll cells. The transport of sucrose, in particular, is thought to be driven by a specific carrier protein which co-transports H^+. Sucrose transport into mesophyll cells may be reversed by perturbing this transport system, either by altering experimentally the extracellular chemical environment in opposition to the gradient (increasing pH, cation levels), or by infiltration of leaf tissue with compatible bacteria (Atkinson & Baker, 1987a). Both of these treatments lead to a decreased uptake of sucrose (determined as cumulative uptake into leaf discs) and an increased loss of sucrose (determined as release of C^{14} sugar). Exactly how pathogenic bacteria are able to draw sucrose from plant cells by altering H^+/K^+ gradients without causing structural damage to the membrane is not clear, but the release of nutrients in the compatible combination may be particularly important in the higher build-up of bacterial population that occurs in this situation.

Changes in plant cell metabolism

During the hypersensitive response, programmed cell death involves a complex sequence of metabolic changes which arise either directly or indirectly (via second messengers) from bacterial challenge. Some of these changes involve the synthesis of novel proteins due to specific activation of resistance response genes, while other more general metabolic changes appear to result from a general breakdown in cell organisation.

Activation of resistance response genes

Daniels *et al.* (1987) have shown that new species of mRNA are transcribed following activation of specific plant genes due to the combined effects of wounding and bacteria. The identification of plant genes that are expressed following infiltration has been investigated by isolating mRNA for *in vitro* translation studies with subsequent analysis by hybridisation of cloned cDNA copies. Within about 4 hours of control (medium only) infiltration, out of a total of about 500 discrete mRNAs, 40 show a marked increase and 10 a decrease. The presence of compatible *Xanthomonas campestris* pv. *campestris* has no effect on gene expression up to 9 hours, but incompatible *X. c.* pv. *vitians* results in the production of eight new gene products within the first 4 hours, and a further five or six by 9 hours. One major role of these resistance gene products is the synthesis of phytoalexins.

The synthesis of new proteins during HR has been investigated by Slusarenko *et al.* (1989) in leaves of *Phaseolus vulgaris*, where the synthesis of enzymes associated with membrane peroxidation is followed by synthesis of a range of polypeptides. The sequence is outlined in Fig. 6.14, and involves:

1. Once lipid peroxidation has been initiated by the activity of LAH and LOX enzymes, further peroxidation continues auto-oxidatively due to subsequent liberation of active oxygen species (Glazener *et al.*, 1991; Sutherland, 1991) which cause general damage to cell constituents, particularly proteins.

2. Levels of superoxide dismutase (SOD) and peroxidase (POX) rise as a specific response to the increase in active oxygen radicals, and may be important in limiting HR necrosis to a localised region of the tissue.

3. The synthesis of isoflavonoid phytoalexins is promoted by the co-ordinated activation of genes for synthesis of phenylalanine ammonia lyase (PAL) and chalcone synthase (CHS), as shown by probing of RNA dot blots and Northern blots with appropriate cDNA probes. Maximum accumulation of transcripts occurred during the latent period, with a subsequent rise in the transcription product (PAL) to a maximum level at the beginning of visible tissue necrosis. No measurable accumulation of these mRNA transcripts or resulting proteins occurred during the compatible interaction.

4. Synthesis of the hydrolytic enzyme chitinase is induced during the HR, with an increase in chitinase mRNA activity during the early stages of the latent period. The induction of this enzyme is not limited to the incompatible reaction, since it also occurs (at a later stage) following inoculation of compatible bacteria.

General metabolic changes

These include overall changes in RNA and DNA synthesis, cell respiration and general levels of hydrolase enzymes.

Autoradiographic studies by Al-Issa and Sigee (1983) on uptake of H^3-uridine following infiltration of tobacco leaves with bacterial suspension have shown that RNA synthesis increases during the induction period, followed by a sharp decrease during the latent and symptom periods. Control infiltration of suspension medium also leads to an initial stimulation of RNA synthesis, and it is difficult to separate the twin aspects of infiltration wound-response and the bacterial cells. In contrast to these results, parallel studies on the synthesis of DNA (uptake of H³-thymidine) showed that neither nuclear nor cytoplasmic synthesis was inhibited until the end of the latent period (Sigee, 1984), by which time mesophyll cells were showing signs of fine structural change.

Other general metabolic changes occurring during the latent period include an increased rate of respiration (in both compatible and incompatible reactions) and increases in the level of various hydrolytic enzymes (for references, see Sigee, 1984). Many of these general metabolic events appear to be a feature of cell breakdown rather than the specific result of incompatible bacterial challenge.

Fine structural changes

During the experimentally induced hypersensitive reaction rapid changes occur in plant cell fine structure, leading to early cell death. These changes initially become apparent during the middle of the latent period, and are the result of metabolic events taking place in the induction and early latent periods. The fine structural changes are of two main types: localised and general, and have been examined in particular detail in infiltrated mesophyll tissue (Sigee, 1984).

Localised response

Localised reaction is shown by a marked thickening of the cell wall in the immediate vicinity of the associated bacteria (Fig. 6.17*b*), and has been observed in both resistant (Al-Issa & Sigee, 1982a) and disease (Sigee & Epton, 1976) situations.

General mesophyll changes

General changes in the fine structure of mesophyll cells can be seen at both light (Fig. 6.17*a*) and electron microscope (Fig. 6.17*b*) levels, and are summarised in Fig. 6.14. These changes do not appear to be spatially related to sites of bacterial association at the cell surface, and involve a well-defined sequence of events that is characterised by:

1. Initial disruption of chloroplast fine structure, with swelling of thylakoids and displacement to one side of the organelle. At the same time, the frequency of polyribosomes and ribosomes in the general cytoplasm typically decreases.

2. Chloroplasts subseqently become spherical and electron-dense, with further degeneration of thylakoid membranes and cell membranes generally. This breakdown is frequently accompanied by the appearance of osmiophilic droplets which result from the release of membrane lipids.

3. Breakdown of the tonoplast results in the intermixing and degradation of all cell contents. Chloroplasts lose their outer envelope and dense matrix, appearing now as large pale structures by transmission light and electron microscopy.

4. The final stages of the sequence involve collapse of mesophyll protoplasts, with the release of cell fluid into the intercellular spaces, followed by tissue desiccation and shrinkage.

Although the above sequence appears to be typical of the hypersensitive response generally (Goodman & Plurad, 1971; Sigee & Al-Issa, 1983), variations in detail may occur between plant species and between cells within the same tissue. Sigee and Epton (1976) reported two types of mesophyll fine structural change (differing in details of chloroplast and ribosome alterations) following infiltration of *Phaseolus vulgaris* cv. Red Mexican with incompatible *Pseudomonas syringae* pv. *phaseolicola* race 1. These were respectively interpreted as representing direct bacterially induced necrosis, and indirect senescence caused by nutrient deprivation of isolated mesophyll cells.

The above sequence of changes is also characteristic of the reaction to compatible bacteria that cause leaf cell necrosis (Sigee & Epton, 1976), though in the latter case the changes are much delayed. It is also closely similar to the fine structural events that occur during normal and induced leaf senescence (Butler & Simon, 1971). These similarities suggest that a common underlying cause (e.g. perturbation of protein synthesis or activation of specific genes) may lead to a common pattern of events (Davies & Sigee, 1984).

Bacterial changes during the hypersensitive reaction

Although the effects of incompatible bacteria on the metabolism and fine structure of plant cells have been well documented, apart from population changes there is relatively little detailed information on the reciprocal effects of the plant cells on the bacteria. However, some aspects of bacterial change after inoculation have been investigated, however, including details of fine-structure, metabolism and flagellation (motility).

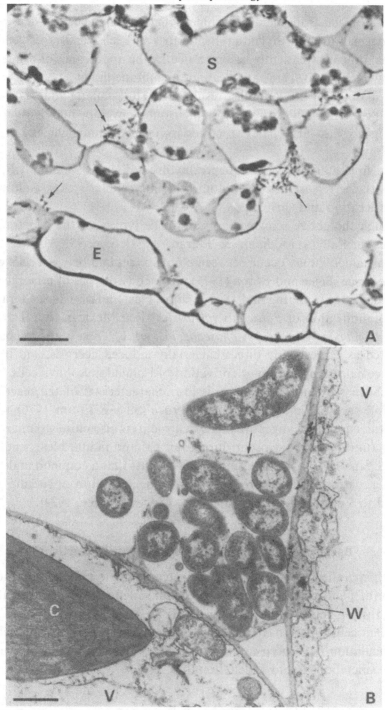

Fig. 6.17 Hypersensitive reaction in tobacco leaf tissue, 5 hours after infiltration with *Pseudomonas syringae* pv. *pisi*. (*a*) Phase-contrast light micrograph of 2 μm section, showing degenerating spongy mesophyll cells (S) and epidermal cells (E) with small groups of associated bacteria (arrows). Bar scale: 15 μm. (*b*) Transmission electron

Fine structural changes

Infiltration of bacteria into plant tissue is followed by well-marked changes in the fine structure of the bacterial cells, and also in the association of the bacteria with the plant cells.

Bacterial cells

Changes in bacterial fine structure have been examined following infiltration of *Pseudomonas syringae* pv. *phaseolicola* into resistant and susceptible leaves of French bean (Sigee & Epton, 1975). During the resistant interaction, which takes place over a relatively long (168-hour) period, the central nucleoid becomes diffuse and dense ribosomal aggregates appear in the peripheral cytoplasm. In susceptible leaves, the bacteria develop a more clearly defined central nucleoid and the whole of the surrounding cytoplasm becomes highly electron-dense. Bacteria in susceptible (but not resistant) plants also develop prominent surface vesicles, some of which become detached, releasing their contents (and membrane fragments) into the intercellular spaces (see Fig. 2.3c).

Bacteria–plant cell associations

During the hypersensitive reaction, bacteria frequently occur within the plant tissue as localised groups (Fig. 6.17a) and are closely associated with the plant cells, either singly or in droplets.

In ultrathin section (Fig. 6.17b), these droplets typically have an electron-dense film over the surface, which various workers have considered to be part of the incompatible resistance mechanism. It has been suggested that entrapment of bacteria within these droplets leads to physical localisation of the pathogen (Politis & Goodman, 1978), limiting the multiplication and spread of bacteria (Sequeira *et al.*, 1977) and promoting bacterial attachment to the plant cell (Bonatti *et al.*, 1979). Other workers have argued that the bacterial droplets arise as a physical aspect of the post-infiltration sequence, and that the surface film is simply a physical interface that develops at this time (Hildebrand *et al.*, 1980; Al-Issa & Sigee, 1982a).

Whether this immobilisation of incompatible bacteria within droplets is part of an active plant defence mechanism or is a simple physical effect remains to be seen, but it should be pointed out that:

Caption for Fig. 6.17 *(cont.)*
micrograph of ultrathin section, showing bacterial droplet at the junction of two mesophyll cells. The droplet has a surface film (arrow) which has ruptured, liberating some of the bacteria into the intercellular space. The mesophyll cells show localised cell wall thickenings (W) and have vacuoles (V) which are limited by intact tonoplast membranes. The chloroplast (C) appears electron-dense, with disorganised internal thylakoid membranes. Bar scale: 1 μm.

1. Bacterial confinement to droplets does not appear to prevent their multiplication or to limit their metabolic activity since they are involved in cell division (Al-Issa & Sigee, 1982a) and are active in DNA and RNA synthesis (Al-Issa & Sigee, 1982b, 1983).

2. Bacterial droplets with a surface film are also seen after leaf infiltration of compatible (Sigee & Epton, 1976) and saprophytic bacteria, so they are not restricted to the incompatible situation.

Bacterial chemistry and metabolism

Relatively little information is available on chemical and metabolic changes in bacterial cells during the hypersensitive reaction. Under experimental conditions, the multiplication of bacteria throughout the induction, latent and early symptom periods (Fig. 6.3) would suggest that they are metabolically active during the major part of the HR and the specific inhibitory effects of a range of antibiotics (Table 6.2) are consistent with a range of metabolic activities being involved in promoting the HR. Specific information has been obtained on changes in bacterial elemental composition and nucleic acid synthesis after infiltration of leaf tissue.

Elemental composition

Changes in the elemental composition of bacterial cells extracted from leaf tissue have been monitored for both the incompatible (Sigee & Hodson, 1993) and compatible (El-Masry & Sigee, 1989) situations, using electron probe X-ray microanalysis (see Chapter 2). In both cases, the elemental composition of bacteria in the plant was closely similar to that of cells grown *in vitro*, with P, S, K and Ca being the major detectable elements present (see Fig. 2.10). In both cases also, the elemental composition remained relatively stable over the time course of the reaction, indicating that the chemical composition of the pathogen was little affected by anti-microbial compounds and other plant metabolites. Only at a very late stage in the sequence, when bacteria could be clearly seen to be undergoing degeneration, was there a major efflux of soluble cell components, with a marked fall in the level of K (El-Masry & Sigee, 1989).

Nucleic acid synthesis

The bacterial synthesis of DNA and RNA during the course of the hypersensitive reaction has been investigated by Al-Issa & Sigee (1982b) and Sigee & Al-Issa (1982) following infiltration of *Pseudomonas syringae* pv. *pisi* into tobacco leaves. Using light and electron microscope autoradiography, these authors showed that synthesis of DNA (uptake of H[3]-thymidine) closely followed the viable count, with a sharp rise in the proportion of labelled bacteria during the period of rapid population increase (6–12 hours after infiltration). Synthesis of RNA was also maximal during the first 12 hours after infiltration, with continued uptake of H^3-uridine into bacterial cells long after mesophyll incorporation had ceased.

Bacterial motility

Although relatively little is known about the mechanism of bacterial movement within infected leaf tissue (see Chapter 5), the presence of flagellate cells *in planta* suggests that active motility may be important in disease development.

Electron microscope examination of leaf macerates during the hypersensitive reaction has shown that the total proportion of cells with flagella falls from 60–20% within the first 12 hours of inoculation (Sigee & Hodson, 1993). This is in marked contrast to the disease situation (see Fig. 5.5), suggesting that an inhibition of bacterial motility may be an important aspect of the HR.

The mechanism of resistance during the hypersensitive reaction

The hypersensitive reaction is one of the most important reactions that occurs between microbes and plant cells, since it represents a general mechanism to avoid invasion by the pathogen and resulting disease. It may also be one of the most common interactions, since chance entry of pathogenic bacteria into plant tissue would normally induce a resistant response rather than a disease reaction.

Exactly how the hypersensitive reaction prevents the development of disease, and which factors are most important in bacterial limitation, have been the subject of some discussion. The reduced multiplication of bacteria at the site of infiltration (Fig. 6.3) and the minimal spread to other parts of the plant are clearly important, and a number of factors which bring this about have been mentioned. The most obvious of these are death of the plant cells and production of phytoalexins (the two major plant responses in HR; see below), but other aspects may also be important: including bacterial immobilisation, pH changes in the intercellular fluid, release of non-phytoalexin antimicrobial compounds and tissue desiccation. Reduction in bacterial motility may be an important factor in limiting the spread of the pathogen.

In the compatible (disease) situation, the prolonged multiplication and spread of the pathogen (Fig. 6.2) largely result from an absence (e.g. no phytoalexins) or delay (e.g. plant cell death) in the above adverse conditions. The ability of some bacteria to cause the specific leakage of plant cell nutrients by the reversal of cell surface pumps (p. 158) may also be important.

Role of plant cell death and phytoalexin production

Plant cell death is usually considered to be one of the most important aspects of the hypersensitive response to invading phytopathogens, since it is assumed to limit the period of plant pathogen growth and also to isolate the pathogen within a region of dead tissue, thus preventing further spread. However, in the case of hypersensitivity to bacteria two experiments particularly suggest that cell death plays a secondary role to phytoalexin production:

1. Pre-infiltration with heat-killed bacteria (Fig. 6.9) leads to phytoalexin production without cell death, but with very effective subsequent limitation of live bacteria.

2. Conversely, infiltration of dark-grown leaf tissue with the incompatible phytopathogen (Fig. 6.13) leads to mesophyll cell death, no phytoalexin production, and no limitation in bacterial spread and multiplication.

In laboratory experiments on the HR, infiltration of very high levels of incompatible bacteria probably represents a very abnormal situation in which the localised production of phytoalexins is quite inadequate in limiting bacterial growth. In this situation, the multiplication of bacteria beyond about 12 hours is probably prevented by limitations in space and nutrient supply and by tissue desiccation. Under natural or field conditions, where relatively low levels of inoculum would be expected to enter the plant tissue, phytoalexin production would be far more effective in limiting bacterial population increase within the infection zone.

References

Al-Issa A. N. & Sigee D. C. (1982a). The hypersensitive reaction in tobacco leaf tissue infiltrated with *Pseudomonas pisi*. 1. Active growth and division in bacteria entrapped at the surface of mesophyll cells. *Phytopathol. Z.* **104**, 104–14.

Al-Issa A. N. & Sigee D. C. (1982b). The hypersensitive reaction in tobacco leaf tissue infiltrated with *Pseudomonas pisi*. 3. Changes in the synthesis of DNA in bacteria and mesophyll cells. *Phytopathol. Z.* **105**, 198–213.

Al-Issa A. N. & Sigee D. C. (1983). The hypersensitive reaction in tobacco leaf tissue infiltrated with *Pseudomonas pisi*. 5. Inhibition of RNA synthesis in mesophyll cells. *Phytopathol. Z.* **106**, 23–34.

Allington W. B. & Chamberlain O. W. (1949). Trends in the multiplication of pathogenic bacteria within the leaf tissues of susceptible and immune plant species. *Phytopathology* **39**, 656–60.

Anderson A. J. & Jasalavich C. (1979). Agglutination of pseudomonad cells by plant products. *Physiol. Plant Pathol.* **15**, 149–59.

Atkinson M. M. & Baker C. J. (1987a). Alteration of plasmalemma sucrose transport in *Phaseolus vulgaris* by *Pseudomonas syringae* pv. *syringae* and its association with K^+/H^+ exchange. *Phytopathology* **77**, 1573–8.

Atkinson M. M. & Baker C. J. (1987b). Association of host plasma membrane K$^+$/H$^+$ exchange with multiplication of *Pseudomonas syringae* pv. *syringae* in *Phaseolus vulgaris*. *Phytopathology* **77**, 1273–9.

Atkinson M. M., Huang J.-S. & Van Dyke C. G. (1981). Absorption of pseudomonads to tobacco cell walls and its significance to bacterium–host interactions. *Physiol. Plant Pathol.* **18**, 1–5.

Baker C. J., Atkinson M. M. & Collmer A. (1987). Concurrent loss of Tn5 mutants of *Pseudomonas syringae* pv. *syringae* of the ability to induce the HR and host plasma membrane K$^+$/H$^+$ exchange in tobacco. *Phytopathology* **77**, 1268–72.

Baker C. J., Mock N., Atkinson M. M. & Hutcheson S. (1990). Inhibition of the hypersensitive response in tobacco by pectate lyase digests of cell wall and polygalacturonic acid. *Physiol. Mol. Plant Pathol.* **37**, 155–67.

Barton-Willis P. A., Wang M. C., Holliday M. J., Long M. R. & Keen N. T. (1984). Purification and composition of lipopolysaccharides from *Pseudomonas syringae* pv. *glycinea*. *Physiol. Plant Pathol.* **25**, 387–98.

Bonatti P. M., Dargeni R. & Mazzuchi U. (1979). Ultrastructure of tobacco leaves protected against *Pseudomonas aptata*. *Phytopathol. Z.* **96**, 302–12.

Brown I. R. & Mansfield J. W. (1988). An ultrastructural study, including cytochemistry and quantitative analyses, of the interactions between pseudomonads and leaves of *Phaseolus vulgaris*. *Physiol. Mol. Plant Pathol.* **33**, 351–76.

Bruegger B. B. & Keen N. T. (1979). Specific elicitors of glyceollin accumulation in the *Pseudomonas glycinea*/soybean host–parasite system. *Physiol. Plant Pathol.* **15**, 43–51.

Butler R. D. & Simon E. W. (1971). Ultrastructural aspects of senescence in plants. *Adv. Gerontol. Res.* **3**, 73–127.

Cook A. A. & Stall R. E. (1968). Effect of *Xanthomonas vesicatoria* on loss of electrolytes from leaves of *Capsicum annum*. *Phytopathology* **58**, 617–19.

Cook A. A. & Stall R. E. (1977). Effects of watersoaking on response to *Xanthomonas vesicatoria* in pepper leaves. *Phytopathology* **67**, 1101–3.

Cramer C. L., Ryder, T. B., Bell J. N. & Lamb C. J. (1985). Rapid switching of plant gene expression induced by fungal elicitor. *Science* **227**, 1240–2.

Croft K. P. C., Voisey C. R. & Slusarenko A. J. (1990). Mechanism of hypersensitive cell collapse: correlation of increased lipoxygenase activity with membrane damage in leaves of *Phaseolus vulgaris* (L.) inoculated with an avirulent race of *Pseudomonas syringae* pv. *phaseolicola*. *Physiol. Mol. Plant Pathol.* **36**, 49–62.

Daniels M. J., Collinge D. B., Maxwell Dow J., Osbourn A. E. & Roberts I. N. (1987). Molecular biology of the interaction of *Xanthomonas campestris* with plants. *Plant Physiol. Biochem.* **25**, 353–9.

Daniels M. J., Dow J. M. & Osborn A. E. (1988). Molecular genetics of pathogenicity in phytopathogenic bacteria. *Ann Rev. Phytopathol.* **20**, 285–312.

Darvill A. G. & Albersheim P. (1984). Phytoalexins and their elicitors – a defence against microbial infection in plants. *Ann. Rev. Plant Physiol.* **35**, 243–75.

Davies I. & Sigee D. C. (1984). Cell ageing and cell death: perspectives. In *Cell Ageing and Cell Death*, ed. I. Davies & D. C. Sigee. Cambridge: Cambridge University Press, pp. 347–50.

Diachun S. & Troutman J. (1954). Multiplication of *Pseudomonas tabaci* in leaves of burley tobacco, *Nicotiana longiflora*, and hybrids. *Phytopathology* **44**, 186–7.

El-Masry M. H. & Sigee D. C. (1989). Electron probe X-ray microanalysis of *Pseudomonas syringae* pv. *tabaci* isolated from inoculated tobacco leaves. *Physiol. Mol. Plant Pathol.* **34**, 557–73.

Ercolani G. L. (1970). Bacterial canker of tomato. IV. The interaction between virulent and avirulent strains of *Corynebacterium michiganense*. *Jens. Phytopathol. Medit.* **9**, 145–50.

Ercolani G. L. (1973). Two hypotheses on the aetiology of response of plants to phytopathogenic bacteria. *J. Gen. Microbiol.* **75**, 83–95.

Fett W. F. & Sequiera L. (1980). A new bacterial agglutinin from soybean. *Plant Physiol.* **66**, 853–8.

Glazener J. A., Orlandi E. W., Harmon G. L. & Baker C. J. (1991). An improved method for monitoring active oxygen in bacteria-treated suspension cells using luminol-dependent chemiluminescence. *Physiol. Mol. Plant Pathol.* **39**, 123–33.

Goodman R. N. (1968). The hypersensitive reaction in tobacco: A reflection of changes in host cell permeability. *Phytopathology* **58**, 872–3.

Goodman R. N. & Plurad S. B. (1971). Ultrastructural changes in tobacco undergoing the hypersensitive reaction caused by plant pathogenic bacteria. *Physiol. Plant Pathol.* **1**, 11–15.

Goodman R. N., Huang P. Y. & White J. A. (1976). Ultrastructural evidence for immobilisation of *Pseudomonas pisi* in tobacco tissue. *Phytopathology* **66**, 754–64.

Harper S., Zewdie N., Brown I. R. & Mansfield J. W. (1987). Histological, physiological and genetical studies of the responses of leaves and pods of *Phaseolus vulgaris* to three races of *Pseudomonas syringae* pv. *phaseolicola* and to *Pseudomonas syringae* pv. *coronafaciens*. *Physiol. Mol. Plant Pathol.* **31**, 153–72.

Hildebrand D. C., Alosi M. C. & Schroth M. N. (1980). Physical entrapment of pseudomonads in bean leaves by films formed at air–water interfaces. *Phytopathology* **70**, 98–109.

Holliday M. J., Keen N. T. & Long M. (1981). Cell death patterns and accumulation of fluorescent material in the hypersensitive response of soybean leaves to *Pseudomonas syringae* pv. *glycinea*. *Physiol. Plant Pathol.* **18**, 279–87.

Keppler L. D., Atkinson M. M. & Baker C. J. (1988). Plasma membrane alteration during bacteria-induced HR in tobacco suspension cells as monitored by intracellular accumulation of fluorescein. *Physiol. Mol. Plant Pathol.* **32**, 209–19.

Keppler L. D. & Novacky A. (1986). Involvement of membrane lipid peroxidation in the development of a bacterially induced hypersensitive reaction. *Phytopathology* **76**, 104–8.

Keppler L. C. & Novacky A. (1987). The initiation of membrane lipid peroxidation during bacteria-induced hypersensitive reaction. *Physiol. Mol. Plant Pathol.* **30**, 233–45.

Klement Z. (1982). Hypersensitivity. In *Phytopathogenic Prokaryotes*, Vol. 2. New York: Academic Press, pp. 149–77.

Klement Z., Farkas G. L. & Lovrekovich L. (1964). Hypersensitive reaction induced by phytopathogenic bacteria in the tobacco leaf. *Phytopathology* **54**, 474–7.

Klement Z. & Goodman R. N. (1967). The hypersensitive reaction to infection by bacterial plant pathogens. *Ann. Rev. Phytopathol.* **5**, 17–44.

Knoche K. K., Clayton M. K. & Fulton R. W. (1987). Comparison of resistance in tobacco to *Pseudomonas syringae* pv. *tabaci* races 0 & 1 by infectivity titrations and bacterial multiplication. *Phytopathology* **77**, 1364–8.

Lelliott R. A., Billing E. & Hayward A. C. (1966). A determinative scheme for the fluorescent plant pathogenic pseudomonads. *J. Appl. Bacteriol.* **29**, 470–89.

Lozano J. C. & Sequeira L. (1970). Prevention of the hypersensitive reaction in tobacco leaves by heat-killed bacterial cells. *Phytopathology* **60**, 875–9.

Lyon G. D. (1989). The biochemical basis of resistance of potatoes to soft rot *Erwinia* spp. – a review. *Plant Pathol.* **38**, 313–39.

Lyon F. & Wood R. K. S. (1976). The hypersensitive reaction and other responses of bean leaves to bacteria. *Ann. Bot.* (London) **40**, 479–91.

Mansfield J. W. (1986). Induced antimicrobial systems in plants. In *Natural Antimicrobial Systems*, FEMS Symposium No. 35, ed. G. W. Gould, M. E. Rhodes-Roberts, A. K. Charnley, R. M. Cooper & R. G. Board. Bath: Bath University Press.

Moesta P., Hahn M. G. & Grisebach H. (1983). Development of a radioimmunoassay for the soybean phytoalexin glyceollin 1. *Plant Physiol.* **73**, 233–7.

Moesta P., Seydell V., Lindner B. & Grisebach H. (1982). Detection of glyceollin on the cellular level in infected soybean by laser microprobe mass analysis. *Z. Naturforsch.* **37C**, 748–51.

Morgham A. T., Richardson P. E., Essenberg M. & Cover E. C. (1988). Effects of continuous dark upon ultrastructure, bacterial populations and accumulation of phytoalexins during interactions between *Xanthomonas campestris* pv. *malvacearum* and bacterial blight-susceptible and resistant cotton. *Physiol. Mol. Plant Pathol.* **32**, 141–62.

Niepold F. & Huber S. J. (1988). Surface antigens of *Pseudomonas syringae* pv. *syringae* are associated with pathogenicity. *Physiol. Mol. Plant Pathol.* **33**, 459–71.

Norelli J. L., Aldwinckle H. S. & Beer S. V. (1988). Virulence of *Erwinia amylovora* strains to *Malus* sp. *Novole* plants grown *in vitro* and in the greenhouse. *Phytopathology* **78**, 1292–7.

Novacky A. & Ullrich-Eberius C. I. (1982). Relationship between membrane potential and ATP level in *Xanthomonas campestris* pv. *malvacearum* infected cotton cotyledons. *Physiol. Mol. Plant Pathol.* **21**, 237–49.

Pavlovkin J., Novacky A. & Ullrich-Eberius C. I. (1986). Membrane changes during bacteria-induced hypersensitive reaction. *Physiol. Mol. Plant Pathol.* **28**, 125–35.

Pierce M. & Essenberg M. (1987). Localization of phytoalexins in fluorescent mesophyll cells isolated from bacterial blight-infected cotyledons and separated from other cells by fluorescence-activated cell sorting. *Physiol. Mol. Plant Pathol.* **31**, 273–90.

Politis D. J. & Goodman R. N. (1978). Localised cell wall appositions: Incompatibility response of tobacco leaf cells to *Pseudomonas pisi*. *Phytopathology* **68**, 309–16.

Sasser M. (1982). Inhibition of antibacterial compounds of the hypersensitive reaction induced by *Pseudomonas pisi* in tobacco. *Phytopathology* **72**, 1513–17.

Sasser M., Andrews A. K. & Doganay Z. V. (1974). Inhibition of photosynthesis diminishes antibacterial action of pepper plants. *Phytopathology* **64**, 770–2.

Sequeira L. (1976). Induction and suppression of the hypersensitive reaction caused by phytopathogenic bacteria: specific and non-specific components. In *Specifity in Plant Diseases*, ed. R. K. S. Wood & A. Graniti. New York: Plenum Press.

Sequeira L. (1978). Lectins and their role in host–pathogen specificity. *Annu. Rev. Phytopathol.* **16**, 453–81.

Sequeira L. (1984). Plant–bacterial interactions. In *Cellular interactions*, Encyclopaedia of Plant Physiology, Vol. 17, ed. H. F. Linskens & J. Heslop-Harrison. Berlin: Springer-Verlag.

Sequeira L., Gaard G. & De Zoeten G. A. (1977). Interaction of bacteria and host cell walls: Its relation to mechanisms of induced resistance. *Physiol. Plant Pathol.* **10**, 43–50.

Sequeira L. & Graham T. L. (1977). Agglutination of avirulent strains of *Pseudomonas solanacearum* by potato lectin. *Physiol. Plant. Pathol.* **11**, 43–54.

Sigee D. C. (1984). Induction of leaf cell death by phytopathogenic bacteria. In *Cell Ageing and Cell Death*, Society for Experimental Biology, Seminar Series Vol. 25. ed. I. Davies & D. C. Sigee. Cambridge: Cambridge University Press, pp. 295–322.

Sigee D. C. & Al-Issa A. N. (1982). The hypersensitive reaction in tobacco leaf tissue infiltrated with *Pseudomonas pisi*. 2. Changes in the population of viable, actively metabolic and total bacteria. *Phytopathol. Z.* **105**, 71–86.

Sigee D. C. & Al-Issa A. N. (1983). The hypersensitive reaction in tobacco leaf tissue infiltrated with *Pseudomonas pisi*. 4. Scanning electron microscope studies on fractured leaf tissue. *Phytopathol. Z.* **106**, 1–15.

Sigee D. C. & Epton H. A. S. (1975). Ultrastructure of *Pseudomonas phaseolicola* in resistant and susceptible leaves of French bean. *Physiol. Plant Pathol.* **6**, 29–34.

Sigee D. C. & Epton H. A. S. (1976). Ultrastructural changes in resistant and susceptible varieties of *Phaseolus vulgaris* following artificial inoculation with *Pseudomonas phaseolicola*. *Physiol. Plant Pathol.* **9**, 1–8.

Sigee D. C. & Hodson N. (1993). Transmission electron microscope studies on incompatible bacteria isolated from inoculated leaf tissue: Changes in elemental composition and degree of flagellation. (In press).

Slusarenko A. J., Croft K. P. & Voisey C. R. (1989). Biochemical and molecular events in the hypersensitive response of bean to *Pseudomonas syringae* pv. *phaseolicola*. In *Biochemistry and Molecular Biology of Plant–Pathogen Interactions*. Proceedings of the International Symposium of the European Phytochemical Society.

Slusarenko A. J. & Wood R. K. S. (1983). Agglutination of *Pseudomonas phaseolicola* by pectin polysaccharide from leaves of *Phaseolus vulgaris*. *Physiol. Mol. Plant Pathol.* **23**, 217–27.

Somlyai G., Holt A., Hevesi M., El-Kady S., Klement Z. & Kari C. (1988). The relationship between the growth rate of *Pseudomonas syringae* pathovars and the hypersensitive reaction in tobacco. *Physiol. Mol. Plant Pathol.* **33**, 473–82.

Stall R. E. & Cook A. A. (1979). Evidence that bacterial contact with the plant cell is necessary for the hypersensitive reaction but not the susceptible reaction. *Physiol. Plant Pathol.* **14**, 77–84.

Sutherland M. (1991). The generation of oxygen radicals during host plant responses to infection. *Physiol. Mol. Plant Pathol.* **39**, 79–93.

Turner J. G. & Novacky A. (1974). The quantitative relationship between plant and bacterial cells involved in the hypersensitive reaction. *Phytopathology* **64**, 885–90.

7

Bacterial virulence and plant disease

Although compatible phytopathogenic bacteria share a common ability to spread and multiply within the host plant, the manner in which they do this and the effect they have on the host plant (disease) vary considerably. This chapter considers general aspects of disease induction, different types of disease that are caused by plant pathogenic bacteria and the range of bacterial characteristics that are important in disease development.

The induction of bacterial disease

The ability of plant pathogenic bacteria to cause disease in a particular host plant depends on many features, including environmental aspects, plant physiology and development, and the expression of pathogenicity and virulence factors by the bacterial cells.

Environmental and physiological factors affecting disease development

Environmental factors are important in the development of plant disease for their direct effects on infection (Chapter 5) and for their indirect effects in determining the physiological status of the plant.

The various aspects of the plant which affect disease development are discussed by Lozano & Zeigler (1990) and include nutritional status, photoperiodic conditioning and stage of maturity and development.

Levels of macronutrients have been shown to be important in plant susceptibility to *Erwinia stewartii*, where elevated levels of N and P increase susceptibility and high levels of Ca and K increase resistance. The role of macronutrients appears to be variable, however, since in some cases fertilisation has been shown to increase resistance (e.g. to *Xanthomonas campestris* pv. *hederae*).

Plant tissues may also differ in their susceptibility to disease according to their stage of development and maturity. Examples of this are provided by the decreasing ability of *Xanthomonas campestris* pv. *manihotis* to infect stem tissues of cassava as lignification proceeds with age, and also by the ability of *Pseudomonas fuscovaginae* to infect only leaf sheaths that envelope the inflorescence, but not sheaths at the seedling stage.

Pathogenicity and virulence factors

The expression of genetically determined pathogenicity and virulence factors is a key aspect in disease induction. Pathogenicity is the fundamental ability of a pathogen to cause disease, while virulence is the degree to which that pathogen affects the health of the plant.

Pathogenicity

One of the most important characteristics of the majority of plant pathogenic bacteria is their ability to perturb the host cell plasmalemma and cause leakage of water, cations and organic nutrients into the intercellular space (see Fig. 6.15), thus creating an essential microenvironment for bacterial growth. Although this pathogenicity factor (with the reciprocal ability to induce a hypersensitive reaction (HR) in incompatible tissue) is possessed by most phytopathogenic bacteria, it does not appear to be a normal feature of those that cause tumour diseases (Table 7.1). The use of the HR as a pathogenicity test is discussed in Chapter 6.

Virulence

The virulence of a plant disease relates to those bacterial characteristics that determine the speed of pathogen growth and spread in the host, and the extent of destruction of host tissue. Some of the major virulence factors are indicated in Table 7.1, but these are not exclusive, and clearly a whole range of bacterial features are involved in the progression of any one disease.

The roles of major virulence factors such as production of toxins, extracellular polysaccharides, cell wall degrading enzymes and plant hormones are considered later in this chapter, and the genetic determination of these characteristics is discussed in Chapter 8.

Table 7.1. *Major types of bacterial disease*

Host range	Symptoms	Compatible response	Major virulence factors	Incompatible response
Necrotic diseases				
Usually narrow	Necrotic leaf spots Stem canker Blossom blight	Delayed cell death (necrosis)	Toxin production	Clear HR
Vascular diseases				
Often wide	Vascular wilts	Long-distance wilting	Extracellular polysaccharide production in xylem	Clear HR
	Yellows	Impaired leaf development	Phloem occlusion	
Soft rot diseases				
Often wide	Soft rots	Tissue maceration	Cell wall degradation	HR at high doses
Tumour diseases				
Often wide	Galls Hyperplasias	Uncontrolled cell division	Hormone disturbance	HR not normally elicited

Major types of plant disease

The symptoms caused by plant pathogenic bacteria include chlorotic (yellow) and necrotic spots of leaves and stems, death of flowers, die-back of shoots and branches, wilts, destruction of root systems, soft rots, gall formation, stunting of organs and whole plants plus alterations in maturation and other physiological changes. Although the compatible interaction at the level of the whole plant thus appears to be highly varied, plant diseases can be grouped into four broad categories in relation to the host tissue response:

1. Necrotic diseases. Where the primary effect of the pathogen is to cause direct cell death, in some cases mediated by specific toxins. The mode of nutrition of the pathogen is typically biotrophic (obtaining nutrients from living cells), since the major phase of growth and division normally occurs between living cells prior to the induction of cell death.

2. Vascular wilt and yellows diseases. Where bacteria multiply within a restricted location (xylem or phloem elements), causing remote disease and death of aerial parts of the plant via the transpiration or phloem stream.

3. Soft rot diseases. Where the pathogen causes separation and destruction

(maceration) of plant tissue by the secretion of wall-degrading enzymes. The mode of nutrition of the pathogen is frequently necrotrophic (obtaining nutrients from dead tissue), since plant cells are normally killed during the maceration process in advance of bacterial colonisation.

4. Tumour diseases. Where the bacterium causes uncontrolled cell proliferation, but which does not result in cell death. In this case, pathogen nutrition is clearly biotrophic, depending on transcription and translation of bacterial genes within living plant cells to generate bacterial nutrients.

These four types of disease not only differ in terms of the tissue response (which relates to virulence factors), but also in relation to host range, symptoms and induction of incompatible response (Table 7.1). In addition to these major types of bacterial disease, naked prokaryotes (Mollicutes) provide a further distinctive category of plant diseases – as discussed previously (p. 29).

The bacteria that induce these diseases (Table 7.1), have been referred to respectively as necrogens (necrotic diseases & wilts), macergens (soft rots) and oncogens (tumours) by Billing (1987)), and show important differences in their pathogenicity and virulence characteristics.

Necrotic diseases (necrotic spots, blights and cankers)

Plant infection with necrogenic bacteria leads initially to localised cell death, which may be apparent as small necrotic lesions (necrotic spots) or more extensive death of whole stems or blossoms (stem and blossom blight).

The majority of phytopathogenic xanthomonads and pseudomonads cause necrotic spots on green parts of susceptible plants, including leaves, stems, fruits and petioles. In some cases the infection remains fairly limited, while in others it spreads throughout the whole plant, causing extensive blight. Areas of necrosis initially appear translucent or water-soaked, due to a massive leakage of water out of cells into the intercellular space. These areas of necrosis may be surrounded by a region of yellowing (chlorotic halo) induced by diffusion of toxin into bacteria-free tissue. Ultimately the area of necrosis becomes desiccated and brown and may disintegrate, leaving holes in the leaf. These features are shown in Fig. 7.1*a*, where halo blight disease of bean has been induced by infiltration of *Pseudomonas syringae* pv. *phaseolicola*. This bacterium multiplies extensively within the intercellular spaces of compatible leaf tissue (Fig. 7.1*b*), producing a toxin (phaseolotoxin, see Table 7.6) which has a major effect on chloroplast metabolism. This leads to rapid disruption of these organelles, with well-defined fine structural changes which are shown in Fig. 7.1*c, d* and described in detail by Sigee & Epton (1976).

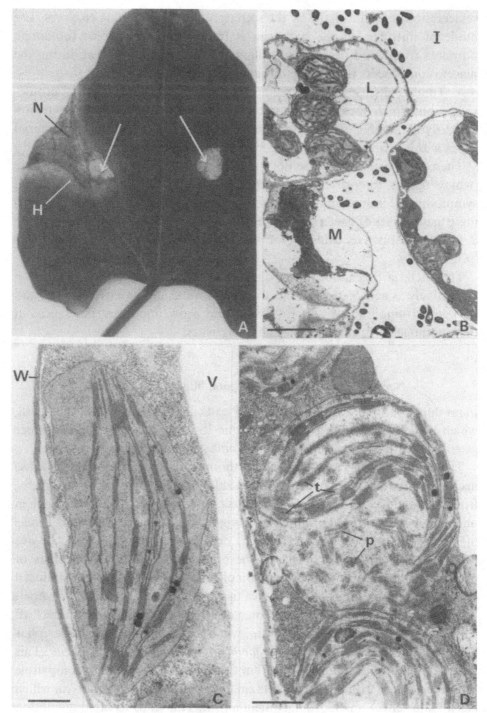

Fig. 7.1 Halo blight of bean caused by *Pseudomonas syringae* pv. *phaseolicola*. (*a*)
Leaf of bean (*Phaseolus vulgaris* cv. Red Mexican), 5 days after spray inoculation

Table 7.2. *Diversity of necrotic diseases*

Symptoms	Bacterium	Disease	Host
Leaf spot	*Pseudomonas syringae* pv. *phaseolicola*	Halo blight	*Phaseolus vulgaris*
	Pseudomonas syringae pv. *tabaci*	Wildfire	*Nicotiana tabacum*
Blossom blight	*Pseudomonas syringae* pv. *syringae*	Blossom blight	*Prunus* sp
Canker & die-back	*Pseudomonas syringae* pv. *mors-prunorum*		
	Erwinia amylovora	Fireblight	Rosaceae

When the region of necrosis occurs in secondary-thickened stems, with infection of bark, cortex and underlying xylem tissue, it is referred to as a canker. If this infection extends around the whole stem (girdling canker), then the entire shoot above this shows quick wilting and drying (die-back). Necrosis, cankers and die-back symptoms are typical of three important tree pathogens of temperate climates: *Pseudomonas syringae* pv. *syringae*, *P.s.* pv. *mors-prunorum* and *Erwinia amylovora* (Table 7.2).

Necrogenic bacteria show some variation in their mode of action. In some cases, host cell death appears to be caused by bacterial toxins, but other factors such as cell wall degrading enzymes may also be important. The host range of these bacteria also shows wide variation (Table 7.2), being limited to

Caption for Fig.7.1 *(cont.)*
with incompatible race 1 (right side) and compatible race 2 (left side) of the pathogen. In both cases, the localised region of bacterial infiltration (arrows) is completely dead and translucent. Compatible bacteria have spread into surrounding tissue to produce a brown region of necrosis (N) which is surrounded by a yellow chlorotic halo (H). (*b*) Transmission electron micrograph (TEM) of diseased tissue (infected with pathogen race 2), 24 hours after infiltration. Bacteria have multiplied in the intercellular spaces (I) and mesophyll cells appear at various stages of degeneration, some still showing details of fine structure (L) while others have completely collapsed contents (M). Bar scale: 5 μm. (*c*) TEM detail from healthy leaf, showing periphery of mesophyll cell with normal chloroplast. W, cell wall; V, vacuole. Bar scale: 0.5 μm. (*d*) TEM detail from diseased leaf (24 hours after bacterial infiltration) showing a degenerating chloroplast. The organelle has become spherical, with displaced thylakoid membranes (t) and localised accumulations of electron-dense particles (p). These are approximately 10 nm in diameter and tend to associate in linear or lattice arrays, with a 10 nm spacing. They are not present in normal (healthy) chloroplasts, but have been observed previously in bean leaves with virus infection and under acute water stress (see Sigee & Epton, 1976). Bar scale: 1 μm. (Photographs in collaboration with H. A. S. Epton.)

Table 7.3. *Some important vascular wilt pathogens*

Bacterium	Disease	Host
Clavibacter michiganensis		
subsp. *michiganensis*	Wilt	Tomato
subsp. *insidiosum*	Wilt	Alfalfa
subsp. *sepedonicum*	Wilt	Potato tuber
Erwinia stewartii	Wilt	*Zea mays*
Erwinia tracheiphila	Wilt	Cucumber
Pseudomonas solanacearum	Wilt	Wide range
Xylella fastidiosa	Pierce's disease	Wide range

single plant species in some cases (particularly foliar diseases caused by *Pseudomonas syringae* pathovars), while in others (e.g. *Erwinia amylovora*) it is quite broad.

Vascular wilt and yellows diseases

The ability of certain pathogens to exclusively invade and colonise the vascular system of herbaceous plants causes two main types of disease, depending on whether the pathogen has occupied the xylem (vascular wilts) or phloem (yellows diseases) tissue.

Xylem invasion and vascular wilt disease

Invasion of the xylem leads to blockage in the transport of water and nutrients, causing wilting and in some cases dwarfing of the entire plant. Although wilt symptoms may not be seen in wet, humid weather, when conditions become dry the effect of the pathogen will become progressively apparent.

A number of important vascular wilt bacteria are shown in Table 7.3. This group of phytopathogens comprises most Coryneform bacteria (including *Clavibacter* and *Curtobacterium*) plus some *Erwinia* sp., pseudomonads, xanthomonads and fastidious prokaryotes. As with other types of bacterial disease, the host range varies from broad to highly specific. With most of these pathogens, wilting is caused partly by occlusion of xylem vessels (reduced water flow) and partly by other factors, including the production of toxins, extracellular polysaccharides (EPS) and cell wall degrading enzymes.

Xylem-limited fastidious bacteria

This group of Gram-negative bacteria constitutes a very distinct group of pathogens, differing from other xylem vascular pathogens in being entirely limited to this part of the plant, requiring complex media for laboratory culture, and requiring insect transmission for infection of new hosts (see Chapter 4). Studies based on *in vitro* growth characteristics, physiology, serology, fatty acid analysis and other molecular studies suggest that these pathogens form a well-defined taxonomic entity (Wells & Raju, 1987), and are generally regarded as forming a single species, *Xylella fastidiosa*.

Although other (non-fastidious) vascular wilt bacteria cause disease by a variety of factors, *Xylella fastidiosa* appears to act entirely by vascular occlusion. This pathogen is particularly important in causing Pierce's disease of grapevine, in which the bacteria block the xylem of petioles and nodes, resulting in a wide range of physical, physiological and biochemical effects (Goodwin et al., 1988). Xylem blockage causes greater resistance to water flow and an increased water stress in the leaves. This is apparent as a reduced water potential (Ψ_w) and lower turgor pressure compared to healthy leaves (Goodwin et al., 1988), resulting further in stomatal closure at various times of day, inhibition of transpiration and photosynthesis plus complex biochemical changes. These include increased levels of Ca^{2+}, Mg^{2+}, glucose, fructose, abscissic acid, proline and chlorophyll, but reduced levels of K^+. The general outcome of these events are accelerated leaf senescence (with marginal chlorosis and necrosis) plus a decline in the general vigour of the plant and an overall stunting.

Xylella fastidiosa has a very broad host range, with known hosts in at least 28 families of dicot and monocot plants. In many of these hosts it produces similar effects to those observed in grapevine (causing almond leaf scorch, alfalfa dwarf and citrus blight), while in other cases the presence of *Xylella* is entirely symptomless (Hopkins, 1988). Other xylem-invading pathogens that cause similar symptoms to *Xylella* include *Xanthomonas albilineans* (sugar cane leaf scald), Sumatra disease bacterium and *Clavibacter xyli* subsp. *xyli* (ratoon stunting disease).

Phloem-invading bacteria and yellows diseases

A number of fastidious prokaryotes (with exacting culture requirements) are able to invade and multiply specifically within phloem tissue, causing a limitation in the flow of photosynthetic products in the plant and resulting in a range of symptoms which include yellowing of leaves, delay in leaf

Table 7.4. *Some major soft rot pathogens and their diseases*

Bacterium	Important diseases	Host	Climatic factors
Xanthomonas campestris pv. *campestris*	Black rot	Crucifers	Cool temperate climate
E. carotovora subsp.			
atroseptica	Blackleg Soft rot	Potato (wide host range)	Cool temperate climate
carotovora	Blackleg Soft rot	Potato (wide host range)	Warm temperate climate
Erwinia chrysanthemi	Blackleg Soft rot	Potato (wide host range)	Subtropics and glasshouses
Clavibacter michiganensis subsp. *sepedonicum*	Bacterial ring rot	Potato (wide host range)	Prevalent in USA, USSR and Scandinavia

development and stunting of growth. The pathogens involved are both true bacteria (with a cell wall) and Mollicutes.

Phloem-inhabiting bacteria (tentatively identified as rickettsia-like in some instances) have been implicated in a number of plant diseases, including potato leaflet stunt, little leaf of *Sida cordifoli*, and club-leaf of clover (Davis, 1990). Mollicutes appear to be of much wider importance and include three species of *Spiroplasma* plus a range of mycoplasma-like organisms (MLOs). The latter are associated with over 200 plant diseases, many of them of major economic importance.

Soft rot diseases

Economically important soft rots of subterranean and aerial plant regions are largely caused by bacteria in the genera *Xanthomonas*, *Erwinia* and *Clavibacter*, major examples of which are noted in Table 7.4. In contrast to other types of bacterial disease, soft rot pathogens cannot readily be identified by their pathogenicity to particular plants since they generally have a wide host range (low specificity).

True pathogenic soft rot bacteria are able to invade and break down the living tissues of roots, tubers, stems, leaves and fruits, where they multiply within the intercellular spaces and secrete large amounts of pectolytic enzymes. These enzymes dissolve the middle lamella, causing cell separation, disintegration and softening of fleshy regions in the infected plant organs. Typical tissue changes and resulting symptoms are shown in Fig. 7.2, where soft rot of potato has been artificially induced by inoculation with *Erwinia carotovora* subsp. *carotovora*. In the region of maceration (Fig. 7.2*b*) cell walls appear pale under phase contrast due to degradation, and cells are plasmolysed and shrunken with large intercellular spaces containing numerous bacteria.

The development of soft rot symptoms in storage tissues such as potato is enhanced by wet, humid conditions, and may be artificially accelerated by enclosing infected tissue under warm conditions in an airtight container. These conditions result in low levels of oxygen due to increased tissue respiration and the presence of a water film which restricts gaseous diffusion. At low oxygen levels, tissue resistance is reduced (with no production of phytoalexins; see Chapter 6) but the growth of facultative anaerobic (*Erwinia* sp.) or obligate anaerobic (*Clostridium* sp.) soft rot bacteria is not adversely affected.

Temperature is a major factor in determining the pathogenicity of particular soft rot bacteria. This is illustrated by the climatic range and experimental behaviour of the three major soft rot pathogens in the genus *Erwinia*: *Erwinia carotovora* subsp. *atroseptica*, *E.c.* subsp. *carotovora* and *E. chrysanthemi* (Table 7.4). If all three erwinias are present below 22°C, *E.c.* subsp. *atroseptica* is the dominant pathogen. However, as temperature rises above 22°C, dominance shifts successively to subsp. *carotovora* and *E. chrysanthemi*. In the field, potato blackleg is caused by *E.c.* subsp. *atroseptica* at temperatures below 25°C, and by *E. chrysanthemi* above this. *Pseudomonas marginalis* and *Bacillus* sp. are more virulent above 30°C than below this temperature.

Although macergenic bacteria act primarily in a different way from necrogenic pathogens, some of the symptoms seen in the field may be closely similar. Thus *Xanthomonas campestris* pv. *campestris*, which produces rotting brown lesions in leaves of cabbage and cauliflower, also generates marginal chlorosis and leaf wilt. Similarly, *Erwinia carotovora* and *E. chrysanthemi* not only cause soft rot of tuber tissue but also blackleg, involving systemic vascular necrosis and wilt. Infection of tubers by *Clavibacter michiganensis* subsp. *sepedonicum* may be symptomless in the early stages, but typically leads to progressive wilting as the disease develops.

Opportunistic pathogens

In addition to the major soft rot pathogens, other bacteria may also cause rotting of tissue in certain situations. Many of these are weak (opportunistic) pathogens, and include *Pseudomonas marginalis*, *Bacillus polymyxa*, *Cytophaga johnsoniae* and *Clostridium* spp. The growth characteristics and diagnosis of these pathogens are discussed by Lelliott & Stead (1987).

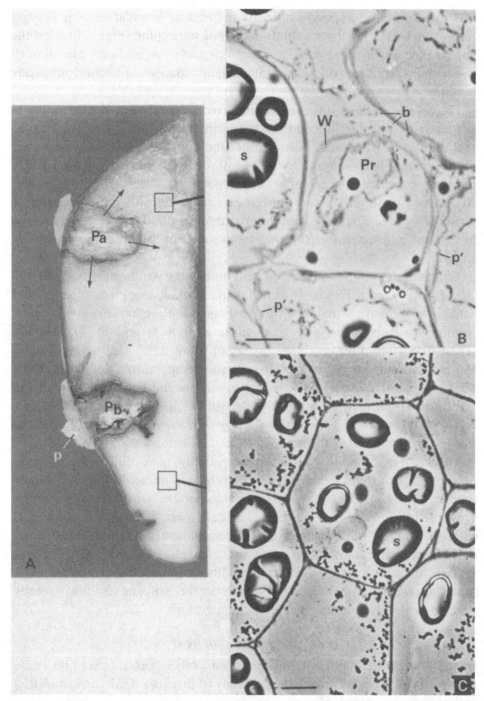

Fig. 7.2 Soft rot of potato caused by *Erwinia carotovora* subsp. *carotovora*. (*a*) Artificial inoculation of potato tuber by infiltration of surface wounds (Pa and Pb) with 5 μl (10⁵ cells/ml) of bacterial suspension. Anaerobic conditions were main-

Some of these weakly pathogenic bacteria are widespread on the plant surface, where they normally occur as saprophytes. *Erwinia rhapontici* and *E. herbicola*, for example, are common inhabitants of the phyllosphere, but have been implicated respectively in crown rot of rhubarb and storage rots of onions. In the soil environment, Campbell *et al.* (1987) reported the isolation of a rhizosphere fluorescent pseudomonad (*Pseudomonas* rp2) that was able to invade roots of cabbage for a limited period shortly after germination, causing host cell wall degradation, destruction of cortical tissue, inhibition of root development and a delay in the maturation of the whole plant.

Tumour diseases

Tumour diseases are caused by plant pathogenic (oncogenic) bacteria which induce uncontrolled cell division (hyperplasia) in infected plants. This proliferation of tissues results in a range of distinct symptoms which are characteristic of the particular pathogen involved. These include:

1. Crown gall and tumours on many dicotyledonous plants, including fruit trees and grapes. Caused by *Agrobacterium tumefaciens* and *A. rubi*.
2. Proliferation of roots (hairy roots) caused by *Agrobacterium rhizogenes*.
3. Proliferation of stems, which may be fused side by side (fasciation), caused by *Rhodococcus fascians*.
4. Galls on olive trees and oleander, caused by *Pseudomonas savastanoi* subsp. *savastanoi*.

In addition to these four major examples of tumour disease, there are also a number of instances where tumours are caused by minor pathogens or where pathogenicity has not been fully characterised. An example of the former is provided by *Erwinia herbicola* pv. *gypsophilae*, which causes crown and root

Caption for Fig.7.2 *(cont.)*
tained by sealing the inoculation plugs with petroleum jelly (p). The tuber has been cut open to show the tissue response, which has been somewhat variable with bacteria spreading from inoculation zone Pa, causing extensive soft rot (arrows), but little disease development from zone Pb. The indicator boxes show regions from which diseased (top) and healthy (bottom) tissue was removed for Fig. 7.2*b* and 7.2*c*. (Photograph in collaboration with N. Walker and H. A. S. Epton.) (*b*) Phase-contrast light micrograph of 2 μm-thick section of soft rot tissue. The cells appear shrunken and plasmolysed, with plasmalemmas (p′) pulled away from the cell wall (W), leading to the formation of detached protoplasts (Pr). Numerous bacteria (b) can be seen in the enlarged intercellular spaces. s, starch grain. Bar scale: 20 μm. (*c*) Light micrograph of healthy tissue, showing turgid, angular cells with little tissue volume occupied by intercellular spaces. Cell walls appear much denser under phase-contrast compared with diseased tissue. Bar scale: 30 μm.

Table 7.5. *Species and biovars of* Agrobacterium

Species	Biovars	Pathogenicity	Disease
Agrobacterium tumefaciens	I, II & III	Tumorigenic strains	Crown gall
Agrobacterium radiobacter	I, II & III	Non-tumorigenic strains	—
Agrobacterium rhizogenes	I & II	Rhizogenic strains	Hairy root
Agrobacterium rubi	–	Tumorigenic	Crown gall

Information taken from Lelliott & Stead (1987).

gall in *Gypsophila*, and represents an unusual case of pathogenicity within a normally saprophytic species (Manulis *et al.*, 1991). Examples of plant tumour diseases caused by pathogenic bacteria which have not been fully characterised include bacterial gall of carrot and various gall diseases of trees (Ogimi & Takikawa, 1988). The development of tumour pathogenicity thus appears to have evolved or been acquired by a variety of taxonomic groups, with at least five species (in three genera) having major oncogenic activity.

Agrobacterium *species*

The taxonomic status of this genus of soil bacteria is confused, since it is split into both species and biovars, with little correlation between the two (Table 7.5).

Pathogenicity in *Agrobacterium* is conferred by the possession of either a root-inducing (Ri) or tumour-inducing (Ti) plasmid, details of which are given in Chapter 8 (see Fig. 8.13). The susceptibility of plant tissues to infection by this pathogen is determined by a number of factors, including age of tissue (young is more susceptible), host specificity, type of tissue (e.g. cambium may be more susceptible than other cell types) and nutritional status of tissue. Plant infection and disease development may also be enhanced by application of exogenous auxins.

Agrobacterium tumefaciens is an opportunistic pathogen that requires tissue damage for the inception of disease and results in the development of crown gall disease. This is characterised by a disorganised growth of the main axis just below the surface of the soil at the point of emergence of lateral roots (the plant 'crown'). Normally, galls are found only on dicotyledonous species; although it has been recently demonstrated that galls can be formed on some monocots. In the case of crown gall, the tumour consists primarily of parenchyma tissue that shows little differentiation, with the inciting bacteria being few in number and confined to the outer regions of the gall.

The extent to which infection with this pathogen affects the overall vigour of the plant varies considerably, and may only result in a localised tumour response. In woody fruit trees such as cherry galls commonly occur in nursery material (on the root system, crown or base of the rootstock) with no consistent difference between infected and non-infected plants in terms of sustained tree growth (Garrett, 1987). In this situation, the disease is important mainly in terms of the adverse appearance of affected nursery material.

The infection process has been investigated in particular detail in this pathogen and involves entry of bacteria into wounded tissue, specific binding of the pathogen to plant surfaces, transfer of the Ti plasmid into plant cells and integration of the molecule into the plant genome. The genetic and molecular events involved in the process of plant transformation are described in Chapter 8 (see Fig. 8.14).

Plant cell transformation results in the high-level synthesis of two major types of compound: phytohormones and opines. The production of plant hormones constitutes an important virulence factor, and is described on p. 206.

The synthesis of opines by transformed plant cells is important for the nutrition of the invading bacteria. These compounds have been divided into two major groups: amino acid derivatives (agropine, octopine and nopaline; Fig. 7.3) and sugar derivatives (the agrocipines). They serve as a unique food supply for the agrobacteria, where they are used as carbon sources specifically by those bacteria that carry the Ti plasmid. They also promote conjugative transfer of the Ti plasmid to plasmidless strains of the bacterium, so that any non-pathogenic cells of *Agrobacterium tumefaciens* present in the gall tissue will rapidly take up the Ti plasmid, assimilate opines and become pathogenic. Host cell changes induced by *Agrobacterium tumefaciens* thus lead to the release of nutrients for the bacterium, which is able to grow and multiply within the microenvironment that the plasmid has created.

Pseudomonas savastanoi subsp. *savastanoi*

Pseudomonas savastanoi subsp. *savastanoi* induces the formation of hyperplastic growths, apparent as knots on young stems and branches (and occasionally on leaves and fruits) in a range of plant species. It commonly occurs as a pathogen on olive (*Olea europaea*) and oleander (*Nerium oleander*), and is also pathogenic on ash (*Fraxinus excelsior*), forsythia (*Forsythia intermedia*), jasmin (*Jasminus* spp.) and privet (*Ligustrum japonicum*).

Infection leading to hyperplasia typically involves entry into host tissue via

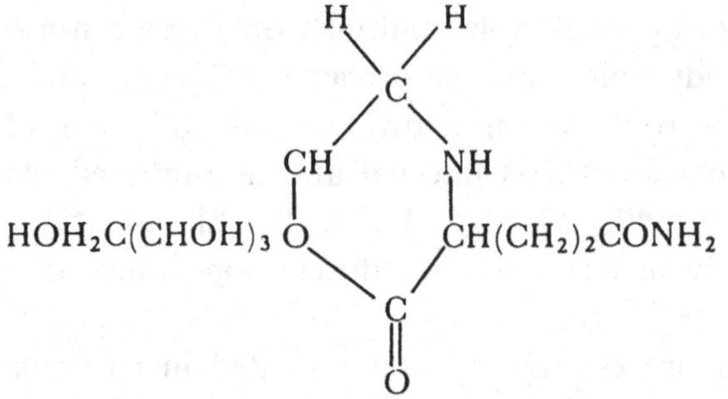

$$\text{HOH}_2\text{C(CHOH)}_3\text{ O}$$

Agropine

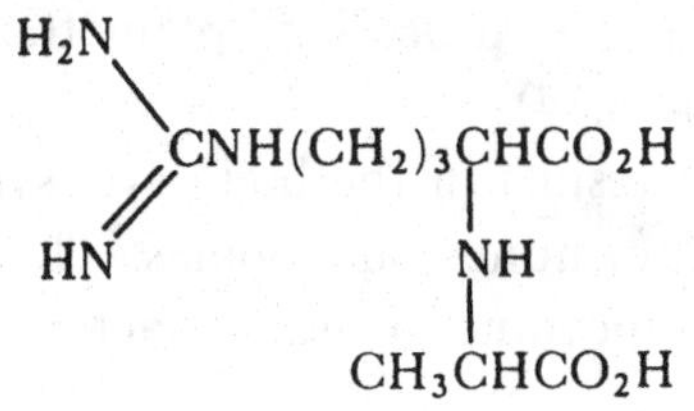

$$\text{H}_2\text{N}$$
$$\text{CNH(CH}_2)_3\text{CHCO}_2\text{H}$$
$$\text{HN}$$
$$\text{NH}$$
$$\text{CH}_3\text{CHCO}_2\text{H}$$

Octopine

$$\text{H}_2\text{N}$$
$$\text{CNH(CH}_2)_3\text{CHCO}_2\text{H}$$
$$\text{HN}$$
$$\text{NH}$$
$$\text{HO}_2\text{C(CH}_2)_2\text{CHCO}_2\text{H}$$

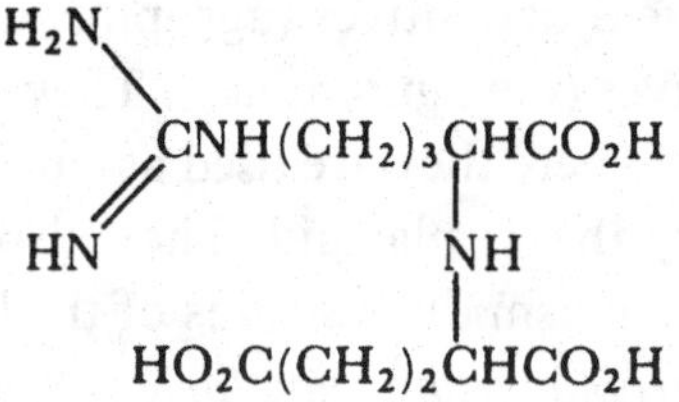

Nopaline

Fig. 7.3 Structures of three opines.

wounds (such as leaf scars and frost injuries) followed by localised accumulation of the bacteria in intercellular spaces and spread of bacteria to the
cambial region. Tumour formation results both from enhanced cambial
activity and proliferation (with enlargement) of host cells around bacterial
cavities (Surico, 1986). Differentiation of vascular elements subsequently
occurs in the tumour. The developmental pattern of hyperplasia, involving
phases of host cell proliferation and differentiation, suggests that at least two
growth factors may be involved in pathogenesis.

The production of plant hormones by this pathogen, and their role in
disease virulence, is described on p. 207.

Virulence factors

As noted previously, virulence factors are characteristics of the bacterial pathogen which are important in the determination of disease severity. Four major virulence factors are considered here: production of toxins, extracellular enzymes, extracellular polysaccharides and plant hormones.

Bacterial toxins

Toxins of phytopathogenic bacteria are generally defined as non-enzymatic metabolic products which are released by the pathogen and cause chemical injury to the host. Toxins have a wide range of physiological and biochemical effects on the host plant, causing symptoms such as chlorosis, water-soaking, necrosis, growth abnormalities and wilting. A general review of bacterial and fungal toxins is presented by Durbin (1983).

Bacterial toxins and disease

Critical evidence for the involvement of particular bacterial toxins in disease would theoretically require isolation of the toxin from diseased (but not healthy) tissue, demonstration of production by bacterial cells *in planta*, purification and chemical identification of the compound, with final experimental testing of the purified extract on a host plant to generate 'disease symptoms'. Although toxin production is considered to be important in many plant diseases, evidence that a particular bacterial metabolite is important in causing a particular effect is often lacking for a number of technical reasons:

1. Toxin isolation from diseased tissue may be difficult due to the complexity of the chemical procedures involved and the low levels of the toxin present. In practice, toxins are often more readily isolated from *in vitro* culture medium prior to purification and experimental testing.

2. Toxins may be chemically unstable, making isolation and purification difficult.

3. Administration of purified compounds to test for symptom development may lead to artefacts, particularly if too high concentrations are used. Toxins are typically active at very low levels (less than 10^{-6}–10^{-8} M), and compounds to be tested should not exceed this concentration. In the case of phaseolotoxin, for example, the development of chlorosis in young leaves may be induced by applying toxin concentrations as low as 30 pmol, and levels of 10–100 nM/g fresh weight of plant tissue resulted in systemic symptoms.

In spite of these problems, a number of toxins have been clearly identified

Table 7.6. *Phytopathogenic bacterial toxins*

Bacterium	Toxin	Target molecule	General reference
Pseudomonas syringae			
Several pathovars	Coronatine	Not known	Bender *et al.* (1989)
pv. *tagetis*	Tagetitoxin	Chloroplast RNA polymerase	Rudolph (1990)
pv. *phaseolicola*	Phaseolotoxin	Ornithine carbamoyl transferase	Hartman *et al.* (1986)
pv. *syringae*	Syringomycin Syringotoxin	Plasmalemma ATPase	Xu & Gross (1988)
pv. *tolaasii*	Tolaasin	Plasmalemma constituent	Rainey *et al.* (1991)
Several pathovars	Tabtoxin (tabtoxinine β-lactam)	Glutamate synthetase	Levi & Durbin (1966)
Pseudomonas andropogonis *Rhizobium japonicum*	Rhizobitoxine	β-cystathionase	Mitchell & Frey (1988)
Xanthomonas campestris pv. *manihotis*	3(methylthio)-proprionic acid	Not known	Perreaux *et al.* (1986)

and characterised, some of which are described on p. 191 and listed in Table 7.6.

Toxin structure and biosynthesis

Toxins produced by bacterial plant pathogens are typically small molecules (MW about 10^3 daltons), except those that are involved in the induction of wilting (MW 10^4–10^6 daltons). Almost all bacterial toxins are products of secondary (non-essential) metabolism, involving well-known metabolic pathways and requiring the translation of novel proteins. Most bacterial toxins appear to be secreted by cells growing both *in vitro* (in synthetic medium) and *in planta*, though certain strains of *Pseudomonas syringae* pv. *tagetis* produce toxins only during disease development (Durbin, 1983).

In common with other secondary metabolites, the production of these compounds is very dependent on nutritional and environmental conditions and occurs at particular phases in the growth cycle. This is shown particularly clearly during *in vitro* cultures, where toxin production may be restricted largely to stationary phase cells.

Bacterial toxins also resemble other secondary metabolites in being produced by a restricted number of organisms (see below), by being resistant to metabolic degradation, and by having no clear function in the normal metabolism of the organism. The resistance of bacterial toxins to inactivation by non-specific degradative enzymes is consistent with a number of chemical features, including the presence of D-amino acids, unusual or modified amino acids, cyclised carbon skeletons and few structural requirements for biological activity.

Toxin specificity and function

Toxin specificity may be considered in relation to the producing organism, occurrence in disease, experimental effects of toxin extract on a range of test organisms and precise biochemical effects. As an example, phaseolotoxin is highly specific in terms of its producing organism (*Pseudomonas syringae* pv. *phaseolicola*), disease occurrence (leaf spot of *Glycine* and haloblight of bean) and biochemical effect (target molecule), but is non-specific in that 'symptoms' are produced following experimental application to a wide range of higher plants and micro-organisms. Toxicity to micro-organisms can be particularly useful in providing a suitable toxin assay. In the case of phaseolotoxin, for example, a quantitative microbiological assay has been developed based on the ability of the toxin to inhibit the growth of *Escherichia coli* and *Salmonella typhimurium* on arginine-deficient medium (Staskawicz and Panopoulos, 1979).

Specificity in toxin production

The origins of some of the major bacterial toxins are shown in Table 7.6. Most toxins are only produced by a single pathovar or restricted group of organisms, and demonstration of production of particular toxins may be an important aid in the identification of bacterial strains (Rudolph, 1990). On a more general level, pseudomonads tend to produce toxins consisting of amino acids or oligopeptides, xanthomonads produce carboxylic acids and related compounds and the production of polysaccharides and peptidosaccharides is a property of all bacterial species.

Disease specificity

Specificity in disease reflects host–pathogen specificity rather than a fundamental property of the toxin, experimental application of which will induce 'symptoms' in a wide range of higher plants and micro-organisms or in higher plants only. Unlike fungal toxins (such as victorin and tentoxin), there do not appear to be any examples of bacterial toxins which are restricted in their action to particular higher plant species or cultivars.

Biochemical target molecules

At the biochemical level, toxins produced by biotrophic pathogens typically have a highly specific metabolic effect, facilitating pathogen invasion and nutrition within the living tissue, with no rapid induction of host cell death. This is in contrast to necrotrophs, where toxins such as fatty acid esters, organic acids and phenolic compounds are biochemically non-specific, resulting in rapid cell death of the host tissue.

Some of the specific biochemical effects of pathogen toxins are listed in Table 7.6. Some toxins (including phaseolotoxin, tabtoxinine-β-lactam and rhizobitoxine) have been shown to bind irreversibly to specific enzymes (target molecules), thereby blocking their catalytic activity. In the case of phaseolotoxin and rhizobitoxin, this has been shown to occur by a mechanism (Kcat) involving specific activation of the toxin by the target molecule, followed by specific alkylation of the target by the activated toxin (Durbin, 1983).

General role of toxins in disease

Although toxins are important for pathogen virulence and the development of disease symptoms, they are not essential for the growth and spread of the pathogen *in planta* – as shown by infiltration of non-toxin producing (TOX⁻) mutants (Chapter 8). Their role in causing remote host cell dysfunction may be to bring about the release of water and nutrients at some distance from the pathogen, creating an environment into which the bacterium can more readily migrate and grow.

Bacterial resistance to toxins

In those situations where the target molecule is common to both the pathogen and higher plant cell, bacterial self-protective mechanisms appear to operate. This has been demonstrated for *Pseudomonas syringae* pv. *phaseolicola* and *P.s.* pv. *tabaci*, where there appears to be a direct relationship between operation of the protection mechanism and toxin production, since TOX⁻ (but not TOX⁺) strains are sensitive to the toxin (see Durbin, 1983). Linkage between twin characters of toxin production and protection will provide a strong selection pressure against the survival of TOX⁻ mutants, which may explain why these cells are rarely encountered under natural conditions.

The adaptive importance of protective mechanisms is also indicated by the fact that they are typically multiple, suggesting evolutionary conservation of all beneficial protective features. In the cases of *Pseudomonas syringae* pv. *tabaci*, for example, the bacterial enzyme glutamate synthetase is sensitive to the toxin, protection against which is achieved by: (1) toxin synthesis within the cell in an inert (precursor) form; (2) activation of the precursor molecule by hydrolysis which is thought to occur outside the plasmalemma (in the periplasm); (3) activation of the precursor only under conditions (high Zn level) typical of the internal plant environment; and (4) inhibition of bacterial uptake of either the toxin or its precursor if they are present in the surrounding medium.

In *Pseudomonas syringae* pv. *phaseolicola*, two types of ornithine carbamoyl transferase are produced: one resistant and one susceptible to toxin. When the bacterium is grown at temperatures which permit toxin production, the resistant type predominates, while at zero toxin production (30 °C) the enzyme is susceptible (Ferguson *et al.*, 1980). Transgenic plants synthesising the resistant enzyme also show resistance to the toxin (Hutziloukas and Panopoulos, 1992).

Characterisation of individual toxins

A number of bacterial toxins have now been characterised in some detail in terms of their chemical structure, biosynthesis and mode of action. Some of these are described below, and are listed in Table 7.6.

Coronatine

Coronatine has a unique composition (Fig. 7.4), consisting of a polyketide structure (coronafacic acid) linked to a cyclopropane component (derived from isoleucine). It is produced by several pathovars of *Pseudomonas syringae*, and is important in disease of ryegrass (pv. *atropurpurea*), soybean (pv. *glycinea*), tomato (pv. *tomato*) and *Prunus* (pv. *mors-prunorum*), where it induces symptoms of chlorosis, stunting and hypertrophy. The importance of coronatine for disease virulence has been demonstrated in both soybean (Leary *et al.*, 1988) and tomato (Bender *et al.*, 1989) by comparing the effects of wild type and nontoxigenic mutants (see Chapter 8).

The toxin is conventionally assayed by induction of chlorosis in soybean leaves, induction of hypertrophy in potato tuber discs or thin layer chromatography of purified culture fluid extracts. More recently, Leary *et al.* (1988) have developed a reliable and rapid method to detect the toxin using antibodies raised to a bovine serum albumin–coronatine conjugate.

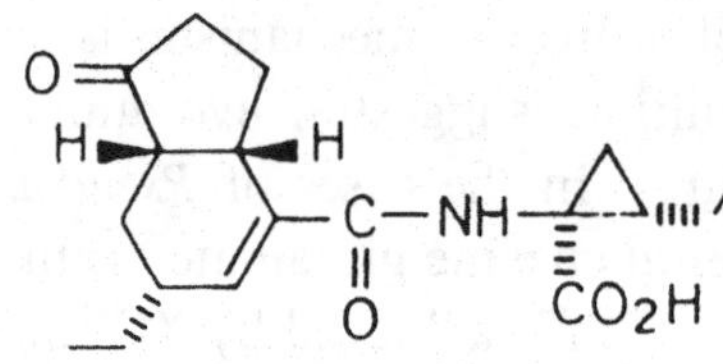

Coronatine

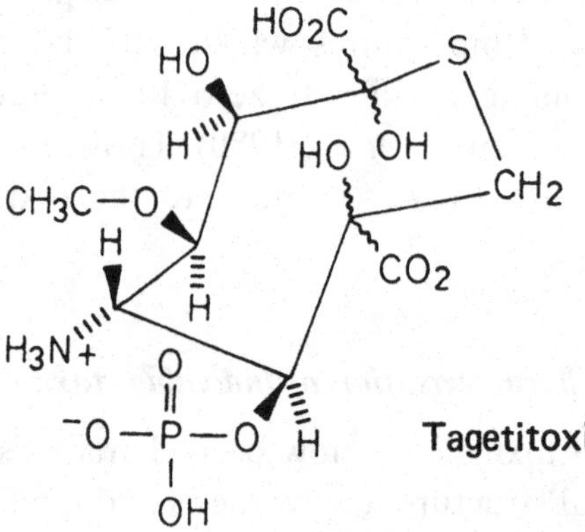

Tagetitoxin

Fig. 7.4 Structure of two cyclic toxins from plant pathogenic bacteria.

Tagetitoxin

This toxin is produced solely by *Pseudomonas syringae* pv. *tagetis* and is a cyclic compound (Fig. 7.4). Most strains of this pathogen do not normally produce the toxin in culture, but do produce it *in planta*, resulting in light-dependent symptoms of apical chlorosis and yellow haloes on leaves. The toxin appears to have a specific effect on chloroplasts by inhibiting chloroplast RNA polymerase, leading to an absence of chloroplast (70s) ribosomes and abnormal thylakoid development.

Phaseolotoxin

The production of phaseolotoxin by the single pathovar *Pseudomonas syringae* pv. *phaseolicola* shows considerable variation between strains, with a direct correlation between the amount of toxin produced and the extent of symptom (chlorotic halo) development.

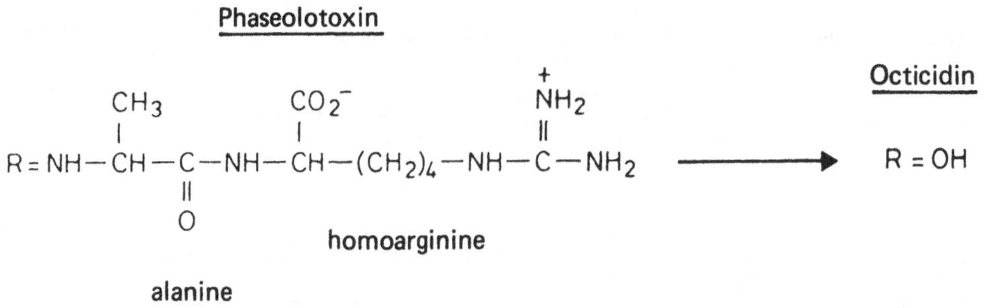

Phaseolotoxin

$$R = NH - CH - C - NH - CH - (CH_2)_4 - NH - C - NH_2 \longrightarrow R = OH$$

Octicidin

alanine homoarginine

Fig. 7.5 Structure of phaseolotoxin and its hydrolysis product. Both the tripeptide (phaseolotoxin) and its hydrolysis product are produced by *Pseudomonas syringae* pv. *phaseolicola* in batch culture, and both toxins are able to induce chlorosis *in planta*.

The chemical structure of phaseolotoxin is shown in Fig. 7.5, hydrolysis of which (by plant peptidases) leads to the formation of octicidin, the main functional toxin *in planta*. Both phaseolotoxin and the peptide-cleavage product act by specific and highly potent inhibition of ornithine carbamoyl transferase (OCTase; Fig. 7.6), an enzyme of the urea (ornithine) cycle. Kinetic studies have shown that the toxin blocks the binding of carbamoyl phosphate (but not ornithine) to the enzyme, resulting in the inhibition of citrulline synthesis and the accumulation of ornithine in the tissue. Carbamoyl phosphate shows little accumulation since it is taken up by nucleic acid synthesis (Durbin, 1983).

OCTase preparations from a variety of sources (several plant species, animals and bacteria) have been shown to be sensitive to phaseolotoxin, the only exception being OCTase from pv. *phaseolicola* itself. In the case of leaf tissue, inhibition of amino acid synthesis by this toxin leads to chlorosis, which can be reversed by application of citrulline. Application of arginine or orotic acid also reverses chlorosis, suggesting that the toxin has a complex effect on carbon flux through the ornithine cycle and related pathways. The induction of chlorosis only occurs in the light, and on young, still-growing leaves, suggesting that the toxin has an inhibitory effect on chlorophyll synthesis without causing chlorophyll degradation. The effect of this toxin on chloroplast fine structure is shown in Fig. 7.1*d*.

Syringomycin and syringotoxin

Syringomycin is a necrosis-inducing toxin produced by a wide spectrum of *Pseudomonas syringae* pv. *syringae* strains, and is implicated in the symptom development of diseases as diverse as bacterial canker of stone trees, bacterial

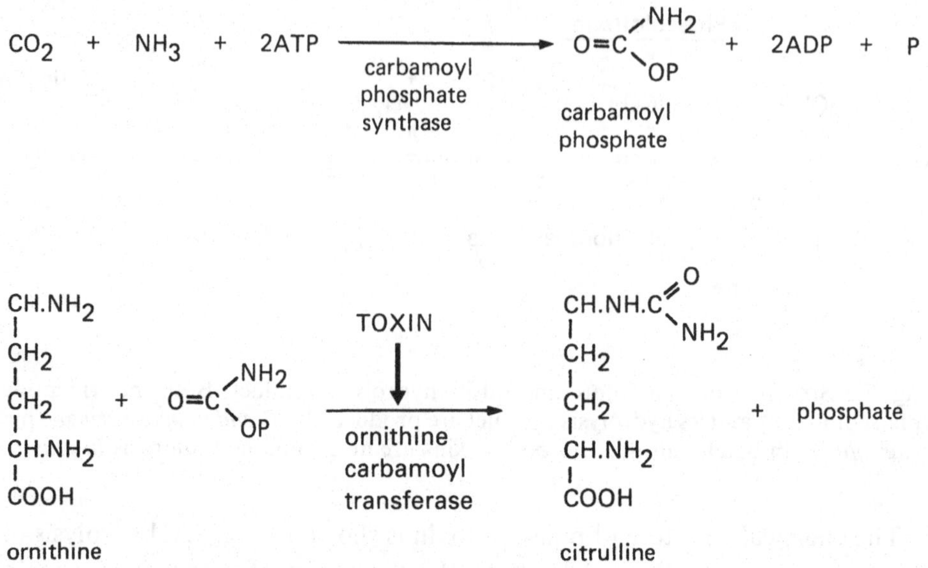

$$CO_2 \; + \; NH_3 \; + \; 2ATP \; \xrightarrow[\substack{\text{carbamoyl} \\ \text{phosphate} \\ \text{synthase}}]{} \; O=C\begin{smallmatrix} \nearrow NH_2 \\ \searrow OP \end{smallmatrix} \; + \; 2ADP \; + \; P$$

carbamoyl
phosphate

$$
\begin{array}{l}
CH.NH_2 \\
| \\
CH_2 \\
| \\
CH_2 \\
| \\
CH.NH_2 \\
| \\
COOH
\end{array}
\; + \;
O=C\begin{smallmatrix} \nearrow NH_2 \\ \searrow OP \end{smallmatrix}
\xrightarrow[\substack{\text{ornithine} \\ \text{carbamoyl} \\ \text{transferase}}]{\text{TOXIN} \; \downarrow}
\begin{array}{l}
CH.NH.C\begin{smallmatrix} \nearrow O \\ \searrow NH_2 \end{smallmatrix} \\
| \\
CH_2 \\
| \\
CH_2 \\
| \\
CH.NH_2 \\
| \\
COOH
\end{array}
\; + \; \text{phosphate}
$$

ornithine   citrulline

Fig. 7.6 Inhibition of ornithinecarbamoyltransferase (OCTase) by phaseolotoxin and octicidin. Octicidin is 20 times more potent than phaseolotoxin in inhibiting OCTase.

brown spot of bean and bacterial blight of wheat. The closely related toxin syringotoxin is unique to citrus isolates. In addition to being phytotoxic, syringomycin and syringotoxin also exhibit broad antibiotic activity, inhibiting the growth of both prokaryotic and eukaryotic micro-organisms.

The toxin acts by disrupting plasmalemma function in plant cells of infected tissue. This occurs by stimulation of a proton pump ATPase, leading to an alteration in the pH gradient across the membrane, increased permeability and ultimately cell necrosis (Xu & Gross, 1988).

Syringomycin is a low MW peptide-containing toxin, purified preparations of which are normally obtained by *n*-butanol extraction followed by ion-exchange chromatography. Recently, an improved purification procedure involving high performance reverse-phase liquid chromatography has revealed a multiplicity of chemically related peptides within the purified sample, suggesting that syringomycin should be regarded as a molecular complex rather than a single compound (Ballio *et al.*, 1988).

Biosynthesis of syringomycin and syringotoxin involves a multiplicity of enzymes, including various peptide synthetases, genetic determination of which is discussed in Chapter 8. In the case of syringotoxin, for example, two synthetases (ST1 and ST2) have so far been identified and shown to be synthesised by cultured cells at the end of exponential phase (Morgan & Chatterjee, 1988). The stationary phase timing of ST1 and ST2 synthesis

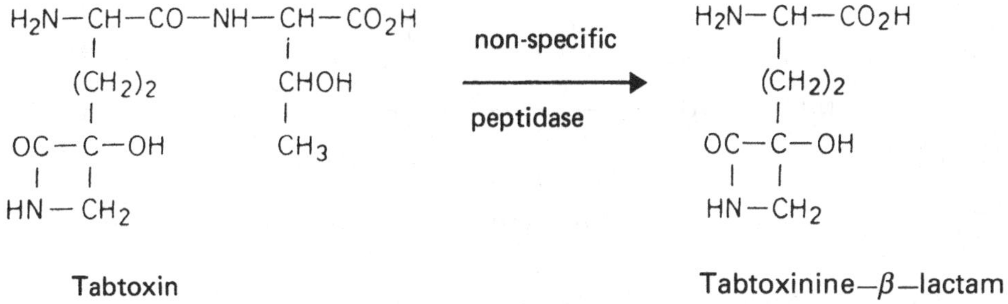

H₂N—CH—CO—NH—CH—CO₂H → H₂N—CH—CO₂H

$$\text{H}_2\text{N}-\text{CH}-\text{CO}-\text{NH}-\text{CH}-\text{CO}_2\text{H}$$

Tabtoxin structure (left):

```
H₂N─CH─CO─NH─CH─CO₂H
    │           │
  (CH₂)₂       CHOH
    │           │
  OC─C─OH      CH₃
    │ │
   HN─CH₂
```

non-specific
───────────►
peptidase

Tabtoxinine–β–lactam structure (right):

```
H₂N─CH─CO₂H
    │
  (CH₂)₂
    │
  OC─C─OH
    │ │
   HN─CH₂
```

Tabtoxin   Tabtoxinine—β—lactam

Fig. 7.7 Structure of tabtoxin and its hydrolysis product.

coincides with the onset of syringotoxin production and is typical of secondary metabolism. Synthesis of these compounds also depends on environmental factors, and appears to be regulated particularly by the concentration of Fe^{3+} ions (within the range 0.2–5.0 μM).

Tolaasin

This toxin is produced by *Pseudomonas syringae* pv. *tolaasii*, a pathogen of mushroom, where it causes disruption of plasmamembranes (Rainey *et al.*, 1991). The structure of tolaasin, a lipodepsipeptide composed of 18 amino acid residues with a β-octanoic acid at the amino terminus, is typical of low molecular weight compounds that are active within membranes. Recent *in vitro* studies on planar lipid bilayers have suggested that this toxin may act by forming voltage-gated, cation selective, ion channels in the plasmalemma of the host plant.

Tabtoxin and tabtoxinine-β-lactam

Tabtoxin is a dipeptide molecule produced in culture by *Pseudomonas syringae* pv. *tabaci* (plus other pathovars such as pv. *atropurpurea*, *coronafaciens* and *garcae*) and is implicated in the induction of chlorosis in host tissue. This is shown by the fact that strains producing tabtoxin in culture always induce chlorosis in infected leaf tissue, while strains that cannot produce the molecule do not cause chlorosis. Furthermore, direct application of tabtoxin induces chlorosis, and both bacterial and direct chemical induction require light sensitisation (Levi & Durbin, 1986).

Tabtoxin itself does not appear to have any biological activity (Turner, 1986), but is readily converted by hydrolases (of bacterial or plant origin) to generate the bioactive compound, tabtoxinine-β-lactam (Fig. 7.7). This binds irreversibly to the enzyme glutamine synthetase (Thomas *et al.*, 1983), which is involved in ammonia assimilation (Fig. 7.8). Inhibition of this enzyme leads to a rapid build-up of ammonia in the plant tissue which causes an uncoupling

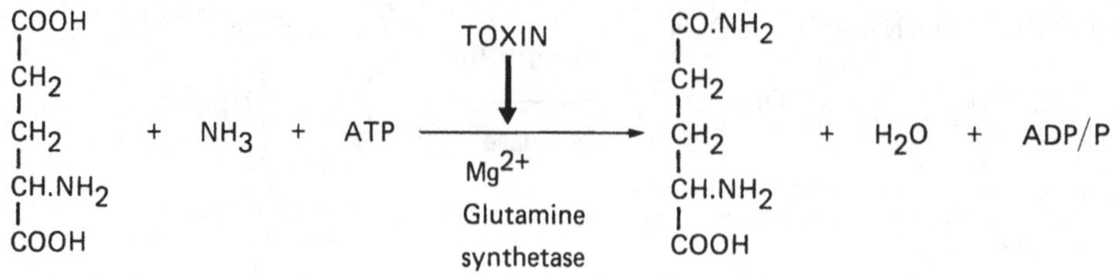

$$
\begin{array}{c}
\text{COOH} \\
| \\
\text{CH}_2 \\
| \\
\text{CH}_2 \\
| \\
\text{CH.NH}_2 \\
| \\
\text{COOH}
\end{array}
\quad + \quad \text{NH}_3 \quad + \quad \text{ATP}
\xrightarrow[\substack{\text{Mg}^{2+} \\ \text{Glutamine} \\ \text{synthetase}}]{\text{TOXIN} \downarrow}
\begin{array}{c}
\text{CO.NH}_2 \\
| \\
\text{CH}_2 \\
| \\
\text{CH}_2 \\
| \\
\text{CH.NH}_2 \\
| \\
\text{COOH}
\end{array}
\quad + \quad \text{H}_2\text{O} \quad + \quad \text{ADP/P}
$$

Fig. 7.8 Inhibition of glutamine synthetase by tabtoxinine-β-lactam.

of the carbon and energy fixation components of photosynthesis, selective destruction of thylakoid membranes and breakdown of chlorophyll, with resulting chlorosis.

Hydrolysis of tabtoxin in culture appears to be mediated by bacterial Zn-activated peptidase activity, since this enzyme is specifically and only present in those strains that produce tabtoxinine-β-lactam *in vitro* (Levi & Durbin, 1986). The bacterial enzyme is also important for hydrolysis of tabtoxin *in planta* since tabtoxinine-β-lactam is only found in intercellular fluid after infection with bacterial strains that have the Zn-activated peptidase; other strains have tabtoxin only in the fluid. Activation of the enzyme *in vitro* may either be carried out by addition of Zn to bacterial cultures, or by adding the cation to purified enzyme extract. In the latter case, the optimal concentration of Zn required for activation is at the same level (25 μM) as that found in tobacco intracellular fluid. The evolution of this system presumably ensures that the bacterially induced formation of the toxic product normally only occurs *in planta* and not elsewhere (e.g. the plant surface).

Rhizobitoxine

Production of rhizobitoxine has been identified for root-nodulating strains of *Rhizobium*, and also more recently (Mitchell & Frey, 1988) for the pathogen *Pseudomonas andropogonis*, which causes bacterial stripe in corn, sorghum and Sudan grass. The chlorosis-inducing compound isolated from this bacterium was shown to be rhizobitoxine on the basis of its ^{1}H and ^{13}C NMR spectra.

The biochemical effect of this toxin is complex and involves inhibition of the synthesis of both homocysteine (by specific inactivation of the enzyme β-cystathionase) and ethylene.

The production of rhizobitoxine by both *Rhizobium* species and pathogenic bacteria of the *Pseudomonas andropogonis* complex is interesting since it implies transfer of genetic information between groups of bacteria that are unrelated and occupy different ecological niches. It is also interesting that the same toxin should be of adaptive significance in two quite different types of bacteria–plant interaction.

Carboxylic acids

Various *Xanthomonas campestris* pathovars have been shown to produce carboxylic acid toxins in culture, which may be important for the induction of symptoms *in planta* (Rudolph, 1990). These pathovars include pv. *oryzeae* (phenylacetic and other acids), pv. *manihot* (3-methylthioproprionic acid) and pv. *campestris* (3-methylthioacrylic acid and 3-methylthioproprionic acid).

Polysaccharide toxins

Bacterial polysaccharides may be important in the induction of wilting and water-soaking symptoms in the host plant. Neither the metabolic effect nor the toxins themselves have been clearly characterised, and it is not even clear whether the effects of the toxin on the plant cells are primarily chemical or physical in nature.

The potential involvement of bacterial polysaccharides as wilt toxins has been mentioned previously in relation to wilt diseases and include products from *Erwinia amylovora* (referred to as 'amylovorin') and *Erwinia stewartii*. The main polysaccharide component produced by *Erwinia amylovora* is a galactan with side-chains from glucuronic acid, glucose and mannose.

Extracellular polysaccharides (EPS) have been implicated in water-soaking, which is an important step in the development of the susceptible reaction for many leaf spot diseases caused by pseudomonads and xanthomonads. With plants kept under high humidity, El-Banoby and Rudolph (1979) were able to directly induce water-soaking in leaf tissue by infiltration of EPS preparations from *Pseudomonas syringae* pv. *phaseolicola*. These EPS preparations contained four macromolecular fractions: levan, alginate, lipopolysaccharide (LPS) and protein.

Extracellular enzymes

The secretion of a wide range of enzymes that are able to degrade higher plant cell walls and other cellular components is an important feature of soft rot bacteria and also bacteria that cause tissue necrosis and vascular wilt diseases. The plant cell is a particularly important target for degradation, consisting of cellulose, hemicellulose, pectins and proteins, all of which are liable to digestion by bacterial enzymes. These can be considered in three major groups: pectolytic enzymes, cellulases and proteases. Table 7.7 indicates the

Table 7.7. *Range of cell wall degrading enzymes produced by bacterial plant pathogens*

Pathogen	Pectolytic enzymes[a]	Cellulases	Proteases
Soft Rot Bacteria			
Erwinia chrysanthemi	Endo-PL (5 isozymes)	Endoglucanase	Protease
Erwinia carotovora subsp. *atroseptica*	Endo-PL (3 isozymes), PG	Endoglucanase	Protease
Xanthomonas campestris pv. *campestris*	PL		Protease
Pseudomonas marginalis	PL		
Pseudomonas cepacia	PG		
Bacterial wilt			
Pseudomonas solanacearum	Endo-PG, PL PME	Endoglucanase	

[a] The key to pectolytic enzyme abbreviations is given in Table 7.8.

occurrence of some of these enzymes that have been identified in important bacterial pathogens.

Recent studies by Pagel and Heitefuss (1990) on the degradation of potato tuber tissue by *Erwinia carotovora* subsp. *atroseptica* have shown that these different enzymes act in a sequential manner: with polygalacturonase (PG), pectate lysase (PL), cellulase, protease and xylanase appearing at discrete times after tissue inoculation.

Differences in enzyme characteristics may also lead to changes in activity with varying environmental conditions. PG differs from PL, for example, in being inhibited by Ca^{2+} and other cations, and by having a low pH optimum. At the beginning of the maceration process PG plays a major role in cell wall degradation, but increasing levels of Ca^{2+} and pH result in increased activity of PL, and within 22 hours of inoculation this enzyme is the major degrading enzyme within the tissue.

Pectolytic enzymes

Pectic substances are a major component of the middle lamella, and comprise both pectic (polygalacturonic) acid and pectin (polymethylgalacturonic acid), as shown in Fig. 7.9. These substances are more exposed than other cell wall constituents to the degrading activity of bacterial pathogens within the intercellular spaces. Pectolytic enzymes produced by these bacteria are of two main types: exo-enzymes (causing terminal degradation of the pectic mole-

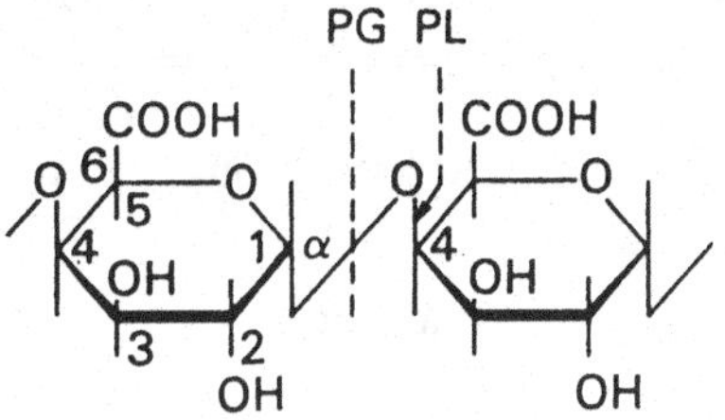

polygalacturonic acid
(pectic acid)

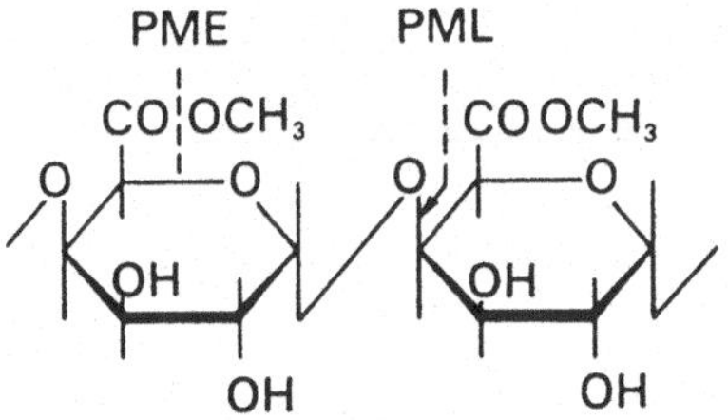

methylated polymer
(pectin)

Fig. 7.9 Sites of action of some major pectic enzymes. PL, pectic lyase; PG, polygalacturonase; PME, pectin methylesterase; PML, pectin lyase.

cule) and endo-enzymes (causing random breakage in the internal regions). The activity of endo-enzymes leads to rapid changes in the viscosity of pectic materials *in vitro*, and is the main enzymatic activity that leads to maceration during disease.

Pectolytic enzymes are of three main types: hydrolases, lyases and esterases (Table 7.8) and include polygalacturonase (PG), pectic lyase (PL), pectin lyase (PML) and pectin methyl esterase (PME), acting as shown in Fig. 7.9.

Endopectate lyases produced by soft rot bacteria are secreted into the surrounding medium, require Ca^{2+} and Mn^{2+} as cofactors, and kill plant cells both within tissue and as protoplasts (Allen *et al.*, 1987). The central role of pectolytic enzymes (particularly endopectate lyases) in tissue maceration caused by soft rot erwinias is shown by a number of experiments:

1. Inoculation of appropriate plant tissue with soft rot bacteria leads to a marked rise in the levels of PG and PL, with a parallel increase in the degree of maceration.
2. Maceration studies with purified endo-PLs (optimal activity at neutral to basic pH) have shown that individual enzymes are able to break down potato tuber cell walls and cause plant cell death.

Table 7.8. *Major pectolytic enzymes produced by bacteria and fungi*

Type of enzyme	Abbreviations	pH optimum	Mainly produced by
Hydrolase			
endopolygalacturonase	Endo-PG	4–5	Fungi
Lyase (transeliminase)			
endopolygalacturonide (pectate) lyase	PGL, PL or endo-PL	8–10	Bacteria
endopolymethyl galacturonide (pectin) lyase	PML, PL or endo-PL	8–10	Fungi
Methylesterase			
pectin methylesterase	PE or PME	7–9	Bacteria
		4–6	Fungi

Adapted from Billing (1987).

3. Cloned *Erwinia* pectolytic enzyme genes confer on non-phytopathogenic *Escherichia coli* the ability to macerate plant tissue.

The role of pectolytic enzymes has been looked at in particular detail in the soft rot bacterium *Erwinia chrysanthemi* (see below) and in the vascular wilt pathogen *Pseudomonas solanacearum*.

Erwinia chrysanthemi

Most strains of *Erwinia chrysanthemi* that have so far been studied produce five different PL iso-enzymes (designated PLa, PLb, PLc, PLd and PLe) coded for by five different *pel* genes. Strain EC16 produces four different iso-enzymes, and has been intensively studied both in terms of genetic determination (see Chapter 8) and in terms of the different characteristics of the particular PL isoenzymes produced by individual genes cloned in *Escherichia coli* (Table 7.9).

The multiplicity of pectolytic enzymes found in this and other soft rot bacteria may reflect differences in the type of substrate occurring in different hosts, and/or a requirement for co-operation between different enzymes for optimal maceration activity. The importance of different enzymes for virulence *in planta* is shown by:

1. The retention and high degree of conservation of multiple enzymes by *Erwinia chrysanthemi* in separate and taxonomically unrelated hosts (such as corn, *Chrysanthemum* and *Dieffenbachia*). Although the same gene products are being produced in different hosts, variation does occur in enzyme

Table 7.9. *Endopectate lyase (PL) isoenzymes produced by* Erwinia chrysanthemi

Enzyme characteristic	PLa	PLb	PLc	PLe
Isoelectric pH	4.6	8.8	9.0	9.8
Effect on tissue:				
a) Maceration	+/−	+ +	+	+ + +
b) Induction of electrolyte loss	+/−	+ +	+	+ + +

The data in this table are taken mainly from Tamaki *et al.* (1988).
The effect of the enzymes on tissue is graded from highly effective (+ + +) to hardly any effect at all (+/−). The ability of the enzymes to macerate tissue was determined (Tamaki *et al.*, 1988) by incubating the enzyme preparations with cucumber slices and assessing the degree of cell separation after 60 minutes at 37°C.

Table 7.10. *Effect of* pel *mutations on the virulence of* Erwinia chrysanthemi *on* Saintpaulia

Mutation	Pathogenicity and virulence
*pel*B-, *pel*C-	Mutants behave like wild-type strains & cause systemic maceration of whole plant
*pel*A-, *pel*D-	Less virulence than wild-type strains, with fewer systemically infected plants and much more limited maceration
*pel*E-	Completely non-invasive, with bacteria and maceration limited to the region of inoculation

Data from Boccara *et al.* (1988).

proportions. In corn stalk rot produced by *E. chrysanthemi*, for example, xylanase activity appears to be particularly prominent (Barras *et al.*, 1987).

2. Specific mutation of particular *pel* genes permits an assessment of the role of individual isoenzymes in virulence and pathogenicity. Using this approach, the importance of PLe but not PLb and PLc for pathogenicity and virulence of *E. chrysanthemi* strain 3937 on *Saintpaulia* has been shown by Boccara *et al.* (1988); see Table 7.10.

Cellulases

Major endoglucanases have been identified in *Erwinia chrysanthemi* and *Pseudomonas solanacearum*, with cloning of the genes concerned in both cases (Chapter 8).

Pseudomonas solanacearum

The importance of endoglucanase as a virulence factor in this pathogen has been shown by specific mutation of the *egl* gene by site directed mutagenesis (Roberts *et al.*, 1988). The resultant mutant was similar to the wild-type strain in the production of extracellular polysaccharide and in the secretion of a general range of cell wall-degrading enzymes (including polygalacturonase) but produced 200 times less endoglucanase. This mutant was significantly less virulent on tomato compared with wild-type. Virulence of the mutant was restored by complementation with the normal *egl* gene, indicating that the production of this enzyme is important but not essential for pathogenicity.

Proteases

Protease secretion by phytopathogenic bacteria has been demonstrated and genetically characterised (see Table 8.10) in *Erwinia carotovora*, *E. chrysanthemi* and *Xanthomonas campestris*. The general role of these enzymes in pathogenesis is not clear, though in *X. c.* pv. *campestris*, *prt*− mutants showed delayed symptoms compared with wild-type bacteria (Tang *et al.*, 1988).

In *Erwinia chrysanthemi*, the mode and timing of protease secretion is quite different from other exoenzymes such as cellulases and pectinases, as noted by Wandersman *et al.* (1987) and summarised in Table 7.11. This distinction reflects fundamental differences in the transport mechanism of these two types of secretory product (see Chapter 2, Fig. 2.11) and suggests that these enzymes may function at different stages of disease development.

Extracellular polysaccharides

The production of extracellular polysaccharides (EPS) and the related possession of a capsule have been shown to be important in pathogenicity and virulence for a number of plant pathogenic bacteria. In this section the role of EPS is considered initially in general terms, then specifically in relation to particular pathogens.

General role in virulence

Various roles for EPS have been suggested, including the induction and maintenance of water-soaking (see previous section on toxins, p. 197),

Table 7.11. *Mode and timing of secretion of different exoenzymes by* Erwinia chrysanthemi

Characteristic	Protease	Cellulase pectinase
Phase of secretion	Growth phase	Stationary phase
Specific transport mechanism	Carboxyl recognition transport system	Amino recognition transport system
a) 'out' mutants	Secretion not affected	Secretion prevented
b) Expression in *E. coli*	Released from cell	Accumulate in periplasmic space
Internal accumulation	No intracellular accumulation	Intracellular build-up. Enzymes occur as bound components prior to release

prevention of bacterial/plant recognition, changes in carbohydrate utilisation and restriction of water movement.

Prevention of bacterial/plant recognition

EPS may be important in promoting the ability of virulent pathogenic bacteria to avoid plant recognition (and thus triggering a hypersensitive reaction (HR)) by masking receptor sites on the bacterial surface. Evidence providing some support for this has come from comparative studies on virulent/avirulent strains of *Erwinia amylovora* where limitation in cell contact by EPS appears to be important for disease development (see below).

Changes in carbohydrate utilisation

The presence of a large and actively growing mass of bacteria within photosynthetic plant tissue creates a physiological sink for products of photosynthesis and leads to nutrient transfer from plant to bacteria. Synthesis of bacterial EPS may be particularly important in this since it will account for a high proportion of the carbohydrates that are diverted from the plant, and their conversion to an insoluble form renders them unavailable for plant metabolism.

Restriction of water movement

The production of EPS is particularly important in vascular wilt diseases, where occlusion of xylem vessels results in reduced water conductivity and associated symptoms. The extensive wilting (without necrosis) that occurs in these diseases results partly from localised accumulations of bacteria, and partly from the release of EPS, which passes into functional vessels, blocking xylem cavities and accumulating on pit membranes. This may be particularly acute at petiole junctions, where pore sizes are too small to allow further EPS migration. Blockage of xylem not only leads to wilting, but may also lead to an internal build-up in pressure, splitting vessels and producing blisters under bark.

EPS may also be important in restricting water flow through primary cell walls of mesophyll tissue, resulting in leaf wilt. This can result where vascular wilt pathogens produce a range of EPS molecular sizes, the smaller ones of which pass through the xylem and cause occlusion in the leaf apoplast (e.g. alfalfa wilt caused by *Clavibacter michiganensis* subsp. *insidiosum*). Localised wilting may also be associated with EPS-producing pathogens that colonise stem cortex or leaf tissue, causing necrosis.

Role of EPS production by different pathogens

The involvement of EPS production in pathogenicity and virulence has been investigated in particular detail in those cases where non-capsulate (EPS−) mutants have been obtained, including *Erwinia amylovora*, *Pseudomonas solanacearum* and *Erwinia stewartii*.

Erwinia amylovora

Isolation and analysis of avirulent mutants by Bennett (1980) and Billing (1984) has shown that EPS production is an important factor in the virulence of this pathogen. Properties of these mutants are summarised in Table 7.12 which shows that although most avirulent mutants (S, L, Q and E) lack EPS, this is not invariably the case (mutant P). Thus, although EPS may be involved in such aspects as masking of receptors or maintenance of water-soaking, the fundamental ability of the pathogen to cause host cell electrolyte leakage (demonstrated by the ability to cause necrosis) is also important.

Although the features shown in Table 7.12 do not initially appear to fit the concept that EPS masking of surface receptors may contribute to virulence in this pathogen, there is a further complexity in terms of population hetero-geneity. Recent studies by Hignett (1988) have shown that culture T is highly variable in its degree of capsulation, and have suggested that virulence depends on some (non-capsulate) cells in the population causing limited cation leakage by making direct contact with the plant cells. Those mutants that lack EPS throughout the whole population cause such extensive cation

Table 7.12. *Extracellular polysaccharide (EPS) production and avirulence mutants of* Erwinia amylovora

Culture	Selection method	EPS production	Pear fruit		Apple shoot infection
			Necrosis	Ooze	
T	Wild-type	+	+	+	+ +
P	Colony variant	+	−	−	−
S,L	Phage-resistant	−	+	−	−
Q	Colony variant	−	+	−	−
E	Colony variant	−	+	−	−
P + S, L, Q or E			+	+	+

Adapted from Mansfield & Brown (1986).

leakage that a full HR develops. In the case of mutant P, where all the cells are capsulate, the absence of any contact-mediated electrolyte leakage prevents the onset of bacterial growth and the development of virulence. Mixing populations of mutant P with mutants S, L, Q or E appears to restore the heterogeneity which is essential for virulence in this organism.

Pseudomonas solanacearum

This vascular wilt pathogen produces large amounts of EPS, which is present as a highly soluble slime rather than as an insoluble capsule. Initial observations (see Sequeira, 1984) had suggested that virulent strains produce EPS and avirulent ones do not. The transition from virulence (strain K60) to avirulence (strain B1) also conveyed the ability to induce an HR on the host plant, suggesting that EPS in this pathogen was acting simply by blocking receptor sites and preventing an incompatible reaction. The situation is not this simple, however, since (1) addition of EPS to strain B1 inoculum still results in an HR; and (2) virulent race 2 strains of the pathogen (from banana) produce large amounts of EPS yet induce an HR in tobacco. This would suggest that other aspects of EPS function, such as water-soaking and vascular dysfunction may be more important for virulence in this pathogen.

Erwinia stewartii

In this bacterium, which is a vascular pathogen of corn, the primary mechanism of virulence is generally considered to be the production of EPS slime, causing occlusion of xylem vessels and associated symptoms. However, other factors are also important in symptom development, such as the ability of the pathogen to induce water-soaking. Recent studies by Dolph

Table 7.13. *Induction of water-soaking and wilting in* Erwinia stewartii

Degree of virulence	Mutants		Symptoms
Normal	wts +	EPS +	Water-soaking and wilting
Partial	wts +	EPS −	Water-soaking only
None	wts −	EPS −	No symptoms
	wts −	EPS +	

Data from Dolph *et al.* (1988).

et al. (1988) have compared virulence in mutants that separately lack EPS production (EPS−) and ability to induce water-soaking (wts−), as shown in Table 7.13.

The results show that ability to induce water-soaking (in leaves) is of primary importance (pathogenicity factor) in disease induction, and that EPS production is of secondary importance (virulence factor) in disease development, where it is involved in the induction of wilting. With this pathogen, water-soaking is presumably caused by some factor other than EPS which increases host cell permeability and leads to leakage of contents.

Plant hormones

The synthesis of plant hormones is of specific importance in the case of tumour diseases, where it leads to a high rate of plant cell division and is the primary cause of hyperplasia. Two major types of hormone are involved: auxins and cytokinins.

Auxins are synthesised from tryptophan by one of two metabolic routes (Manulis *et al.*, 1991):

1. The indoleacetamide pathway. This is summarised in Fig. 8.10, and involves two key enzymes: tryptophan monoxygenase and indolacetamide hydrolase.
2. The indolepyruvate pathway, with indole-3-pyruvate and indole-3-acetaldehyde as intermediates.

The indolacetamide pathway is the major route for auxin production by plant pathogenic bacteria, and the synthesis of key enzymes for this pathway is generally important for the virulence of oncogenic organisms. In contrast, the indolepyruvate pathway is not regarded as important for the induction of hyperplasia, but it is involved in auxin synthesis by higher plant cells and has been shown to occur in non-pathogenic strains of *Erwinia herbicola*, where it may be important in saprophytic survival.

An example of cytokinin synthesis is shown in Fig. 7.10, which illustrates the

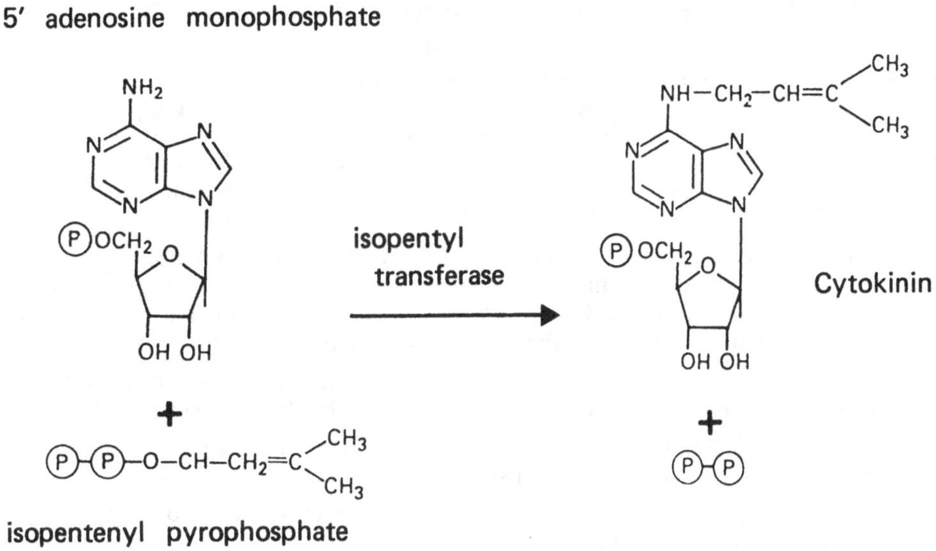

5′ adenosine monophosphate

Fig. 7.10 Biosynthesis of cytokinin during plant infection by *Agrobacterium tumefaciens*. Cytokinin synthesis involves the attachment of an isopentenyl group to 5′AMP by isopentyl transferase. This enzyme is encoded by gene 4 of T DNA.

conversion of adenosine monophosphate to isopentyl adenosine by the enzyme isopentyl transferase.

Synthesis of these hormones has been investigated in particular detail in two tumour-inducing bacteria, *Agrobacterium tumefaciens* and *Pseudomonas savastanoi* subsp. *savastanoi*. These important pathogens differ in that *A. tumefaciens* induces hormone synthesis directly in the plant cells by transformation, while *P.s.* subsp. *savastanoi* produces the hormone (or hormone precursor) as a secretory product of the bacterial cell. The process of transformation by *A. tumefaciens* is described in detail in Chapter 8 (see Fig. 8.14), where details of the genetic determination of hormone production by both pathogens are also given. Recent studies by Manulis *et al.* (1991) have shown that the acquisition of *P.s.* subsp. *savastanoi*-type indole acetic acid genes by *Erwinia herbicola* pv. *gypsophilae* also confers oncogenic activity on this pathogen.

Plant hormone production by Pseudomonas savastanoi *subsp.* savastanoi

Plant hormones are secreted by this phytopathogen both in culture and inside olive and oleander host plants, where they determine the pathogenic and virulence characteristics of the bacterium. Comparative aspects of auxin and cytokinin production, and the role of these hormones in disease, are summarised in Table 7.14.

Table 7.14. *Production of plant hormones by* Pseudomonas savastanoi
subsp. savastanoi

	Auxin	Cytokinin
Compound	IAA	Zeatin, zeatin riboside, 1′-methylzeatin, 1″-methylzeatin riboside
Synthesis	Secondary metabolism Main pathway has indoleacetamide as intermediate metabolite	Secondary metabolism Different cytokinins synthesised at different times during culture cycle
Strain variation	All strains synthesise IAA	Different cytokinins produced by different strains
Effect of mutants	IAA-deficient mutants do not induce normal symptoms	Not yet isolated
Level of synthesis	Determines disease incubation time	Determines size of tumour
Role in disease	Pathogenicity and virulence	Virulence

IAA, indoleacetic acid.
Information taken from Surico (1986).

The general conclusions from studies involving mutant strains or strains producing different levels of hormone are that pathogenicity is determined largely by indoleacetic acid (IAA) production (probably through the promotion of cambial activity) and that virulence is determined by both IAA (speed of tumour induction) and cytokinin (tumour size).

References

Allen C., Georg H., Yang Z., Lacy G. & Mount M. (1987). Molecular cloning of an endo-pectate lyase gene from *Erwinia carotovora* subsp. *atroseptica*. *Physiol. Mol. Plant Pathol.* **31**, 325–35.

Ballio A., Barra D., Bossa F., DeVay J. E., Grguvina I., Iacobellis N. S., Marino G., Pucci P., Simmaco M. & Surico G. (1988). Multiple forms of syringomycin. *Physiol. Mol. Plant Pathol.* **33**, 493–6.

Barras F., Thurn K. K. & Chatterjee A. K. (1987). Resolution of four pectate lyase structural genes of *Erwinia chrysanthemi* (EC16) and characterisation of the enzymes produced in *Escherichia coli*. *Mol. Gen. Genet.* **209**, 319–25.

Bender C. L., Malvick D. K. & Mitchell R. E. (1989). Plasmid-mediated production of the phytotoxin coronatine in *Pseudomonas syringae* pv. *tomato* *J. Bacteriol.* **71**, 807–12.

Bennett R. A. (1980). Evidence for two virulence determinants in the fireblight pathogen *Erwinia amylovora*. *J. Gen. Microbiol.* **116**, 351–6.

Billing E. (1984). Studies on avirulent strains of *Erwinia amylovora*. *Acta Hort.* **151**, 249–53.

Billing E. (1987). *Bacteria as Plant Pathogens*. Aspects of Microbiology No. 14. Wokingham: Van Nostrand Reinhold.

Boccara M., Diolez A., Rouve M. & Kotoujansky A. (1988). The role of individual pectate lyases of *Erwinia chrysanthemi* strain 3937 in pathogenicity on *Saintpaulia*. *Physiol. Mol. Plant Pathol.* **33**, 95–104.

Campbell J. N., Cass D. D. & Peteya D. J. (1987). Colonisation and penetration of intact canola seedling roots by an opportunistic fluorescent *Pseudomonas* sp. and the response of host tissue. *Phytopathology* **77**, 1166–73.

Davis M. J. (1990). Fastidious prokaryotes. In *Methods in Phytobacteriology*, ed. Z. Klement, K. Rudolph, and D. C. Sands. Budapest: Akademiai Kiado.

Dolph P. J., Majerczak D. R. & Coplin D. L. (1988). Characterisation of a gene cluster for exopolysaccharide biosynthesis and virulence in *Erwinia stewartii*. *J. Bacteriol.* **170**, 865–71.

Durbin R. D. (1983). The biochemistry of fungal and bacterial toxins and their modes of action. In *Biochemical Plant Pathology*, ed. J. A. Callow, pp. 137–62. Chichester: Wiley.

El-Banoby F. E. & Rudolph K. (1979). A polysaccharide from liquid cultures of *Pseudomonas phaseolicola* which specifically induces water-soaking in bean leaves (*Phaseolus vulgaris* L.) *Phytopathol. Z.* **95**, 38–50.

Ferguson A. R., Johnson J. S. & Mitchell R. E. (1980). Resistance of *Pseudomonas syringae* pv. *phaseolicola* to its own toxin, phaseolotoxin. *F.E.M.S. Lett.* **7**, 123–5.

Garrett C. M. (1987). The effect of crown gall on growth of cherry trees. *Plant Pathol.* **36**, 339–45.

Goodwin P. H., DeVay J. E. & Meredith C. P. (1988). Roles of water stress and phytotoxins in the development of Pierce's disease of the grapevine. *Physiol. Mol. Plant Pathol.* **32**, 1–15.

Hartman C. L., Secor G. A., Venette J. R. & Albaugh D. A. (1986). Response of bean calli to filtrate from *Pseudomonas syringae* pv. *phaseolicola* and correlation with whole plant disease reaction. *Physiol. Mol. Plant Pathol.* **28**, 353–8.

Hignett R. C. (1988). Effects of growth conditions on the surface structures and extracellular products of virulent and avirulent forms of *Erwinia amylovora*. *Physiol. Mol. Plant Pathol.* **32**, 387–94.

Hopkins D. L. (1988). Natural hosts of *Xylella fastidiosa* in Florida. *Plant Dis.* **72**, 429–31.

Hutziloukas E. & Panopoulos N. J. (1992). Origin, structure and regulation of *argK*, encoding the phaseolotoxin-resistant ornithine carbomoyltransferase in *Pseudomonas syringae* pv. *phaseolicola*, and functional expression of *argK* in transgenic tobacco. *J. Bacteriol.* **174**, 5895–909.

Leary J. V., Roberts S. & Willis J. W. (1988). Detection of coronatine toxin of *Pseudomonas syringae* pv. *glycinea* with an enzyme-linked immunosorbent assay. *Phytopathology* **78**, 1498–500.

Lelliott R. A. & Stead D. E. (1987). *Methods for the diagnosis of bacterial diseases of plants*. Oxford: Blackwell Scientific Publications.

Levi C. & Durbin R. D. (1986). The isolation and properties of a tabtoxin-hydrolysing aminopeptidase from the periplasm of *Pseudomonas syringae* pv. *tabaci*. *Physiol. Mol. Plant Pathol.* **28**, 345–52.

Lozano J. C. & Zeigler R. (1990). Screening for resistance. In *Methods in Phytobacteriology*, pp. 334–8. ed. Z. Klement, K. Rudolph, and D. C. Sands. Budapest: Akademiai Kiado.

Mansfield J. W. & Brown I. R. (1986). The biology of interactions between plants and bacteria. In *Biology and Molecular Biology of Plant–Pathogen Interactions*, ed. J. A. Bailey, pp. 71–98. Berlin: Springer-Verlag.

Manulis S., Gafni Y., Clark E., Zutra D., Ophir Y. & Barash I. (1991). Identification of a plasmid DNA probe for detection of strains of *Erwinia herbicola* pathogenic on *Gypsophila paniculata*. *Phytopathology* **81**, 54–7.

Mitchell R. E. & Frey E. J. (1988). Rhizobitoxine and hydroxythreonine production by *Pseudomonas andropogonis* strains, and the implications to plant disease. *Physiol. Mol. Plant Pathol.* **32**, 335–41.

Morgan M. K. & Chatterjee A. K. (1988). Genetic organisation and regulation of proteins associated with production of syringotoxin by *Pseudomonas syringae* pv. *syringae*. *J. Bacteriol.* **170**, 5689–97.

Ogimi C. & Takikawa Y. (1988). Bacterial gall diseases of trees. Proc. 5th Int. Congr. Plant Pathol., Kyoto, 96.

Pagel W. & Heitefuss R. (1990). Enzyme activities in soft rot pathogenesis of potato tubers: Effects of calcium, pH and degree of pectin esterification on the activities of polygalacturonase and pectate lyase. *Physiol. Mol. Plant Pathol.* **37**, 9–25.

Perreaux D., Maraite H. & Meyer J. A. (1986). Detection of 3(methylthio)proprionic acid in cassava leaves infected by *Xanthomonas campestris* pv. *manihotis*. *Physiol. Mol. Plant Pathol.* **28**, 323–8.

Rainey P. B., Brodey C. L. & Johnstone K. (1991). Biological properties and spectrum of activity of tolaasin, a lipodepsipeptide toxin produced by the mushroom pathogen *Pseudomonas tolaasii*. *Physiol. Mol. Plant Pathol.* **39**, 57–70.

Roberts D. P., Denny T. P. & Schell M. A. (1988). Cloning the *egl* gene of *Pseudomonas solanacearum* and analysis of its role in phytopathogenicity. *J. Bacteriol.* **170**, 1445–51.

Rogers H. J., Perkins H. R. & Ward J. B. (1980). Microbial cell walls and membranes. London: Chapman & Hall.

Rudolph K. (1990). Toxins as taxonomic features. In *Methods in Phytobacteriology*, ed. Z. Klement, K. Rudolph and D. C. Sands, pp. 251–67. Budapest: Akademiai Kiado.

Sequeira L. (1984). Plant–bacteria interactions. In *Cellular Interactions*, ed H. F. Linskens and J. Heslop-Harrison, pp. 187–211. Berlin: Springer-Verlag.

Sigee D. C. & Epton H. A. S. (1976). Ultrastructural changes in resistant and susceptible varieties of *Phaseolus vulgaris* following artificial inoculation with *Pseudomonas phaseolicola*. *Physiol. Plant Pathol.* **9**, 1–8.

Surico G. (1986). Indoleacetic acid and cytokinins in the olive knot disease. An overview of their role and their genetic determinants. In *Biology and Molecular Biology of Plant–Pathogen Interactions*, ed. J. A. Bailey. Berlin: Springer-Verlag.

Staskawicz B. J. & Panopoulos N. J. (1979). A rapid and sensitive microbiological assay of phaseolotoxin. *Phytopathology* **69**, 663–6.

Tamaki S. J., Gold S., Robseon M., Manulis S. & Keen N. T. (1988). Structure and organisation of the *pel* genes from *Erwinia chrysanthemi* EC16. *J. Bacteriol.* **170**, 3468–78.

Tang J. L., Clough C. L., Barber C. E., Dow J. M. & Daniels M. J. (1988). Molecular cloning of protease gene(s) from *Xanthomonas campestris* pv. *campestris*: Expression in *Escherichia coli* and role in pathogenicity. *Mol. Gen. Genet.* **210**, 443–8.

Thomas M. D., Langston-Unkefer P. J., Uchytil T. F. & Durbin R. D. (1983). Inhibition of glutamine synthetase from pea by tabtoxinine-β-lactam. *Plant Physiol.* **71**, 912–15.

Turner J. G. (1986). Activities of ribulose-1,5-bisphosphate carboxylase and glutamine synthetase in isolated mesophyll cells exposed to tabtoxin. *Physiol. Mol. Plant Pathol.* **29**, 59–68.

Wandersman C., Delepelaire P., Letoffe S. & Schwartz M. (1987). Characterisation of *Erwinia chrysanthemi* extracellular proteases: Cloning and expression of the protease genes in *Escherichia coli. J. Bacteriol.* **169**, 5046–53.

Wells J. M. & Raju B. C. (1987). Biology and physiology of fastidious xylem-limited bacteria from plants. In Plant Pathogenic Bacteria, Proceedings of the 6th International Conference on Plant Pathogenic Bacteria, ed. E. L. Civerolo, A. Collmer, R. E. Davis and A. G. Gillaspie, pp. 321–6. Dordrecht: Martinus Nijhoff.

Xu G.-W. & Gross D. C. (1988). Physical and functional analysis of the *syr*A and *syr*B genes involved in syringomycin production by *Pseudomonas syringae* pv. *syringae. J. Bacteriol.* **170**, 5680–8.

8

Genetical analysis of plant pathogenic bacteria

The ability of plant pathogenic bacteria to survive and multiply outside and inside plants, and to cause disease, is determined to a large extent by their genetic constitution. The genetic analysis of plant pathogenic bacteria currently involves the application of molecular techniques for the identification and investigation of bacterial genes that are important in all of these aspects, and will be considered first. Following sections discuss the role of specific genes and gene systems in the activity of plant pathogenic bacteria in relation to the determination of compatibility and incompatibility, disease virulence, and non-pathogenic characteristics. The final part of this chapter deals with the occurrence and role of plasmids in these bacterial cells.

Molecular genetics: identification and investigation of bacterial genes

Bacterial genes, occurring on either chromosomal or plasmid DNA, are involved in the determination of a wide range of phenotypic characteristics. In recent years new techniques of molecular biology have been particularly successful in the genetic analysis of plant pathogenic bacteria (Daniels *et al.*, 1988), and have been described in detail in a number of recent texts (e.g. Brown, 1986; Sambrook *et al.*, 1989). The major objectives of molecular genetics are:

1. Identification and isolation (cloning) of specific genes with defined functions.
2. Gene mapping and determination of the nucleotide sequence, which gives information on gene functioning and homologies.
3. Determination of the chemical nature of the primary gene product, which may give information on the location and function of the translated polypeptide.

A general scheme for the identification and investigation of bacterial genes

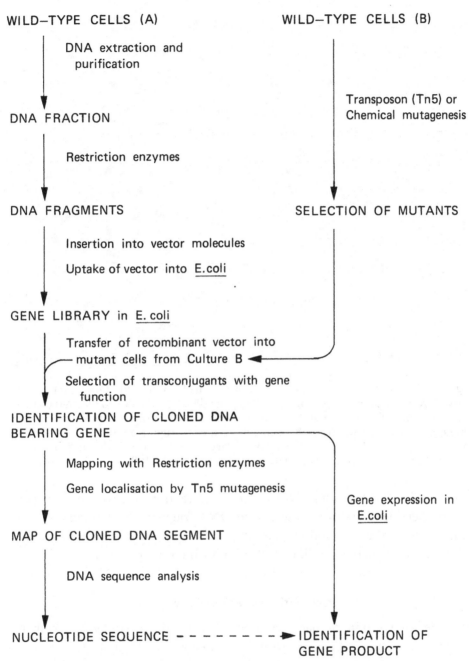

Fig. 8.1 General scheme for the identification and investigation of cloned bacterial genes.

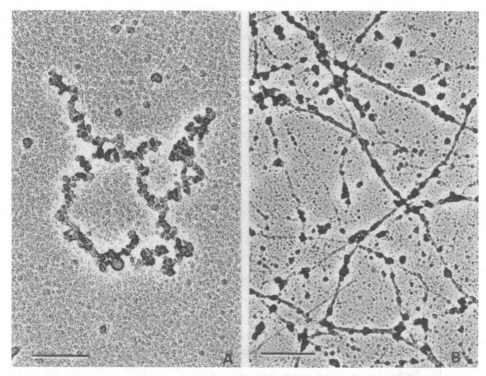

Fig. 8.2 Electron microscope appearance of regions from a crude DNA preparation derived from *Pseudomonas syringae*, showing details of a supercoiled plasmid (A) and the reticulum of linear (mainly chromosomal) DNA (B). The preparation was obtained from the supernatant that remains after centrifuging a lysed cell preparation for 30 minutes at 48 000 × g, followed by air-drying and rotary shadowing. Bar scale: 1.0 μm. (Taken, with permission, from El-Masry & Sigee, 1986.)

is shown in Fig. 8.1, and involves DNA extraction, formation of a gene library then identification of the cloned DNA fragment containing a specific gene. This may be followed by mapping of the gene within the cloned DNA, DNA sequence analysis and identification of the gene product.

Preparation of a DNA fraction

A crude preparation of bacterial DNA can be prepared by breaking up bacterial cells in suspension with lysozyme and detergent, followed by removal of cell debris (and some chromosomal DNA) by centrifugation. Electron microscope examination of the crude extract reveals the presence of linear strands of genomic (chromosomal) DNA plus plasmids (Fig. 8.2).

For molecular experiments involving DNA manipulation, the polynucleotide sample must be in a highly purified state, with minimal RNA and protein contamination. A typical extraction sequence is summarised in Fig. 8.3 and is based on the technique described by Gillis *et al.* (1990). This involves three main stages:

1. Breakdown of bacterial cells and release of cell contents. Most Gram-negative bacteria can be lysed with treatment with 2% sodium dodecylsulphate, but Gram-positive bacteria require further degradation using enzymes such as lysozyme, proteinase and *N*-acetylmuramidase. Physical treatments such as mild osmotic shock, sonication and freeze–thawing may also be useful in achieving cell breakdown, either directly or by assisting enzyme penetration.

2. Preparation of crude DNA extract, in which most protein has been removed by phenol (or phenol-chloroform) extraction. Impure DNA may be removed from the aqueous layer above the phenol (Fig. 8.3) by the addition of 95% ethanol and collecting the DNA on a glass rod by gentle stirring.

3. Purification of the DNA can then be carried out by dissolving the sample in buffer, removal of RNA (by RNase) followed by further deproteinisation. As a final step, DNA may be purified by centrifugation in a caesium chloride gradient, with dialysis of the pure fraction against buffer over a 36-hour period.

Formation of a gene library

This involves cutting the DNA at specific sites with restriction endonuclease enzymes into fragments of defined length, then cloning each size category as separate populations within a host bacterium (typically *Escherichia coli*) for subsequent identification of the DNA clone containing the gene of interest. Individual fragments are cloned by insertion into a vector molecule, which can be introduced into a recipient host cell and replicated within the dividing bacterial population.

Vector molecules

DNA fragments from the restriction enzyme digest can be inserted into vector molecules, as shown in Fig. 8.4. In this example, DNA circle pJB8 is cut by the restriction endonuclease BamHI at a single specific site (with the nucleotide sequence 5′.GGATCC.3′), and a fragment of insert DNA (also cut by BamHI) is joined by the enzyme DNA ligase to form a recombinant molecule.

The value of recombinant DNA technology is that sequences of foreign DNA can be incorporated into circular molecules which act as vectors

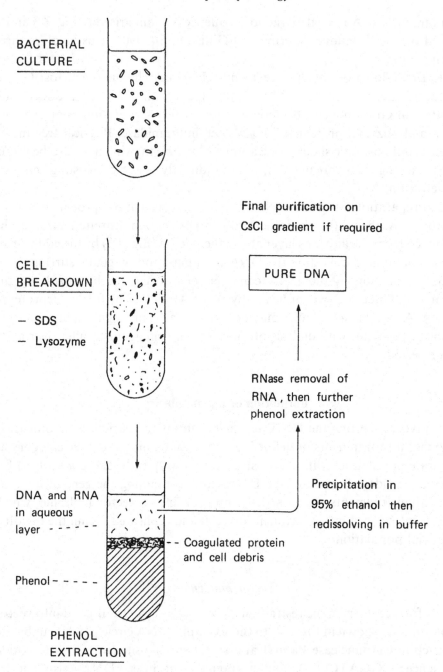

BACTERIAL
CULTURE

Final purification on
CsCl gradient if required

CELL
BREAKDOWN

PURE DNA

— SDS
— Lysozyme

RNase removal of
RNA, then further
phenol extraction

DNA and RNA
in aqueous
layer - - - - -

Precipitation in
95% ethanol then
redissolving in buffer

- - Coagulated protein
and cell debris

Phenol - - - - -

PHENOL
EXTRACTION

Fig. 8.3 Extraction and purification of bacterial DNA. SDS, sodium dodecyl sulphate.

Table 8.1. *DNA vector molecules*

Vector	Maximum length of DNA insert[a]	Example	Mode of transfer into recipient
Plasmid	5 kb	pBR322	Transformation
Phage DNA	3 kb	M13 }	{ Transfection
	23 kb	EMBL }	{ Transduction
Cosmid	40 kb	pJB8 }	{ Transformation
		pLAFR1 }	{ Transduction
			{ Conjugation[b]

[a] Typical values for the vector category, taken from Brown (1986).
[b] By triparental mating.

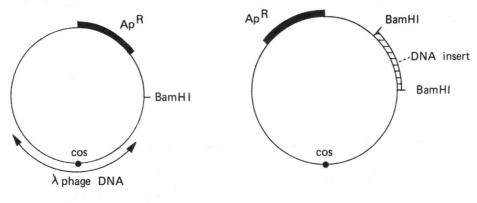

(a) Cosmid pJB8   (b) Recombinant molecule

Fig. 8.4 Formation of a recombinant cosmid vector molecule. (*a*) General diagram of cosmid pJB8, showing the phage DNA region (containing a cos site) indicated by the arrow, and the plasmid DNA with ampicillin resistance region (Ap[R]) and BamHI endonuclease cutting site occupying the rest of the molecule. (*b*) Recombinant molecule. A region of foreign DNA has been inserted into the cosmid by opening up the DNA circle with BamHI and joining the cut ends to the insert DNA with ligase.

(carriers), thus facilitating transfer of the foreign DNA into recipient bacterial cells, with subsequent survival, replication and gene expression in the new cellular environment. Vector molecules are of three main types: plasmids, phage DNA (e.g. M13) and phage DNA/plasmid hybrids (cosmids). These vectors differ in the maximum length of DNA insert that they can carry, and also in the means by which they are transferred into recipient bacterial cells (Table 8.1).

Under normal conditions, molecules of DNA can transfer from one bacterial cell to another in three main ways: transformation (including

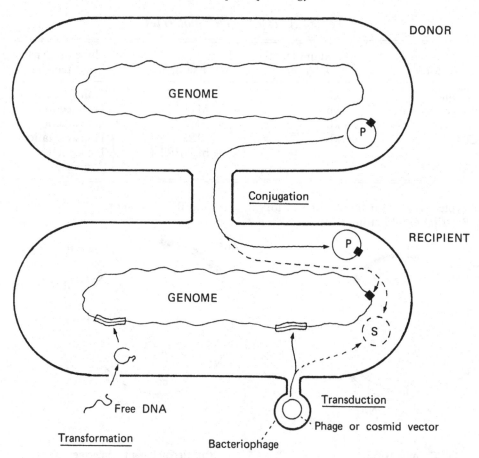

Fig. 8.5 Uptake of foreign DNA into a recipient bacterial cell by transformation, transduction or during conjugation. Conjugative transfer of DNA is illustrated by passage of a plasmid (P) bearing a transposon (filled square). This may subsequently become incorporated into the recipient cell genome, with the remaining part of the vector molecule becoming lost due to lack of replication (suicide vector, S). The vector molecule involved in transduction may also follow a similar fate.

transfection), transduction and conjugation, as summarised in Fig. 8.5. These naturally occurring processes may each be used experimentally to introduce foreign DNA (within recombinant molecules) into recipient cells.

Transformation

Transformation involves the direct incorporation of small fragments or molecules of DNA such as plasmids into bacterial cells from the surrounding environment, and employs sophisticated mechanisms of binding and uptake. Direct uptake of phage (or phage-derived) DNA by this process is specifically referred to as transfection. Transformation is particularly important for the transfer of recombinant plasmids, as illustrated in Fig. 8.7, where the recombinant cosmid derived from pLAFR1 is introduced into *Escherichia coli*, prior to triparental mating.

Bacteria which are able to take up DNA from the surrounding medium are referred to as 'competent'. Most populations of bacteria only have a small proportion of competent cells, so there is only limited uptake of DNA under normal conditions. For experimental work, it is essential to increase competency to enhance the transformation process. This may be achieved in a variety of ways, depending on the particular bacterial species concerned, and can involve both chemical (e.g. 50 mM calcium chloride) and physical (e.g. heat shock) treatment.

Successful uptake of DNA and its survival in the host cell may be determined by selection for specific gene expression. Uptake and survival, for example, of the plasmid pBR322 (commonly used in cloning experiments) which carries genes for the detoxification of the antibiotics ampicillin and tetracycline, may be demonstrated in transformed cells by selection for ampicillin and tetracycline resistance.

Transduction and conjugation

The experimental transfer of DNA may involve the use of bacteriophage or bacterial vectors.

Transduction In the case of bacteriophage vectors (such as M13 and λ) the experimental transfer of phage DNA combined with other DNA of interest may be directly achieved by the normal process of infection, and is referred to as transduction.

The experimental use of λ-phage vectors involves packaging the recombinant DNA with phage protein to form complete viral particles in the test tube (*in vitro* packaging) prior to transduction. The enzymes required to assemble the linear λ DNA into phage particles require the presence of specific recognition sites ('*cos* sites') for endonuclease cleavage prior to packaging in the viral protein. Cosmids have proved particularly useful in DNA transfer by transduction since they are able to carry larger fragments of insert DNA than phage DNA vectors (see Table 8.1). These hybrid molecules comprise:

1. A region of λ-phage DNA, including a *cos* site for *in vitro* packaging.
2. A region of plasmid, containing an antibiotic marker site to identify the recipient bacterium, as shown in Fig. 8.4 for the cosmid pJB8.

Although transduction is generally important in bacterial genetics, it has not been widely used with phytopathogenic bacteria.

Conjugation The use of bacterial cells as vectors involves the transfer of plasmids or cosmids (e.g., pLAFR1, pLAFR3) during the process of conjugation, with contact between donor and recipient bacteria to form a mating aggregate. DNA transfer leads to the recipient acquiring genetic information and becoming a transconjugant. The transfer of plasmids (and derived cosmids) by conjugation requires the activity of two types of gene: one for establishing the mating aggregate (*tra* gene) and one for mobilising DNA transfer (*mob* gene).

Conjugative plasmids carry both of these genes (*tra*$^+$, *mob*$^+$) and are able to promote their own transfer from donor to recipient cell. Non-conjugative plasmids lack the *tra* gene and are not able to promote their own transfer. This can only occur if they carry the *mob* gene and if *tra* functions are provided in trans (i.e. from other DNA in the cell), either on another plasmid (helper plasmid) or on the chromosome of the donor.

The use of helper plasmids to promote transfer of non-conjugative vector molecules such as cosmids can be carried out by triparental mating between recipient (phytopathogen) cells and separate cells containing the cosmid and helper plasmids. Both cosmid and helper plasmids are normally contained in cells of *Escherichia coli*, into which they have either been artificially introduced by transfection (cosmid) or are naturally occurring (helper plasmid). During the triparental mating the helper plasmids are transferred at high frequency into the cosmid-containing *E. coli* cells, where they subsequently promote transfer of both cosmid and helper plasmids into the recipient bacterial cell.

An example of triparental mating is shown in Fig. 8.7, where the recombinant cosmid derived from plAFR1 is transferred into *Pseudomonas syringae* using the helper plasmid pRK2013. These particular vector and helper molecules have been widely used in molecular plant pathology, since the cosmid has a broad host range and can be accepted by a wide range of recipient cells, while the helper plasmid is only able to replicate in enteric (e.g. *Escherichia coli*) bacteria, ensuring that it is not inherited in the phytopathogenic recipient.

One specific and widely used example of plasmid transfer of DNA is provided by transposon mutagenesis.

Identification of cloned genes by complementation

This is normally carried out by introducing cloned DNA into bacteria that are lacking in specific gene activity, and determining which clone can produce or restore gene function in the transconjugant (complementation). The essential absence of gene function in the recipient bacterium may either be naturally occurring, or may be induced by mutation. An example of the former situation is shown in Fig. 8.7, where an incompatibility gene from one race of bacterium is introduced into another race where it is naturally absent.

Where mutagenesis is involved, the recipient bacterium can be treated with a mutagen and the loss of particular gene function identified by the loss of specific and defined phenotypic characteristics. The gene on the cloned DNA that restores this characteristic is identified as the genetic determinant of that particular feature. This approach is shown in Fig. 8.1, where cloned DNA fragments from wild-type cells (culture A) are inserted for screening into defined mutants derived from wild-type cells (culture B).

Mutagenesis of the recipient bacteria is normally carried out either by chemical mutagens or by transposons.

Transposon mutagenesis

Transposons are short lengths of DNA which are able to transfer from a vector molecule and insert into target DNA with which they have little or no base sequence homology. This insertion may occur randomly or at particular sites, and leads to a local discontinuity in the existing base sequence, resulting in the loss of gene function. Insertion of transposons thus represents a specific and very useful type of mutagenesis.

One particular transposon, Tn5, has been widely used for the generation of phytopathogenic bacterial mutants. This transposon has a number of useful characteristics, including: (1) high rates of transposition from vector molecules, resulting in large yields of mutant colonies; (2) low insertion-site specificity, so that mutations are randomly induced in the target DNA; (3) relatively stable mutants (with low rates of reversion); and (4) various restriction sites for the activity of different restriction endonucleases. These sites are important for the location of the Tn5 segment during mapping and sequencing of the target DNA. The Tn5 transposon has a length of 5.7 kb, with a central gene for kanamycin resistance. Mutant reversion, with loss of the Tn5 segment, will thus be accompanied by susceptibility to this antibiotic.

Tn5 mutagenesis is typically carried out by introduction of the transposon on a plasmid vector during conjugation (see Fig. 8.5). The plasmid used for this is typically selected for its inability to survive and replicate in the host cell ('suicide plasmid', e.g. pJB4JI, used in *Rhizobium leguminosarum*) so that only the transposon remains, incorporated within the genomic DNA.

Mapping of cloned DNA

Once a segment of cloned DNA has been identified as containing the gene of particular interest, then the DNA can be cut into smaller fragments (sub-cloned) and mapped by a variety of restriction enzymes (restriction mapping). The position of the gene within this segment can be located by the position at which Tn5 mutagenesis causes loss of gene function. This is illustrated in Fig. 8.8, where the location of *avr*A gene is mapped in a cloned fragment.

DNA sequence analysis and homology

Segments of DNA containing one or more genes can be analysed in terms of their nucleotide sequence. This provides information on the transcriptional activity of the DNA (e.g. whether the genes are transcribed as a group or as individual units) and also allows comparison between closely related genes in terms of their homology.

DNA homology can also be identified by standard molecular hybridisation procedures such as Southern blotting. For example Todd *et al.* (1990) demonstrated a high degree of gene homology between pathovars of *Xanthomonas campestris* by probing *Eco*RI digests of DNA from pv. *oryzae* with a 7.5 kb DNA fragment (cloned virulence gene) from pv. *campestris*.

Identification of the gene product

Information on the primary gene product (i.e. the translated protein) may be obtained both from the DNA nucleotide sequence and from *in vitro* expression in non-phytopathogenic host cells (e.g. *Escherichia coli*). The amino acid sequence provides useful information on the nature of the polypeptide, and may indicate, for example, whether the protein is membrane-bound or not (see section on avirulence genes, p. 226).

In vitro expression of particular genes also permits the isolation and chemical analysis of pure gene product. The use of *in vitro* gene product for biochemical characterisation is particularly important where the gene product is difficult to separate from closely related proteins *in vivo*, such as individual pectate lyase enzymes of *Erwinia chrysanthemi* (see later). *In vitro* gene product also has potential in other ways, including its use as highly pure antigen for the production of specific antibody probes.

Genes for compatibility or incompatibility

When a bacterial cell enters a plant, subsequent events depend primarily on whether the bacterium has the ability to induce a pathogenic reaction or not, and if such a reaction occurs, whether it will lead to disease (compatibility) or resistance (incompatibility).

Evidence suggests that the genetic control of compatibility and incompatibility are closely related, and are largely determined by single Mendelian-type genes. This evidence comes from two main lines of investigation:

1. Studies on interrelationships between pathogen races and host cultivars.
2. Mutation studies in single pathogens in host and non-host combinations.

Race–cultivar interactions

Gene-for-gene hypothesis

Much of the early work on interactions between pathogens and plants was carried out with fungal pathogens. Studies by Flor in particular have been carried out on interactions within a narrow pathogen/host range, and have led to the proposal that these interactions between whole organisms may be related to specific interactions between individual pathogen/host genes, referred to as the 'gene-for-gene hypothesis'. This gene-for-gene interaction was initially shown in studies by Flor (see Flor, 1955) on the fungal pathogen *Melampsora lini*, which causes rust disease in flax. Race 1 of the pathogen is avirulent on flax variety Kyoto. X-ray induced mutations (deletions) led to race 1 of the pathogen becoming virulent, leading Flor to conclude that resistance usually results when complementary or reciprocal genes in pathogen and host are dominant, and that susceptibility results when dominance is experimentally eliminated or when either gene is recessive. This situation is shown in Fig. 8.6, which summarises hypothetical interactions between pathogen avirulence genes (A, a) and host resistance genes (R, r).

The gene-for-gene interaction implies that incompatibility results from the active interaction of the pathogen avirulence and plant resistance genes, and that absence of the gene product from either results in a compatible situation. Since the early work of Flor, gene-for-gene interactions have been identified in various fungal race–cultivar systems, including mildews of wheat and barley, potato late blight and stem rust of wheat. Evidence has also been obtained for similar systems involving bacterial plant pathogens, as shown in the interaction between races of *Pseudomonas syringae* pv. *glycinea* and cultivars of soybean. Incompatibility in this host/pathogen system is deter-

Table 8.2. *Interactions between four races of* Pseudomonas syringae *pv.* glycinea *and four cultivars of soybean*

Pathogen race	Soybean cultivars			
	(W) r1,R2,r3	(X) R1,r2,R3	(Y) r1,R2,R3	(Z) R1,r2,r3
(1) a1,a2,A3	−	+	+	−
(4) a1,a2,a3	−	−	−	−
(5) a1,A2,a3	+	−	+	−
(6) A1,a2,a3	−	+	−	+

+, incompatible interaction.
−, compatible interaction.
Postulated genotypes are given in relation to three avirulence genes (A1–A3) and three resistance genes (R1–R3). The soybean cultivars are: (W), Flambeau; (X), Harasoy; (Y), Norchief; and (Z), Peking.
(Data from Ellingboe, 1985).

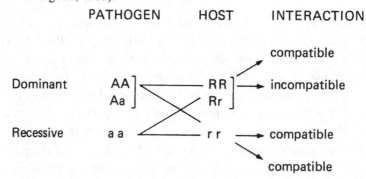

Fig. 8.6 Genetic determination of incompatibility. An incompatible interaction occurs between pathogen and plant when one or more dominant avirulence (A) genes and resistance (R) genes are both present in the respective partners.

mined by a number of avirulence and resistance genes, and is summarised for four pathogen races (1, 4, 5 and 6) and four plant cultivars (A–D) in Table 8.2.

In Table 8.2, an incompatible reaction occurs when any two complementary genes combine in the dominant state. Thus race 6 will induce an incompatible reaction in cultivar Z (interaction of genes A1 and R1) but race 5 will not.

Molecular genetics of avirulence genes

The successful employment of molecular genetics to study the avirulence genes in these pathogens is possible for two main reasons. Firstly, as already

Table 8.3. *Cloning of some chromosomal and plasmid avirulence genes*

Pathogen	Host	Genes	Reference
CHROMOSOMAL			
Pseudomonas syringae pv. *glycinea*	Soybean	$avrA$ $avrB$	Staskawicz *et al.* (1984) Tamaki *et al.* (1988a, 1991)
Pseudomonas syringae pv. *phaseolicola*	Bean	avrAsph1	Hitchin *et al.* (1989)
Pseudomonas syringae pv. *pisi*	Pea	avrAspi1	Vivian *et al.* (1989)
Xanthomonas campestris pv. *malvacearum*	Cotton	$avrB_2$ $avrB_3$ $avrB_6$ $avrB_N$ $avrB_{1n}$	Gabriel *et al.* (1986)
Xanthomonas campestris pv. *vesicatoria*	Pepper	avrRxv $avrBs_2$	Whalen *et al.* (1988)
PLASMID			
Pseudomonas syringae pv. *glycinea*	Soybean	$avrC$	Tamaki *et al.* (1988a, 1991).
Xanthomonas campestris pv. *vesicatoria*	Pepper	$avrBs_1$ $avrBs_3$	Stall *et al.* (1986) Whalen *et al.* (1988)

Table adapted partly from Vivian (1990), from which gene codes have been taken.

noted, bacteria (unlike fungal pathogens) are particularly suitable in terms of genome size and established techniques for this type of analysis. Secondly, race–cultivar relationships have been characterised in detail for a number of bacterial phytopathogen–plant systems (see Chapter 3). Modern techniques in molecular genetics have resulted both in the cloning of a range of avirulence genes and analysis of the DNA sequences involved.

Cloning of avirulence genes

Avirulence (*avr*) genes have now been cloned in a wide range of phytopathogenic bacteria, some of which are summarised in Table 8.3. These genes are present on both chromosomal and plasmid DNA, in some cases within the same organism. The earliest molecular studies on an *avr* gene were carried out by Staskawicz *et al.* (1984) on *Pseudomonas syringae* pv. *glycinea* and will be described by way of example.

Cloning and analysis of the avr*A gene from* Pseudomonas syringae *pv.* glycinea Staskawicz *et al.* (1984) carried out an initial cloning of the *avrA* gene (Fig. 8.7) followed by further analysis of the cloned DNA (Fig. 8.8).

Gene cloning was carried out by preparation of a genomic library of race 6 using the cosmid vector pLAFR1, transfer of the library into race 5 by conjugation, then testing of the transconjugants by inoculation into the four cultivars shown in Table 8.2. One single transconjugant, containing a 27.2 kb fragment of DNA from race 6, changed the compatible reaction of race 5 on cultivars X and Z to incompatible. This result was consistent with the race 6 DNA fragment containing an avirulence gene (*avr*A) which had been cloned in the *Escherichia coli* cells.

The 27.2 kb fragment of race 6 DNA bearing avirulence gene *avr*A was further analysed to determine:

1. The size and location of the *avr*A gene, by restriction mapping and Tn5 mutagenesis. The 27.2 kb insert DNA was cut into seven fragments using *Eco*R1 (Fig. 8.8) and particular fragments responsible for avirulence identified by Tn5 mutagenesis. All of the Tn5 insertions which induced loss of the race 6 phenotype occurred in the two adjacent small *Eco*RI fragments (6,7), identifying these as the probable major site of the *avr*A gene.

2. The specificity of the *avr*A gene to race 6. The unique nature of *avr*A gene DNA to race 6 was tested by determining the degree of hybridisation (Southern blot analysis) with DNA of race 6 and other races of the pathogen, using *Eco*RI fragments 6 and 7 as probes. Clear hybridisation only occurred with the DNA of race 6, suggesting that at least part of the A gene was unique to this race.

Avirulence gene sequencing and gene products

The isolation of cloned avirulence genes provides the opportunity to gain information on the gene product, either indirectly by determination of the nucleotide sequence, or directly by analysis of the protein produced in transformed *Escherichia coli* cells. This gene product is of particular interest, since it is presumably involved in the initial interaction between the plant and the incompatible bacterium, possibly in terms of a surface recognition molecule or the production of an elicitor of the HR (see Chapter 7).

Surface recognition The complementary interaction between bacterial avirulence genes and cultivar resistance genes appears to involve some kind of recognition event (see page 140). If this recognition occurs at the pathogen–plant cell surfaces, it might be expected that the avirulence gene protein would be present in the bacterial surface membrane.

Avirulence genes *avr*B and *avr*C from *Pseudomonas syringae* pv. *glycinea* have now been sequenced (Tamaki *et al.*, 1988a) and shown to encode single gene products of 36 and 39 kDa respectively. Computer analysis of the resulting amino acid sequence has failed to identify any features in the resulting polypeptide which would be expected in a membrane-associated protein, such as recognisable signal peptide sequences or stretches of hydrophobic amino acids. This would suggest that the primary gene product is not directly involved in cell surface recognition, which is further indicated by the fact that protein produced from cloned gene (*avr*C) in *Escherichia coli* cells did not elicit an HR when infiltrated into resistant soybean cultivars.

Elicitor production Recent studies by Keen *et al.* (1990) have indicated that expression of the *avr*D gene cloned from *Pseudomonas syringae* pv. *tomato* leads to the production of a diffusible elicitor which initiates cultivar-specific induction of the HR. The 34 kDa *avr*D-encoded protein appears to possess an enzymatic function which converts a normal bacterial metabolite to the low molecular weight *avr*D elicitor. Evidence for this is provided by the fact that specific induction of an HR in incompatible soybean leaves could be caused by infiltration of *Escherichia coli* and *P. syringae* pathovars containing the cloned *avr*D gene, and by culture fluids from these bacteria, but not by direct infiltration of the purified *avr*D protein.

DNA sequence studies of the range of *avr* genes now available suggest important differences in gene operation, variation in the size and chemical characteristics of the gene products, plus major similarities and dissimilarities in the degree of homology of the amino acid sequence. Sequencing of the *avr*A gene from *Pseudomonas syringae* pv. *glycinea*, for example, indicates a single open reading frame (ORF) specifying a much larger (100 kDa) protein than the other *avr* genes from this pathogen (Napoli & Staskawicz, 1987). Furthermore, avirulence gene *avr*Bs1 from *Xanthomonas campestris* pv. *vesicatoria* comprises two ORFs, of which ORF1 probably provides transcriptional readthrough for expression of ORF2, and ORF2 encodes a 50 kDa protein (Ronald & Staskawicz, 1988). This 50 kDa protein shows amino acid sequence homology with *avr*A (*P.s.* pv. *glycinea*), as do the predicted gene products of *avr*B and *avr*C, suggesting gene derivation from common ancestral sequences.

Although similarities in gene homology imply a close evolutionary relationship, the resulting phenotype is quite distinct. Thus closely related genes *avr*C and *avr*D from *Pseudomonas syringae* pv. *glycinea*, which share a common homology, interact with quite different soybean resistance genes to elicit the HR (Tamaki *et al.*, 1991). This shows that even minor differences in their amino acid sequence can affect their race specificity, and is consistent with an enzymatic or catalytic role of *avr* gene products. Future cloning and sequenc-

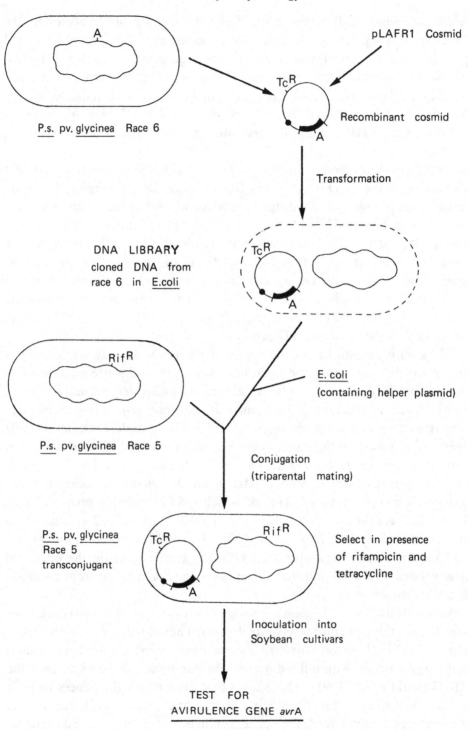

Fig. 8.7. For legend see p. 229

ing of other *avr* genes will clarify the rather fragmentary picture that occurs at the present time.

Mutation studies on single pathogens in host/non-host situations

The induction of mutations in a particular bacterial phytopathogen, by transposon or chemical mutagenesis, leads to a range of defects in the ability of the pathogen to incite a plant reaction. The genes involved can be divided into two main types on the basis of their phenotypic effects:

1. Hypersensitivity/pathogenicity genes – referred to as *hrp* genes by Lindgren *et al.* (1986) – are required for both the development of the hypersensitive reaction on non-host plants and the expression of disease symptoms in host plants.
2. Virulence genes – also referred to as disease specific (*dsp*) genes by Boucher *et al.* (1987) – are concerned only with disease development and will be discussed later in the section on genetic determination of virulence.

hrp *genes*

These genes are involved in the induction of both compatible and incompatible plant reactions. Mutation of *hrp* genes results in a partial or complete inability of the pathogen to cause disease in the host, coupled with an absence of HR when injected into non-host plants (or resistant cultivars). The dual effect of these genes supports the idea that hypersensitivity and pathogenicity share common aspects, such as damage to the plant cell plasmalemma, controlled by the bacterial genome.

In a number of pathogens *hrp* genes have now been identified by isolating transposon (Tn5) mutants that have lost both pathogenicity (path⁻) and the ability to induce HR (HR⁻), and identifying the cloned fragment of wild-type DNA that is able to restore these functions (Table 8.4). Very few mutants show loss of hypersensitivity, and the screening procedure may be very laborious. Bauer and Beer (1987), for example, screened 5200 Tn5-containing strains of *Erwinia amylovora* for lack of HR induction by tobacco leaf

Fig. 8.7 Cloning of *Pseudomonas syringae* pv. *glycinea* avirulence gene *avr*A and its expression in transconjugant cells. Summary of experiment carried out by Staskawicz *et al.* (1984). Genomic DNA is shown as an undulating circle. Gene symbols are: A, avirulence gene *avr*A; Tc^R, tetracycline resistance; and Rif^R, rifampicin resistance. Transfer of recombinant plasmid pLAFR1 containing a fragment of race 6 DNA involved triparental mating between *Pseudomonas syringae* pv. *glycinea* race 5, *Escherichia coli* containing the cosmid, and *E. coli* containing helper plasmid pRK2013. The cell of *E. coli* is shown with a broken outline.

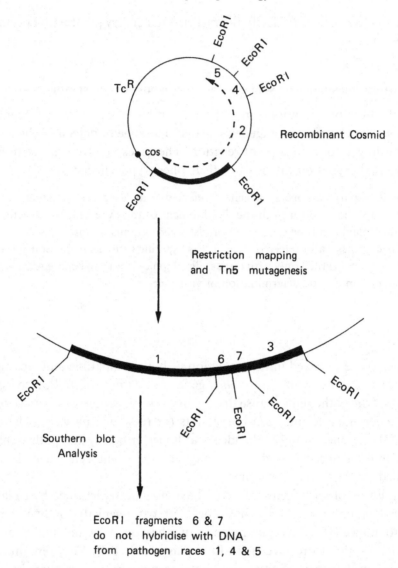

Fig. 8.8 Localisation of avirulence gene *avr*A in the cloned DNA fragment. The figure shows an *Eco*RI restriction map of cosmid pLAFR1 containing cloned DNA (broken line), with the overall recombinant cosmid (top figure) and part of the cloned DNA (bottom figure) presented separately. DNA fragments cut by the restriction enzyme are labelled 1–7 in decreasing size. Insertion of Tn5 transposons into segments 6 and 7 led to loss of race-specific incompatibility, identifying these regions as the probable site of the *avr*A gene. (Adapted from Staskawicz *et al.*, 1984.)

Table 8.4.. hrp *genes in plant pathogenic bacteria*

Host	Pathogen	Experimental details	Reference
Bean	*Pseudomonas syringae* pv. *phaseolicola*	8 *hrp*⁻ mutants *hrp* cluster composed of 7 loci on 22 kb fragment of chromosomal DNA	Lindgren *et al.* (1986) Rahme *et al.* (1992)
Bean	*Pseudomonas syringae* pv. *syringae*	Cloning/sequencing of *hrp* gene cluster on 3.9 kb DNA fragment	Mukhopadhyay *et al.* (1988)
Tomato	*Pseudomonas solanacearum*	9 hrp⁻ mutants in 17.5 kb region on megaplasmid	Boucher *et al.* (1987)
Apple/ Pear	*Erwinia amylovora*	3 hrp⁻ mutants in 17 kb DNA fragment. 20 other mutants were hrp⁺ path⁻	Bauer & Beer (1987)

N.B. All *hrp*⁻ mutants are unable either to elicit a hypersensitive reaction (hr⁻) or induce a disease response (path⁻).

infiltration before they obtained just three HR⁻ mutants. In the case of *Pseudomonas syringae* pv. *phaseolicola*, which causes halo blight of beans, Lindgren *et al.* (1986) were able to isolate eight independent *hrp* transposon mutants, six of which caused loss of pathogenicity on bean and HR on tobacco, and two caused reduced pathogenicity and variable HR responses.

Typically, *hrp* genes occur as a gene cluster. This was initially shown by the work of Lindgren *et al.* (1986) on *Pseudomonas syringae* pv. *phaseolicola*, where a single recombinant plasmid containing a short insert of wild-type DNA was able to restore pathogenicity and HR induction to almost all Tn5 mutants by complementation. In this pathogen, the *hrp* cluster is chromosomal, with an estimated length of 22 kb, and, as determined by Rahme *et al.* (1991), consists of seven loci (*hrp*L, *hrp*AB, *hrp*C, *hrp*D, *hrp*E, *hrp*F, and *hrp*SR). A particular feature of this gene cluster is the transcriptional convergence, with gene sets in the two halves of the cluster forming two transcriptional units which are separately transcribed towards the centre of the locus.

DNA sequences within the *hrp* gene cluster of *Pseudomonas syringae* pathovars are closely similar, suggesting a high degree of phylogenetic conservation within this group of organisms. Hybridisation data obtained by Boucher *et al.* (1987) also points to a close structural homology between the *hrp* regions of *Pseudomonas solanacearum* and *Xanthomonas campestris* pathovars (but not *Pseudomonas syringae* pathovars) – providing further evidence of common pathogenicity genes in different phytopathogens and *hrp* gene conservation during evolution.

There is some evidence from DNA sequence studies that *hrp* genes, unlike *avr* genes, code for membrane proteins. This has been obtained by Mukho-padhyay *et al.* (1988), who have shown that the *hrp* locus in *Pseudomonas syringae* pv. *syringae* is contained on a 3.9 kb DNA fragment which has two ORFs and which codes for an 83 kDa polypeptide with a deduced amino acid sequence typical of integral membrane proteins.

Comparison of *hrp* and *avr* genes in the determination of incompatibility

Both pathogenicity cluster (*hrp*) and race specificity (*avr*) genes are involved in the determination of incompatibility, appearing at the present time to represent two distinct but interrelated genetic systems. These two gene systems seems to differ in a number of important respects:

1. *hrp* genes are required for pathogenicity, *avr* genes are not.
2. *avr* genes determine avirulence on resistant plants that carry complementary resistance genes. No such complementarity has yet been demonstrated with *hrp* genes.
3. The activity of *avr* genes has typically been demonstrated in relation to race-specific incompatibility, while the activity of *hrp* genes is characteristically observed at a more general level.

Although the activity of *avr* genes has normally been demonstrated within a particular race–cultivar situation, the activity of these genes is not restricted to determining race-specific resistance. This is shown by the fact that *avr* genes can be transferred from one pathovar to another, conferring incompatibility against a non-host plant (Whalen *et al.*, 1988; Fillingham *et al.*, 1992). Examples are also known where *avr* genes in different pathovars (e.g. *Pseudomonas syringae* pv. *tomato* and pv. *glycinea*) have been shown to be genetically identical (Heath, 1991).

Interactions between avr *and* hrp *genes*

There is increasing evidence that interaction occurs between *avr* and *hrp* genes in the determination of incompatibility:

1. Studies by Lindgren *et al.* (1988) have shown that the determination of incompatibility by *avr* genes depends upon a completely operational *hrp* system. This was demonstrated in *Pseudomonas syringae* pv. *tabaci* and pv.

glycinea by transfer of *hrp*⁻ mutant DNA into the genome from *P.s.* pv. *phaseolicola* by marker exchange. The resulting mutants, with normal (and different) *avr* genes but defective *hrp* genes, were non-pathogenic on their host plants and did not induce HR on normally incompatible host cultivars. These mutants were restored to the hrp⁺ phenotype by introduction of a plasmid containing a functional *hrp* locus from *P.s.* pv. *phaseolicola*.

2. In some cases, *hrp* genes appear to control either the expression or the successful functioning of *avr* genes (Knoop *et al.*, 1991). An example of the former is provided by *Pseudomonas syringae* pv. *glycinea*, where the induction of *avr*B *in vitro* or *in planta* is controlled by the *hrp* locus.

Hrp gene control of *avr* gene function rather than induction has been demonstrated in *Xanthomonas campestris* pv. *vesicatoria*, where *avr*Bs₃ is expressed constitutively *in planta* but only functions in the presence of a fully operational *hrp* locus. In this example, the *hrp* locus itself is inducible *in planta*, and it has been suggested that *hrp* gene products may control passage of elicitor (*avr* gene product) at the plant cell surface.

The role of virulence genes in the determination of plant disease

As noted previously, the ability of a particular pathogen to cause disease depends on two sorts of genes: pathogenicity (*hrp*) genes and virulence genes. Pathogenicity genes are a fundamental requirement for disease to be induced, while virulence genes determine the type of disease and its severity. The general importance of virulence genes in disease is shown by the effects of mutagenesis, while the specific identity and role of individual genes is revealed by cloning experiments.

Loss of virulence by mutagenesis

The multiplicity of genes involved in the determination of virulence is shown by the variety and range of mutants that limit or prevent the development of disease *in planta*. Although naturally occurring mutants provide some information on this, the use of transposon or chemical mutagens to produce large numbers of mutants is potentially more useful, particularly where the mutations occur randomly throughout the genome. Some recent examples where this has been carried out are summarised in Table 8.5.

Where Tn5 mutagenesis has been carried out, the transconjugants (selected on the basis of kanamycin resistance) have been screened for two major types of mutant: auxotrophs and avirulent prototrophs.

Table 8.5. *Mutagenesis of plant pathogenic bacteria*

Bacterium	Mutagen	Auxotrophs[a]	Avirulent prototrophs[a]	Reference
Xanthamonas campestris	Tn5		0.5%	Daniels et al. (1984)
Pseudomonas syringae pv. *phaseolicola*	Tn5	1.3%	0.3%	Somlyai et al. (1986)
Pseudomonas syringae pv. *phaseolicola* race 3	Nitrosoguanidine	7%	6 genes 0.3%	Harper et al. (1987)
Pseudomonas solanacearum	Tn5	0.35%	0.14% 10 genes	Boucher et al. (1985)

[a] Frequencies of auxotroph and prototroph mutants are expressed as a percentage of the total number of clones.

Table 8.6. *Cloning of genes involved in toxin production*

Bacterium	Gene product	Details	Reference
Pseudomonas syringae pv. *tomato* (str PT 23.2)	Coronatine	Gene(s) carried on 101 kb plasmid	Bender et al. (1989)
Pseudomonas syringae pv. *phaseolicola*	Phaseolotoxin	Clustered genes for toxin production on 25 kb fragment	Peet et al. (1986)
Pseudomonas syringae pv. *syringae*	Syringomycin	*syrA* and *syrB* genes encode synthetase proteins	Xu & Gross (1988)
	Syringotoxin	Two genes identified encoding synthetase proteins	Morgan & Chatterjee (1988)
Pseudomonas syringae pv. *tabaci*	Tabtoxin	Gene cloned and expressed in non-toxin producing *Pseudomonas syringae*	Kinscherf et al. (1991)

Auxotrophs

These mutants differ from the original (prototrophic) wild-type bacteria in requiring a growth factor supplement for *in vitro* culture. These typically have a metabolic block which renders them unable to synthesise a particular amino acid, which must, therefore, be externally supplied for growth to occur. The large number of auxotrophs generated is of interest, since it indicates the widespread and random insertion of the Tn5 transposon throughout the bacterial genome. As might be expected, a relatively high proportion of the auxotrophs are also typically avirulent. This avirulence may be regarded as a secondary effect, since the gene concerned does not have a primary role in causing disease, and presumably arises because the host internal environment lacks the required supplement for bacterial growth.

Avirulent prototrophs

These mutants do not require any supplement for *in vitro* culture, but differ from wild-type bacteria in having a reduced ability to promote disease in host plants. The prototrophic nature of these avirulent mutants suggests that the avirulence does not arise from some general metabolic defect, but concerns genes that have a more specific effect on virulence.

One of the major technical problems encountered in this type of experimental study lies in the large number of transconjugant clones that have to be screened. For example, in the work by Boucher *et al.* (1985) on *Pseudomonas solanacearum*, over 8000 clones were screened for virulence by inoculating individual tomato seedlings with bacterial suspension. This led to the identification of 12 avirulent prototrophs, of which Southern blot analysis demonstrated that the Tn5 insertion had occurred in at least 10 different *Eco*RI fragments (i.e. at least 10 different genes).

As noted earlier, these avirulent mutants are of two main types, depending on whether the mutation also prevents the induction of a hypersensitive reaction in the non-host (*hrp* genes) or not (virulence genes). In the studies by Boucher *et al.* (1985) 5 out of the 12 avirulence mutants had no effect on the induction of HR, indicating that up to five virulence genes had been identified in these experiments.

In general, mutagenesis of bacterial phytopathogens has shown that numerous genes are involved in the ability of the pathogen to cause disease in the host. These experiments give no indication of the phenotypic effects of these genes, elucidation of which requires the use of gene cloning.

Cloning of virulence genes

In an increasing number of cases, the use of gene cloning has enabled the precise identification of virulence gene products, and DNA sequencing has provided valuable information on gene homologies and function.

Genes that have been clearly characterised are involved in the determination of various features that are important in virulence (see Chapter 7), including the production of toxins, extracellular enzymes, extracellular polysaccharides and plant hormones.

Toxins

As noted in Chapter 7, bacterial toxins generally have a low molecular weight and are not specific to a particular host. The non-specific and wide-ranging effects of these toxins is particularly useful in the genetic analysis of toxin production since it allows the use of toxin-sensitive microbes as an alternative bioassay to the employment of host plants, resulting in a more convenient and rapid screening procedure.

Genetic analysis has so far been carried out on the production of four bacterial toxins: coronatine, phaseolotoxin, syringomycin (plus syringotoxin) and tabtoxin (Table 8.6).

In most cases, the production of toxins appears to involve a number of genes, some of which are directly involved in the toxin biosynthesis. For example, the production of syringomycin by *Pseudomonas syringae* pv. *syringae*, requires the synthesis of at least 5 proteins (SR1–SR5), ranging in size from 130 to 470 kDa. Two genes, *syr*A and *syr*B, have so far been identified from an EcoR1 cosmid library on the basis of their ability to restore toxin production to TOX⁻ Tn5 mutants. These genes, occurring on separate cosmids, respectively encode proteins SR4 (350 kDa) and SR5 (130 kDa) which are believed to be components of the syringomycin synthetase complex. Subcloning, followed by restriction and Tn5 insertion analysis, has localised these genes to regions of DNA approximately 3 kb in length.

The molecular determination of toxin activity not only involves toxin biosynthesis, but also complex regulatory activity. In the case of phaseolotoxin, for example, there is temperature regulation of toxin production and also regulation of relative rates of synthesis of resistant and susceptible bacterial ornithine carbamoyl transferase (the target molecule; Peet *et al.*, 1986).

Table 8.7. *TOX⁻ mutants and the assessment of virulence*

Pathovar[a]/ toxin	TOX⁻ strain	Virulence[b]	Reference
pv. *tomato* coronatine	Tn5 mutants	Only small lesions and limited population increase in tomato leaves	Bender *et al.* (1987)
pv. *phaseolicola* phaseolotoxin	UV-induced	Normal population increase at inoculation site but no systemic spread	Patil *et al.* (1974)
pv. *syringae* syringomycin	Tn5 mutants	Reduced virulence or completely non-pathogenic on cherry	Xu & Gross (1988)
pv. *tabaci* tabtoxin	Spontaneous mutants	Growth curve *in planta* differs from wild-type after 3 days	Turner & Taha (1984)

[a] Pathovars of *Pseudomonas syringae*.
[b] Virulence: comparison of mutant with wild-type.

Avirulence mutants

The importance of toxin production to pathogen virulence is shown by comparison of the growth of wild-type bacteria and non-toxin producing (TOX⁻) mutants *in planta*, and the relative development of symptoms in these two situations. In general, TOX⁻ cells reach lower population levels in host tissue compared with wild-type bacteria and produce less-pronounced symptoms, suggesting that toxins are important but not essential in disease development. The production of toxins is required for normal symptom development. Some examples of TOX⁻ mutants and the effects of loss in toxin production are given in Table 8.7.

The pathogenic effect of TOX⁻ mutants has been investigated in particular detail in *Pseudomonas syringae* pv. *tabaci*, where the absence of tabtoxin production by leaf-infiltrated bacteria leads to symptomless infection (Turner & Taha, 1984) and the following major changes in bacteria–plant cell activities and interactions:

1. Effects of bacteria on plant cells: infected tobacco leaf tissue exhibited none of the physiological effects (accumulation of ammonia, inactivation of glutamine synthetase, loss of chlorophyll) that were induced by wild-type bacteria.

2. Bacterial activity: some aspects of bacterial activity are not altered by the absence of toxin production, including early growth and invasion of

inoculated leaves, and bacterial transmission from infected to healthy leaves by rain splash. Growth and spread of TOX⁻ bacteria in inoculated tissue after 3 days is considerably less than wild-type, possibly due to lower levels of nutrient leakage into the apoplast. TOX⁻ bacteria also differ in that their growth is markedly inhibited by tabtoxin, while the growth of wild-type cells is unaffected or even stimulated by quite high levels of the toxin.

The above results suggest that tabtoxin is a major determinant of virulence in *Pseudomonas syringae* pv. *tabaci*, and that TOX⁻ mutants behave essentially as non-virulent parasites. The importance of other (e.g. *hrp*) genes in pathogenicity is shown by the fact that cloning and expression of tabtoxin genes in a non-pathogenic strain of *P. syringae* does not lead to pathogenicity (Kinscherf *et al.*, 1991).

Where toxin production is determined by a number of genes, the effect of mutation on pathogenicity and virulence may depend on which particular genes are inactivated. The syringomycin Tn5 TOX⁻ strains raised by Xu and Gross (1988) fell into three main groups:

1. Path⁺, HR⁺
2. Path⁻, HR⁺ and
3. Path⁻, HR⁻,

where Path⁺ signifies pathogenicity in cherry (but attenuated in virulence) and HR⁺ signifies ability to elicit an HR in an incompatible plant.

The occurrence of Path⁻ strains in groups 2 and 3 suggests a linkage between toxigenicity and pathogenicity, even though syringomycin production is not essential for pathogenicity (group 1). Group 3 may have arisen by insertion of the transposon into *hrp* genes.

Extracellular enzymes

Plant pathogenic bacteria produce a range of extracellular enzymes, including pectolytic enzymes, cellulases and proteinases. In some cases, the production of these enzymes has been clearly linked to the virulence of the pathogen (see Chapter 7). The secretion of extracellular enzymes is under genetic control, and is discussed in Chapter 2.

Pectolytic enzymes

The production of pectolytic enzymes is particularly important in soft rot bacteria (e.g. *Erwinia chrysanthemi*) and in some bacteria that cause wilt diseases (e.g. *Pseudomonas solanacearum*). Some recent work on cloning of genes that encode these enzymes is summarised in Table 8.8, followed by more detailed accounts of studies carried out with *Erwinia chrysanthemi* and *Pseudomonas solanacearum*.

Erwinia chrysanthemi Strain EC16 of *Erwinia chrysanthemi* produces four separate pectate lyase enzymes (PelA, PelB, PelC and PelE). These enzymes are difficult to separate on the basis of molecular weight or isoelectric point, and the cloning of their respective genes with the production of individual enzymes in transconjugant *Escherichia coli* has been particularly important in elucidating the enzyme characteristics.

The *pel* genes of *Erwinia chrysanthemi* have now been sequenced and the gene products analysed by expression in *Escherichia coli* cells. These genes are organised in two clusters:

1. *pel*A–*pel*E, which also contains the remains of gene *pel*D (no longer operating in this strain).
2. *pel*B–*pel*C.

Comparison of the nucleotide sequences and the similarity of the synthesised proteins suggest that each gene cluster has arisen by duplication of an existing gene. This is particularly evident for the *pel*B–*pel*C cluster, where there is 84% amino acid identity in the sequencing data from the two genes. The *pel*A–*pel*E cluster appears to show greater divergence, since there is less amino acid similarity, the protein products have very different isoelectric points, and the *pel*D gene is no longer functional.

Despite their structural homology and tandem organisation, these genes appear to be independently regulated, as the sequence data shows that they each have an efficient transcriptional terminator after the translational stop at the end of the open reading frame.

Pseudomonas solanacearum The *pgl*A gene involved in the production of a single subunit major polygalacturonase is important, but apparently not absolutely necessary for pathogenesis. Mutant strains lacking the gene (*pgl*A$^-$) have reduced virulence but are still pathogenic on tomato plants.

Cosmid clones containing the gene were obtained by screening for polygalacturonase activity on pg plates. The location of the gene was subsequently determined by subclone and mutation analysis.

Cellulases

Genes that encode cellulase (endoglucanase) enzymes have been cloned in a number of genera of plant pathogenic bacteria (Table 8.9).

Table 8.8. Cloning of genes that encode pectolytic enzymes

Bacterium	Gene product	Gene	Details	Reference
Soft rot bacteria				
Erwinia chrysanthemi strain EC16	Pectate lyase	4 functional *pel* genes	Two gene clusters	Barras *et al.* (1987)
	PelA (45 kDa)	*pelA*	Gene cluster *pelA*–E	Tamaki *et al.* (1988b)
	PelE (47 kDa)	*pelE*		
	PelB	*pelB*	Gene cluster *pelB*–C	
	PelC (39 kDa)	*pelC*		
Erwinia carotovora subsp. *carotovora*	Pectin lyase	*pnlA*	Gene within 2.2 kb DNA fragment	Chatterjee *et al.* (1991)
Pseudomonas fluorescens	Pectate lyase	Single *pel* gene	Gene within 4.3 kb DNA fragment	Liao (1991)
Xanthomonas campestris pv. *campestris*	Pectate lyase 3 isozymes	3 *pel* genes	Genes cloned in non-pecto lytic *X.c.* pv. *translucens*	Daniels *et al.* (1987)
Bacterial wilt				
Pseudomonas solanacearum	Endopoly galacturonase (52 kDa)	*pg1A*	Gene contained within 1.8 kb DNA fragment	Schell *et al.* (1988)

Table 8.9. *Cloning of genes that encode cellulase enzymes*

Bacterium	Gene product	Gene	Reference
Soft rot bacteria			
Erwinia chrysanthemi	Major endoglucanases	*celZ* *celY*	Boyer *et al.* (1987)
Pseudomonas solanacearum	Major 43 kDa endoglucanase	*egl*	Roberts *et al.* (1988)

Pseudomonas solanacearum The *egl* gene in this bacterium encodes the major secreted endoglucanase. A cosmid bearing the *egl* gene was cloned in *Escherichia coli*, with subsequent subclone analysis and restriction enzyme mapping. Bacteria lacking the *egl* gene were generated by site marker exchange mutagenesis. This *egl⁻* mutant had normal production of extracellular polysaccharide and polygalacturonase, but 200 times less endoglucanase, and was significantly less virulent on tomato compared with wild-type strains. Virulence was restored by complementation with the normal *egl* gene.

Proteases

The secretion of proteases appears to be widespread amongst phytopathogenic bacteria, including many pathovars of *Xanthomonas campestris* (Tang *et al.*, 1987), and recent examples of protease gene cloning in *X.c.* pv. *campestris* and other phytopathogens are given in Table 8.10.

Although the general occurrence of protease enzymes might suggest that they have some major importance in plant cell degradation, the experimental evidence suggests that they have only a minimal role in the host–pathogen interaction. Tn5 and marker exchange protease⁻ mutants of *Xanthomonas campestris* pv. *campestris*, for example, showed no difference from wild-type (protease secreting) bacteria in terms of pathogenicity in turnip plants, although differences in symptom development did suggest that proteases may play a minor role in the progression of disease (Tang *et al.*, 1987).

In *Erwinia chrysanthemi*, three antigenically and structurally distinct proteases, plus a protease inhibitor, are encoded by a single cluster of genes. The protease inhibitor is a low MW, heat-stable protein which remains inside the cell and is able to bind specifically to the proteases. It is not required for either synthesis or secretion of the proteases, and is thought to act as a safety inactivation mechanism should the proteases become activated inside the cell (Wandersman *et al.*, 1987).

Table 8.10. *Cloning of genes that encode extracellular protease enzymes and their secretion*

Bacterium	Gene product	Gene	Details	Reference
Erwinia carotovora subsp. *carotovora*	Protease	Single *prt* gene	Cloned and expressed in *E. coli*	Allen *et al.* (1986)
Erwinia chrysanthemi	Proteases A,B,C	*prt*A,B,C	4 genes on 40 kb DNA cloned and expressed in *E. coli*	Wandersman *et al.* (1987)
	Protease inhibitor	*inh*		
	Transport proteins	*prt*D,E,F	4 genes on 5.5 kb DNA sequenced	Letoffe *et al.* (1990)
Xanthomonas campestris pv. *campestris*	Protease	Structural gene(s)	On 10 kb DNA Cloned and expressed in *E. coli*	Tang *et al.* (1987)
			Sequenced and identified as metalloproteins	Kyostio *et al.* (1991)

Table 8.11. *Cloning of genes that encode extracellular polysaccharide production*

Bacterium	Genes	Details	Reference
Pseudomonas solanacearum		At least 3 gene clusters: 1. EPS production and virulence	Cook & Sequeira (1991)
	ops genes	2. EPS and LPS biosynthesis	Kao & Sequeira (1991)
Erwinia stewartii	*cps* genes	Identification of 5 genes *cps*A–E involved in EPS production & pathogenicity	Dolph *et al.* (1988)
Xanthomonas campestris pv. *campestris*	*xps* genes	Genes clustered in 22 kb region of DNA that restores EPS (xanthan) production to naturally occurring mutants	Harding *et al.* (1987)

Extracellular polysaccharides

Extracellular (capsular) polysaccharides have been implicated in the disease virulence of a number of plant pathogenic bacteria, including *Pseudomonas solanacearum, Erwinia amylovora, Erwinia stewartii* and *Xanthomonas campestris* pv. *campestris* (see Chapter 7).

The genetic analysis of extracellular polysaccharide (EPS) production in these pathogens has involved extensive use of naturally occurring non-capsulate (EPS⁻) mutants as well as the cloning of genes for EPS production. The assessment of virulence in such naturally occurring mutants in *Erwinia amylovora* (see Table 7.12), *Pseudomonas solanacearum* and *Erwinia stewartii* (see Table 7.13) is discussed in chapter 7.

Some recent work on experimental induction of EPS⁻ mutants and cloning of EPS genes is summarised in Table 8.11. Molecular studies on *Erwinia stewartii* and *Xanthomonas campestris* have been facilitated by the availability of the naturally occurring EPS⁻ mutants, and have involved analysis of cloned wild-type DNA that restores EPS production to these cells.

In the case of *E. stewartii*, which causes vascular wilt and leaf blight of corn, EPS production is restored to 14 naturally occurring EPS⁻ deletion mutants by a single fragment of wild-type DNA cloned in cosmid pES2144. Subcloning, restriction analysis and Tn5 mutagenesis of this cosmid have demonstrated that its ability to restore casule production resides in 5 regions (genes), designated *cps*A–E. These genes appear to be organised in three operons and

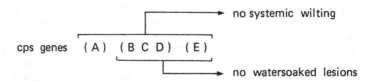

Fig. 8.9 Genetic determination of extracellular polysaccharide production and pathogenicity in *Erwinia stewartii*. The 5 genes involved in EPS production are arranged in three operons. Tn5 mutagenesis of any of these genes prevents the systemic induction of wilting in the host plant, but only two operons are involved in the induction of water-soaked lesions. (After Dolph *et al.*, 1988.)

are important in pathogenicity, both in the induction of systemic wilting and the production of water-soaked lesions in leaf tissue (Fig. 8.9). The importance of bacterial ability to cause water-soaking and EPS production in the general severity of wilt disease is summarised in Table 7.13.

In *Pseudomonas solanacearum*, at least three gene clusters important for EPS biosynthesis have been identified, one of which (*ops* genes) is also required for LPS biosynthesis (Kao & Sequeira, 1991).

Plant hormones

The induction of hypertrophic growth of plant tissue by oncogenic bacteria involves the promotion or production of plant growth substance synthesis by the bacterial cells (Chapter 7). The genetic determination of this has been investigated in two particular cases: *Pseudomonas savastanoi* subsp. *savastanoi* and *Agrobacterium tumefaciens*, both of which carry chromosome or plasmid-borne genes involved in the synthesis of 3-indoleacetic acid (IAA) and various cytokinins (Morris, 1986), as summarised in Table 8.12.

Synthesis of IAA

In *Pseudomonas savastanoi* subsp. *savastanoi*, studies by Kosuge and his co-workers (Glass & Kosuge, 1988) have shown that IAA is produced from tryptophan by the action of two enzymes: tryptophan monoxygenase and indoleacetamide hydrolase, as illustrated in Fig. 8.10. These enzymes are encoded by genes *iaa*M and *iaa*H respectively, which constitute the IAA operon, and function as one transcriptional unit. In strains of the pathogen isolated from olive galls, this operon is located on the main chromosome, while in strains derived from *Oleander* galls it is located on a plasmid, designated pIAA, which occurs in three main sizes: 52 kb(pIAA1),

Table 8.12. *Synthesis of IAA and cytokinin by* Pseudomonas savastanoi *subsp.* savastanoi *and* Agrobacterium tumefaciens

Gene product	Gene	Details	Reference
IAA SYNTHESIS			
Pseudomonas syringae subsp. *savastanoi* (Oleander isolate)			
Tryptophan monoxygenase	*iaa*M	Carried on	Glass & Kosuge
Indoleacetamide hydrolase	*iaa*H	plasmid	(1988)
IAA lysine synthetase	*iaa*L	pIAA	
		Gene sequencing	Yamada *et al.* (1985)
		In vitro transcription/ translation	Palm *et al.* (1989)
Agrobacterium tumefaciens			
Tryptophan monoxygenase	*iaa*M	Carried on Ti	Schroder *et al.*
Indoleacetamide hydrolase	*iaa*H	plasmid	(1984)
CYTOKININ SYNTHESIS			
Pseudomonas syringae subsp. *savastanoi*			
26.8 kDa protein	*ptz*	Carried on pcK1 plasmid	Powell & Morris
		Cloned and sequenced	(1986)
Agrobacterium tumefaciens			
Isopentyl transferase	*tmr*	Located in T-region of Ti plasmid	Buchmann *et al.* (1985)
Enzyme involved in transzeatin synthesis	*tzs*	Located in *vir* region of Ti plasmid	Beaty *et al.* (1986)

73 kb(pIAA2) and 90 kb(pIAA3). Transcription and translation of the IAA operon in an *Escherichia coli in vitro* system leads to the production of two proteins with molecular weights of 62 and 47 kDa. These correspond quite closely to the protein sizes predicted from the DNA sequence analysis, where the *iaa*M locus has an ORF encoding 557 amino acids (61.8 kDa) and the *iaa*H locus 455 amino acids (48.5 kDa).

In *Oleander* isolates of the pathogen, IAA is further converted to an amino acid conjugate (by the enzyme IAA lysine synthetase) which enters the plant tissue and is subsequently broken down by a plant hydrolase. The importance of this conjugate to bacterial virulence in *Oleander* tissue is shown by Tn5 inactivation of the *iaa*L locus, which leads to a loss of ability to induce gall formation, and much reduced bacterial populations compared with wild-type cells.

Genes *iaa*M and *iaa*H in *Agrobacterium tumefaciens* have an analogous function to the corresponding genes in *Pseudomonas savastanoi*, with close homologies in their nucleotide sequence. This close homology provides strong

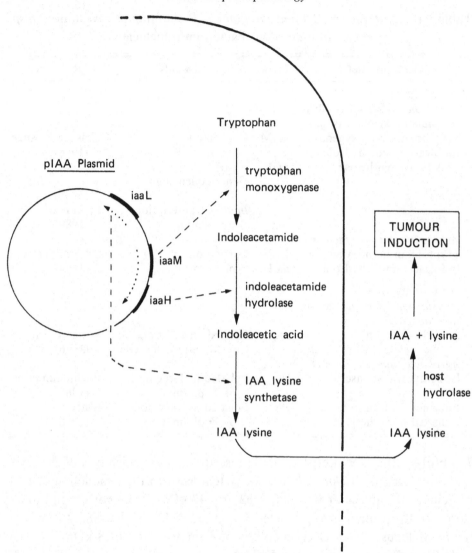

Fig. 8.10 Genetic determination of IAA lysine production by *Pseudomonas savastanoi* subsp. *savastanoi* in oleander galls.

evidence that the two IAA operons have a common ancestor, and suggests a strong similarity in the pathways of IAA synthesis that occur in the two types of bacteria.

Genes that are homologous to *iaa*M and *iaa*H of *Pseudomonas savastanoi* have also been specifically located by DNA probes in *Erwinia herbicola* pv. *gypsophilae*, where they confer oncogenic activity (Manulis *et al.* 1991).

Synthesis of cytokinins

The pathways of cytokinin synthesis are less clearly defined than those for IAA, but appear to comprise two main synthetic steps (Powell & Morris, 1986). These involve addition of an isoprenoid side-chain to 5'AMP by isopentyl transferase (Ipt) yielding isopentenyladenine and isopentenyladenosine (Fig. 7.10), followed by hydroxylation to produce transzeatin and transribosylzeatin.

A number of genes involved in the synthesis of cytokinins have been cloned and characterised in both *Pseudomonas savastanoi* subsp. *savastanoi* and *Agrobacterium tumefaciens* (Table 8.12). As with genes involved in IAA synthesis, a high degree of homology exists between the cytokinin genes of the two pathogens, providing further evidence of common gene ancestry. Unlike *A. tumefaciens*, the location of cytokinin genes in *P.s. subsp. savastanoi* occurs on a separate plasmid from the IAA determinants, and IAA-deficient mutants lacking the (pIAA) plasmid produce cytokinins at the same level as wild-type strains (Surico, 1986).

When the *tzs* gene is cloned in *Escherichia coli* it expresses Ipt activity and also leads to secretion of zeatin and ribosylzeatin into culture medium.

Evolutionary conservation of pathogenicity and virulence genes

There is increasing evidence that evolutionary pathways leading to pathogenicity and virulence are highly conserved and that fundamental homologies exist between pathogens which are otherwise very different in terms of disease symptoms and host plant.

Phylogenetic conservation of *hrp* genes has already been mentioned, with two major homology groups (*Pseudomonas solanacearum*/*Xanthomonas campestris* and *Pseudomonas syringae* pathovars) identified so far.

Similar conservation appears to apply also to some at least of the virulence genes. Todd *et al.* (1990) have shown, for example, that genes required for the virulence of *Xanthomonas campestris* pv. *campestris* (which causes soft rot of crucifers) have conserved functional analogies with *X.c.* pv. *oryzae* (causing a wilting disease of rice without any associated rotting) and also with *X. campestris* pvs. *graminis*, *poae*, *phlei* and *arrhenatheri*. These genes, which are involved in the export of extracellular proteins, are thus common to pathogens which differ in host identity, type of disease and geographic location.

Control of gene activity: plant regulation of bacterial gene expression

The control of bacterial gene activity during the plant association is an area of increasing interest, and is important in understanding the interaction that occurs between the two types of organism. In some cases (constitutive genes), transcriptional activity occurs on a continuous basis, irrespective of whether the bacterium is growing *in vitro* or *in planta*. Expression of avirulence gene *avr*Bs$_3$, for example, occurs in both minimal and complex media, and is not influenced by the plant environment (Knoop *et al.*, 1991).

In other cases (inducible genes) bacterial gene expression may be switched on inside the plant or near the plant surface. Such induction of gene activity may result either due to the presence of high substrate levels (e.g. induction of pectolytic enzymes during tissue maceration) or due to the presence of plant signal molecules (e.g. induction of *hrp* genes).

Plant signal molecules

Various plant molecules have now been isolated that specifically activate particular bacterial genes at critical points in the infection process (Mo & Gross, 1991), including:

1. Phenolic compounds. These are of particular importance and are involved in the activation of *vir* genes in *Agrobacterium tumefaciens* (e.g. acetosyringone). Phenolic glycosides have also been shown to be important in the induction of *syr*B gene in *Pseudomonas syringae* pv. *syringae*.
2. Flavonoids. In *Rhizobium*, specific plant flavonoids are required for induction of nodulation.
3. Sugar molecules. A second class of *vir*-inducing compounds has recently been identified in *Agrobacterium tumefaciens*, comprising the simple pyranose sugars D-fucose, D-galactose, D-glucose and D-mannose. These saccharides markedly enhance *vir* gene induction when only low levels of the phenolic signal molecule are present.

Plant induction of *hrp* genes has recently been shown for a range of pathogens, including *Pseudomonas syringae* pv. *phaseolicola* (Rahme *et al.*, 1992) and *Xanthomonas campestris* pv. *vesicatoria* (Schulte & Bonas, 1992). Signalling mechanisms for these genes are as yet undefined, but appear to be complex.

Genes which determine non-pathogenic characteristics

Although the majority of molecular biological studies on plant pathogenic bacteria have concentrated on genes that determine host specifity or the degree of virulence *in planta*, other phenotypic aspects are also important.

Table 8.13. *Genetic determination of ice nucleation in* Pseudomonas

Bacterium	Gene product /Gene	Details	Reference
Pseudomonas syringae	InaZ protein /*ina*Z	Gene cloning of 3.5–4 kb DNA and expression in *E. coli*	Orser *et al.* (1985)
		Gene sequencing	Green & Warren (1985)
Pseudomonas fluorescens	InaW protein (180 kDa) /*ina*W	Gene cloning of 4.6 kb DNA and expression in *E. coli*	Corotto *et al.* (1986)
		Gene sequencing	Warren *et al.* (1986)
		Homologies between genes and gene products	Deininger *et al.* (1988)

These non-pathogenic aspects are relevant, for example, to the growth and survival of bacteria outside the host and will determine the response of the bacterium to different physical and biological environmental situations. Three such non-pathogenic characteristics which have been investigated in detail are the ability of particular phylloplane bacteria to cause ice nucleation at subzero temperatures, siderophore metabolism and antibiotic resistance (see later section on plasmids).

Genetic determination of ice nucleation activity

The ability of ice nucleating active bacteria to bring about ice nucleation at temperatures of −5 to 0°C (Ina$^+$ phenotype), while other bacteria supercool without ice crystal formation (Ina$^-$ phenotype) is discussed in Chapter 4. The genetic determination of ice nucleation activity has been investigated in various strains of *Pseudomonas syringae*, *Pseudomonas fluorescens* and *Erwinia herbicola*.

Pseudomonas syringae *and* Pseudomonas fluorescens

In these bacteria the Ina$^+$ phenotype is imparted by single genes, referred to respectively as *ina*Z and *ina*W, which encode the ice nucleating proteins InaZ and InaW (Table 8.13).

Expression of the *ina*Z gene of *Pseudomonas syringae* and the *ina*W gene of *Pseudomonas fluorescens* in *Escherichia coli* (normally devoid of ice nucleation activity) leads in each case to the INA$^+$ phenotype. The InaW protein is incorporated into the inner and outer membrane fractions of transformed *E. coli* cells, consistent with ice nucleation being a cell surface phenomenon.

DNA sequencing of the two ice nucleating genes has shown close homologies, suggesting a common ancestry. In each case, the polypeptide produced is composed of repeating units based on the consensus octapeptide sequence Ala-Gly-Tyr-Gly-Ser-Thr-Leu-Thr (Green & Warren, 1985). Such conservation of repetitive sequences leads to ice nucleating proteins that have a regular and well-defined structure. These proteins are believed to bind water molecules in an orderly array, resembling an ice crystal, thus providing a suitable template for ice nucleation. In the case of InaW protein, Green and Warren (1985) showed that 68 out of the 122 octapeptide repeats could be deleted without completely destroying ice nucleation activity, suggesting that it is the overall pattern rather than individual polypeptide sequences that are important in ice nucleation.

Homologies between ice nucleating proteins are further demonstrated by a common antibody response, since anti-InaW has been shown to react with InaW (the original antigen) plus InaZ and the ice nucleating protein from *Erwinia herbicola*.

Although the *ina*Z and *ina*W genes are closely related, they differ in the timing of expression during *in vitro* culture of bacteria. Transcription of *ina*Z in *Pseudomonas syringae* appears to commence at stationary phase, coincident with the onset of ice nucleation activity. In contrast to this, *ina*W activity in *Pseudomonas fluorescens* remains constant throughout the period of bacterial growth, in parallel with the continuous ice nucleating activity seen throughout the growth cycle in this organism.

Erwinia herbicola

Although the restriction site map of the ice nucleating gene (*ice*E) in *Erwinia herbicola* is very different from that of *Pseudomonas syringae* (in line with their taxonomic separation), Southern blot analysis indicates substantial conservation, giving further support to a common gene ancestry (Orser *et al.*, 1985). The *ice*E gene has now been sequenced, and a consensus sequence with other ice nucleation genes determined (see Turner *et al.*, 1991).

Genetic control of siderophore metabolism

The production of extracellular Fe-chelating compounds (siderophores) by bacteria at the plant surface is an important aspect of the growth and ecology of both pathogenic and saprophytic organisms (see Chapter 4), and has major implications for biological control (Chapter 9). In view of this, there is considerable interest in the molecular biology (including the genetic determi-

Table 8.14. *Genetic determination of siderophore metabolism*[a]

Organism	Phenotypic effect	Genetic or molecular analysis
SIDEROPHORE PRODUCTION		
Growth-promoting bacteria		
Pseudomonas fluorescens strain B-10	Pseudobactin synthesis and uptake	At least 12 genes involved, arranged in 4 clusters. 1 gene determines ferric-pseudobactin uptake
strain TR21	Iron-independent growth	At least 10 genes involved
Pseudomonas putida	Production of fluorescent pigment	Minimum of 8 genes reported
Phylloplane bacteria and phytopathogens		
Pseudomonas syringae pv. *syringae*	Presence (Flu+) or absence (Flu−) of pyoverdin production	At least 4 different regions of DNA involved
Erwinia chrysanthemi	Chrysobactin production	Cluster of 4 *cbs* genes[b] Cloned & expressed in *E. coli*
SIDEROPHORE RECEPTORS (Outer membrane proteins)		
Pseudomonas syringae pv. *syringae*	70 kDa protein Fe-pyoverdin receptor	Synthesis repressed by iron. Absent from mutant which cannot take up Fe-complex
Erwinia chrysanthemi	Fct protein	Production controlled by *fct* gene[b]

[a] Data taken mainly from Panopoulos & Peet (1985).
[b] Franza & Expert (1991).

nation) of siderophore production and retrieval (Panopoulos and Peet, 1985), with information contributed from three main directions (Table 8.14) at the present time:

1. Siderophore production by plant growth-promoting soil bacteria (where deleterious micro-organisms are out-competed).
2. Siderophore production by phylloplane and phytopathogenic bacteria (*in vitro* studies).
3. Studies on outer membrane siderophore receptors (see Fig. 4.5).

Studies on siderophore synthesis and metabolism have so far concentrated particularly on fluorescent siderophores, reflecting the major role played by fluorescent pseudomonads at the plant surface. The results suggest that this aspect of cell metabolism is controlled by a multiplicity of genes occurring at different sites on the genome. These genes encode enzymes involved in siderophore synthesis as well as outer membrane proteins that are important in the uptake of the iron-siderophore complex, and their activity is essential for microbial growth and competition at low external iron levels.

Non-fluorescent siderophore systems have also been investigated, including that of *Erwinia chrysanthemi*. In this bacterium a series of genes have been identified for encoding the surface receptor protein Fct (*fct* gene) and the siderophore chrysobactin (4 *cbs* genes). These genes all occur within a single 8 kb cluster (in the order *fct*, *cbs*A, *cbs*B, *cbs*C and *cbs*E) and have been cloned and expressed in *Escherichia coli* (Franza & Expert, 1991).

The occurrence and genetic importance of plasmids in plant pathogenic bacteria

Plasmids are important and widely occurring constituents of plant pathogenic bacteria (Coplin, 1982). They are typically present as circular covalently closed molecules of DNA (cccDNA), and are transmitted from one generation to another during cycles of growth and cell division as autonomous self-replicating units. Plasmids may also pass between bacterial cells during the process of conjugation, and some (conjugative) plasmids specifically promote bacterial conjugation and plasmid transmission between cells in addition to their other genetic effects.

In this section, plasmids will be considered initially in terms of their isolation and characterisation, followed by discussion of their genetic importance in the different activities of plant pathogenic bacteria. In recent years it has been shown that plasmids encode a wide range of functions that are important in bacterial–plant interactions (including pathogenicity and virulence factors) and that they are also important in the determination of certain non-pathogenic features. In addition to naturally occurring plasmids, recombinant plasmids have considerable potential for future developments in producing genetically engineered organisms.

Plasmid isolation and characterisation

The plasmid composition of a bacterial strain is normally determined by a bulk preparation and analytical method, which involves growing a population of the cells in liquid culture, collecting a sample of bacteria and extracting the constituent plasmids for subsequent analysis.

Different species of plasmid may be separated and characterised on the

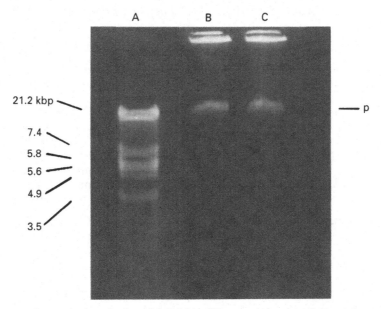

Fig. 8.11 Bulk analysis of plasmid DNA. Electrophoresis of *Erwinia amylovora* plasmid DNA on a 0.5% agarose gel, showing a single plasmid band at 24 kbp (channels B,C). Channel A shows electrophoresis of linear marker DNA for approximate comparison of molecular sizes.

basis of their distinct sizes. Plasmid size categories are normally determined from bulk population samples by four main experimental steps (Broda, 1979): bacterial lysis, removal of chromosomal DNA and cell debris, concentration of remaining (largely plasmid) DNA, followed by separation of plasmid species by centrifugation or electrophoresis. An example of this is shown in Fig. 8.11, where agarose gel electrophoresis of a plasmid sample from *Erwinia amylovora* reveals a single plasmid species of approximately 20 kb in size.

Although bulk extraction followed by size separation normally gives a clear resolution of plasmid species, some categories of plasmid may be lost during preparation. This is particularly true for large (less stable) plasmids above 150–200 kb (Mogen *et al.*, 1988), where their size renders them liable to linearisation by shearing, and may also be the case for relaxed (non-supercoiled) molecules. The differential loss of larger plasmids may be reduced if bacterial lysis is carried out directly in electrophoresis wells (Eckhardt, 1978), resulting in the separation of molecules immediately after their release from bacterial cells (thus avoiding shearing).

Bulk analysis has been widely used in the investigation of plasmids in bacterial phytopathogens (Table 8.15a), providing information on a wide range of species and pathovars. In some cases, the plasmid composition of a

Table 8.15 *Analysis of bacterial plasmids*

	Bacterium	Results	Reference
(a) Bulk analysis	*Pseudomonas syringae* pv. *papulans*	Variation in plasmid occurrence	Burr *et al.* (1988)
	Pseudomonas syringae pv. *tomato*	5 size classes of plasmid, 29–101 kb	Bender & Cooksey (1986)
	Xanthomonas campestris pv. *pruni*	10 size classes of plasmid, 4.2–35.1 MDa 1–4 per strain	Randhawa & Civerolo (1987)
	Clavibacter michiganensis subsp. *sepedonicum*	Single 50.6 kb plasmid copy no. 1–5/cell	Mogen *et al.* (1988)
(b) Micro-scale	*Pseudomonas syringae* pv. *tabaci*	EM visualisation of relaxed and super-coiled plasmids	El-Masry & Sigee (1986)
analysis	*Erwinia amylovora*	As above	Sigee & El-Masry (1987)

particular pathovar is highly variable, with considerable differences in size range and number of plasmid species between strains. This is shown, for example, by the work of Burr *et al.* (1988) on *Pseudomonas syringae* pv. *papulans*, and Randhawa & Civerolo (1987) on *Xanthomonas campestris* pv. *pruni* (Table 8.15a). In the former case, great differences in plasmid composition occurred between some strains isolated within a single orchard, while other strains isolated from separate orchards had a closely similar plasmid composition.

Other plant pathogenic bacteria are much less variable in their plasmid content. In *Clavibacter michiganense*, Mogen *et al.* (1988) have demonstrated the presence of a single, highly conserved plasmid in half the strains analysed, while in the other strains, Southern hybridisation showed the plasmid to be integrated into the chromosome. The high degree of conservation and retention of this plasmid suggests that it encodes important metabolic functions.

Electron microscope examination of bacterial plasmids

Transmission electron microscopy provides a powerful tool in the analysis of bacterial plasmids, with the potential to determine plasmid sizes (contour lengths), functional state of the molecule (e.g. replicating or non-replicating), nucleotide sequences (by *in situ* hybridisation) and general fine structural aspects.

Electron microscope images of individual plasmids may be obtained either from plasmid preparations derived from bulk analysis (see above) or from single bacterial cells that have been lysed on an electron microscope grid (microscale analysis).

Microscale analysis of bacterial plasmids by in situ *lysis*

In this approach, plasmids are liberated from individual bacterial cells and observed immediately by transmission electron microscopy (El-Masry & Sigee, 1986). The technique, originally developed by Griffith (1976) to examine the genomic DNA of *Escherichia coli*, involves:

1. Lysozyme treatment of a suspension of bacteria, causing partial degradation of the cell envelope.
2. Sedimentation of bacterial cells onto a formvar-coated electron microscope grid.
3. Final *in situ* lysis on the EM grid by treatment with detergent.
4. Air-drying and contrast enhancement.

In situ lysis leads to a sequential extrusion of bacterial cell contents, with an initial liberation of plasmids followed by the extrusion of genomic DNA. Within a single EM grid, bacteria can be seen at various stages of disruption, including cells surrounded by 5 or 6 plasmids (Fig. 8.12a) presumably derived from the single bacterial cell.

The technique has so far been used with *Pseudomonas syringae* pv. *tabaci* (El-Masry & Sigee, 1986) and *Erwinia amylovora* (Sigee & El-Masry, 1987), resulting in the visualisation of both supercoiled and extended circular (relaxed) molecules in a state that should be close to the situation in the living cell (Fig. 8.12). The extended circular plasmids differed from the supercoiled molecules in their frequent association with membrane material and in the presence of a clear protein sheath, protease-removal of which left a residual circle of naked DNA. In some cases, the DNA circle appeared to have a clear replication fork, suggesting that the relaxed plasmids may differ from the supercoiled molecules in being in a state of replication.

The role of plasmids in the determination of pathogenicity and virulence

Although the majority of genes involved in the induction and development of disease are probably located on the main chromosome, some pathogenicity and virulence genes are present on bacterial plasmids, including *avr* genes (Table 8.3), *hrp* genes (Table 8.4) and genes that encode the production of toxins (e.g. coronatine, Table 8.6) and plant hormones. The genetic determi-

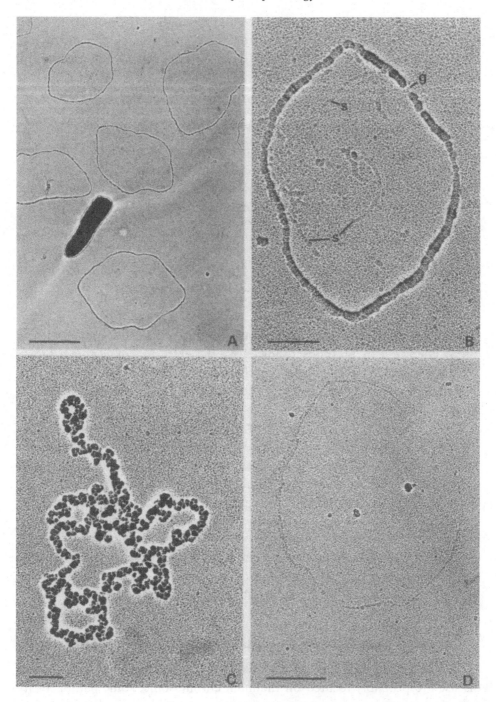

Fig. 8.12 Fine structure of *Pseudomonas syringae* plasmids. Plasmids have been freshly-released from cells lysed on electron microscope grids and are rotary-

nation of indoleacetic acid and cytokinin production by *Pseudomonas savastanoi* subs. *savastanoi* and *Agrobacterium* species (Table 8.12) has already been mentioned, and constitutes a major aspect of tumour induction by these organisms.

Tumour induction by Agrobacterium *species: role of the Ti and Ri plasmids*

The role of plasmids in the pathogenicity of *Agrobacterium tumefaciens* and *Agrobacterium rhizogenes* is now well established, and detailed accounts have been given by a number of authors (e.g. Kahl & Schell, 1982; Walden, 1988) on the respective roles of the tumour inducing (Ti) and root-inducing (Ri) plasmids in these organisms. Critical evidence for the involvement of plasmids as the causal agent of disease has been provided by growing bacteria at high temperature, where loss of pathogenicity is correlated with loss of the plasmid from the bacterial cells.

The Ti and Ri plasmids are large (200 kb), double-stranded DNA molecules that encode information for both hyperplastic plant growth and the production and catabolism of specific aminosugar derivatives, collectively referred to as opines. These compounds include nopaline, octopine, agrocinopine and agropine (see Fig. 7.3) and are important both in promoting plasmid transfer between bacteria and as a carbon/nitrogen source for the pathogen. The plasmid-induced production of opines by the transformed plant cells constitutes the major benefit of the association to the bacterium.

The induction of tumour formation (*Agrobacterium tumefaciens*) or root hair disease (*Agrobacterium rhizogenes*) involves a permanent genetic change (transformation) of the plant cell, in which a specific segment of the bacterial plasmid (transfer or T-DNA) is transferred and integrated into the plant genome.

Functional organisation of the Ti and Ri plasmids

Ti plasmids found in different strains of *Agrobacterium tumefaciens* have four major regions of homology, as summarised in Fig. 8.13. Two of these regions

Caption for Fig. 8.12 *(cont.)*
shadowed. (*a*) Single bacterium, surrounded by relaxed-circle plasmids. Bar scale: 2.0 μm. (*b*) Relaxed-circle plasmid, showing a 10–15 nm thick protein sheath around most of the DNA contour. The sheath has a banded appearance, with some gaps (g) and with associated strands of material (s) lying within the plasmid circle. Bar scale: 0.25 μm. (*c*) Supercoiled plasmid. Bar scale: 0.25 μm. (*d*) Relaxed-circle plasmid, showing complete loss of sheath and internal strands after protein extraction by treatment with sodium dodecyl sulphate. Bar scale: 0.25 μm. (Taken, with permission, from El-Masry & Sigee, 1986.)

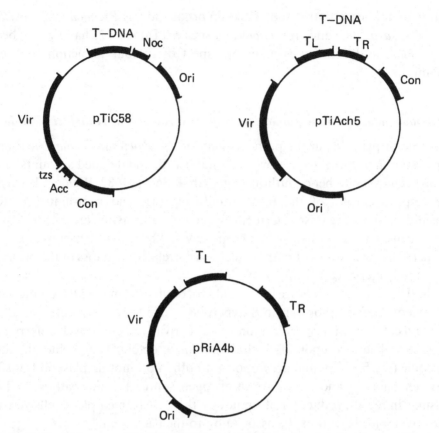

Fig. 8.13 General maps of the Ti and Ri plasmids. Showing details of a nopaline-type plasmid (pTiC58), an octopine-type plasmid (pTiAch5) and an Ri plasmid (pRiA4b). Major functional regions are: T-DNA, transfer DNA; Noc, nopaline catabolism; Ori, origin of replication; Con, conjugative transfer of the plasmid; Acc, agrocinopine catabolism; tzs, transzeatin synthesis; Vir, virulence. In two of the plasmids, the T-DNA region is split into left (T_L) and right (T_R) halves. (Redrawn from Walden, 1988.)

(T-DNA and the '*Vir*' region) are directly involved in tumour induction, while the other ('*Ori*' and '*Con*') regions encode information for plasmid replication and conjugative transfer. Variations in plasmid structure occur in relation to type of opine synthesis (nopaline or octopine), as detailed below.

General organisation of the Ri plasmid in *Agrobacterium rhizogenes* is closely similar to the Ti plasmid (Fig. 8.13), with a homologous *vir* region and a segment of T-DNA that also transfers to the plant genome. As with the Ti plasmid, variations in the organisation of the T-DNA occur in relation to the type of opine synthesis, with agropine plasmids having two separate T-DNA regions and mannopine and cucumopine plasmids having only one. The role of the Ri plasmid in bacterial pathogenicity and the induction of root hair

Table 8.16. *Genes present in the T-DNA and* vir *region of the Ti plasmid*

	Gene	Gene product
T-DNA		
Opine production	*nos*	Nopaline synthase
	ocs	Octopine synthase
	acs	Agrocinopine synthase
	ons	Nopaline or octopine secretion
Indoleacetic acid synthesis	*iaa*M	Tryptophan monoxygenase
	*iaa*H	Indoleacetamide hydrolase
Cytokinin synthesis	*tmr*	Isopentyl transferase
Control of tumour size	*tml*	(Unknown product)
vir-DNA		
Sensory transduction	*vir*A[a]	Sensory membrane protein
	*vir*G[a]	Induction of *vir* and *pin* loci
Excision of T-DNA	*vir*D[a]	Single-stranded endonuclease
Unknown functions	*vir*B[a]	
	*vir*C	
	*vir*E	Acts extracellularly
	*vir*F	Acts extracellularly

[a] These genes are required for pathogenicity (tumour formation is abolished by mutation).
Information taken from Walden (1988).

disease by *Agrobacterium rhizogenes* closely resembles the Ti-mediated pathogenicity of *Agrobacterium tumefaciens*.

Pathogenicity and virulence genes in Agrobacterium tumefaciens

The role of specific genes in the T-DNA and *vir*-DNA is summarised in Table 8.16. Mutation studies show that some of these genes (*vir*A, B, D & G) are absolutely essential for tumour formation and qualify as pathogenicity genes. Loss of function in other genes (*ons, iaa*M, *iaa*H, *tmr*) still permits disease development, but the tumour is now slow-growing, so these genes should be regarded as virulence genes.

T-DNA Nopaline-encoding strains of bacterium have a single 24 kb T-DNA region, while in octopine strains this is split into two regions: T_L (14 kb) and T_R (7 kb). The T-DNA of nopaline strains and T_L DNA of octopine bacteria share regions of close homology (referred to as the 'core' DNA) that contain genes involved with hormone synthesis, opine secretion and the determination of tumour size.

The borders of the T-DNA in both nopaline and octopine Ti plasmids are flanked by a repeat sequence of 25 base pairs. These sequences are carried through into the plant cell during transformation and are integrated into the plant DNA. The right (but not left) border repeat is essential for efficient transfer of the DNA into the plant cell, and can be substituted for this by either a left border or synthetic sequence, so long as they are in the correct orientation.

The sequence of genes in the T-DNA and the borders that flank this region show considerable resemblance to the situation seen in eukaryote DNA, and the possible eukaryote origin of this part of the prokaryote Ti plasmid is of considerable interest.

vir DNA In nopaline bacteria, this region of DNA contains six complementation groups: *vir*A, B, C, D, E and G, with one additional gene (*vir*F) in octopine strains. The genes of the *vir* region encode proteins that are important in sensing the presence of wounded plant cells, and in the molecular transduction of the chemotactic response.

Mechanism of plant cell transformation by Agrobacterium tumefaciens

Transfer of DNA from *Agrobacterium tumefaciens* to recipient plant cells involves a cascade of events which require the active participation of both pathogen and plant cell (Walden, 1988). Although much is now known about the early stages in this process, certain aspects, such as the mechanisms involved in the passage of DNA into the plant cell and its integration into the plant genome, are still not clear. The general sequence of tumour induction is summarised in Fig. 8.14, and occurs as a series of distinct events:

Chemotaxis and plant cell conditioning The earliest events in tumour induction involve a reciprocal interaction between the bacterium and plant cell, with the pathogen responding to plant cell exudate (chemotaxis, see Chapters 2 and 5) and the plant cell responding to bacterial growth factor secretion (conditioning).

Wounded plant tissue releases various phenolic compounds into the rhizosphere, one of which (acetosyringone) has been shown to act as a potent chemotactic attractant at low concentration. The release of growth factor (transzeatin) has been demonstrated for nopaline bacteria, and appears to condition the plant cells for the transformation process, possibly by inducing cell division.

Sensory transduction, and activation of vir *genes* Sensing the presence of wounded host tissue is a primary step in the transformation process, and requires the products of two particular *vir* genes, *vir*A and *vir*G, which are expressed constitutively at substantial levels. The *vir*A gene produces a membrane-bound sensory protein molecule which is able to detect the presence of acetosyringone and subsequently reacts with *vir*G product to induce transcription of both *vir* loci and *pin* loci (the latter are plant-inducible loci which do not appear to be directly involved in virulence).

Expression of the vir *loci* Interaction between *vir*G protein and acetosyringone-activated *vir*A product leads to a sudden increase in the transcription of *vir*B, C, D, E and G. These genes are now expressed at high level, encoding a range of polypeptides (Table 8.16).

Excision of T-DNA Induction of the *vir*D locus results in the production of a site- and strand-specific endonuclease, which cuts the segment of T-DNA on either side of the 25 bp-repeat flanking regions. This T-DNA strand acts as the T-DNA transfer molecule which passes into the plant cell and effects transformation.

Transfer of DNA into the plant cell The passage of the T-DNA strand into the plant cell is thought to be analogous to DNA transfer during bacterial conjugation, with the 5′ end leading and the whole molecule being incorporated into a DNA–protein complex. Recent studies suggest that a single bacterium can infect more than one plant cell, with multiple copies of the T-DNA being inserted into the plant genome. This suggests that replication of the T-DNA occurs prior to final integration into the plant genome, either in the bacterial cell or the plant.

Integration of T-DNA into the plant genome Southern blot analysis of transformed plant DNA shows that this integration occurs randomly within the genome, and that no rearrangement of the T-DNA nucleotide sequence occurs during the transformation process.

Comparison of the plant ('target') DNA before and after incorporation of the T-DNA shows that complex rearrangements can take place (Gheysen *et al.* 1987), and a four-step sequence of integration has been proposed involving sequential introduction of two nicks in the plant DNA, ligation and copying of the T-DNA and repair and replication of the staggered nicks in the plant DNA.

Expression of the T-DNA in planta Transcription of the T-DNA in the plant cell leads to the synthesis of indoleacetic acid and cytokinins, disrupting the

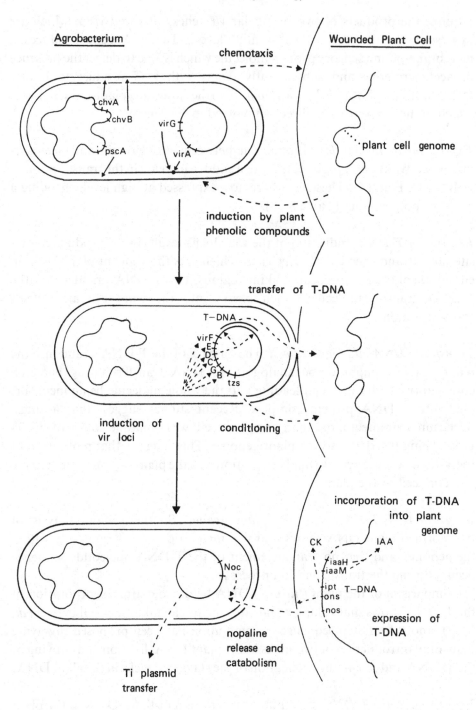

Fig. 8.14 For legend see p. 263.

hormonal balance of the cell and leading to active cell division and dis-organised growth. The establishment of the transformed phenotype is also characterised by the expression of genes that promote opine synthesis. These compounds are taken up into the *Agrobacterium* cells and catabolised by enzymes encoded by the Ti plasmid. The *Noc* region of the nopaline plasmid encodes nopaline oxidase, arginase and ornithine cyclodeaminase, converting nopaline to proline (Sans *et al.*, 1987).

Plasmid determination of non-pathogenic characteristics

Plasmids appear to play an important role in the interactions that occur between pathogenic bacteria and surrounding micro-organisms, since they have been shown to encode production of, and resistance to, at least some of the anti-microbial compounds (particularly bacteriocins) that are involved. The genetic determination of antibiotic resistance is also of considerable importance in relation to the use of these compounds as chemical control agents.

Plasmid determination of bacteriocin production

Bacteriocins constitute one particular group of anti-bacterial substances that appear to be of major importance for microbial interactions in the rhizosphere. This has been demonstrated particularly clearly in the case of *Agrobacterium radiobacter* strain 84, which is a non-pathogenic rhizosphere organism that antagonises *Agrobacterium tumefaciens* by the production of a bacteriocin (agrocin 84) and is an important biocontrol agent.

The production of agrocin 84 in *Agrobacterium radiobacter* is mediated by a small plasmid (Ellis *et al.*, 1979), which can be transferred to, and expressed in, *Escherichia coli*. The agrocin 84 plasmid is non-conjugative, and is always transferred together with a nopaline catabolic plasmid which promotes conjugation and mobilises the agrocin plasmid for transmission. The agrocin plasmid can also transfer to *Agrobacterium tumefaciens* during conjugation (at high nopaline levels) giving rise to a range of transconjugants, including pathogenic bacteria that synthesise agrocin and are also resistant to it.

Fig. 8.14 Major events in the transformation of plant cells by *Agrobacterium tumefaciens*. Details and abbreviations of genes in the T-DNA and *vir* loci are given in Table 8.16. *Chv*A, *chv*B and *psc*A are all chromosomal genes that are thought to be involved in the synthesis of cell wall components and are important for bacterial attachment to the plant cell. (Adapted from Walden, 1988.)

Complete inhibition of agrocin plasmid transfer has now been achieved by genetic modification of the plasmid, as detailed in Chapter 9.

Plasmids and antibiotic resistance

Recent studies have demonstrated a number of cases involving plant pathogenic bacteria where plasmids have the ability to confer resistance to antibacterial chemical agents (including both organic and inorganic molecules) and to be readily transmitted between bacterial cells. This has important implications for the chemical control of these organisms. The potential role of plasmids in the determination of resistance to chemical agents is shown by two main lines of research: studies on plasmid-associated resistance in the field, and studies on the transfer of antibiotic resistance between phytopathogenic and other bacteria in the laboratory.

Antibiotic resistance in the field

A number of examples are known where antibiotic resistance in phytopathogenic bacteria in the field is specifically associated with the presence of conjugative plasmids (Table 8.17a).

In the case of *Pseudomonas syringae* pv. *papulans*, resistance to streptomycin is associated with a 68 MDa conjugative plasmid, which appears to be present in all resistant strains (in the New York area, USA) but not present in sensitive strains. The origin of this plasmid, which now occurs quite commonly within the bacterial population, is obscure. Plasmid characterisation of field strains has shown that considerable diversity occurs within this pathogen, and the plasmid may have occurred indigenously at a low background level prior to its selection by field application of streptomycin. An alternative possibility is that the plasmid may have recently transferred from other phylloplane bacteria, prior to selection. Laboratory studies have shown that the plasmid can transfer at low frequency (approximately 10^{-7} per recipient cell) from *Pseudomonas syringae* pv. *papulans* to *P.s.* pv. *syringae*, and the reverse may also occur.

In some cases, antibiotic resistance appears to be encoded by genomic rather than plasmid DNA. This has been demonstrated in *Erwinia amylovora* (Schroth *et al.*, 1979), where high level resistance to streptomycin was demonstrated in isolates from Californian pear orchards. Non-plasmid determination of streptomycin-resistance in this pathogen was also demonstrated by Panopoulos (1978), who showed that resistance was not transmissible by conjugation, and that resistant and sensitive strains showed no differences in plasmid composition.

Table 8.17. *Field studies on resistance to antibiotics and inorganic control agents*

Bacterium	Resistance	Plasmid determinant(s)	Reference
(a) Plasmid-mediated			
Pseudomonas syringae pv. *glycinea*	Trimethoprim	40 kb plasmid	Leary & Trollinger (1985)
Pseudomonas syringae pv. *papulans*	Streptomycin	68 MDa plasmid	Burr *et al.* (1988)
Xanthomonas campestris pv. *vesicatoria*	Cu	Single 193 kb conjugative plasmid[a]	Stall *et al.* (1986)
	Streptomycin	68 kb plasmid	Minsavage *et al.* (1990)
Pseudomonas syringae pv. *tomato*	Cu	2 conjugative plasmids (67 & 101 kb)	Bender & Cooksey (1986)
Curtobacterium flaccumfaciens	Arsenite, arsenate and antimony	Single 46 MDa conjugative plasmid	Hendrick *et al.* (1984)
(b) Genome-mediated			
E. amylovora	Streptomycin	—	Schroth *et al.* (1979)

[a] Plasmid also carries to an avirulence gene.

The distinction between chromosomal and plasmid-mediated resistance reflects fundamental differences in the nature of the resistance mechanism involved. In the case of resistance to aminoglycoside antibiotics (such as streptomycin), plasmid-mediated resistance is typically at a medium to low level and is due to the synthesis of inactivating enzymes. Chromosomal resistance is normally at a higher level and involves single-step mutations which code for altered ribosomes (Minsavage *et al.*, 1990).

Laboratory studies on plasmid-mediated transfer of resistance

There have been several laboratory studies demonstrating the transfer of plasmids encoding antibiotic resistance (originally referred to as R factors) between plant pathogenic bacteria and other bacteria associated with animals or plants (Table 8.18).

Conjugative plasmid transfer occurs in mixed populations of the donor and recipient cells, and can take place either *in vitro* or *in planta* (Lacy & Leary, 1975). Transmission of antibiotic resistance is thus independent of the host, and is unrelated to the pathogenic nature of the bacterium. Transfer of

Table 8.18. *Laboratory transfer of antibiotic resistance*

Donor bacterium	Recipient cell	Plasmid	Reference
Pseudomonas aeruginosa	*Erwinia herbicola* *Erwinia stewartii*	RP1	Gibbins *et al.* (1976)
Pseudomonas aeruginosa	*Pseudomonas syringae* pathovars	R91-1, R18-1 RP 1	Panopoulos *et al.* (1975)
Pseudomonas syringae pv. *papulans*	*Pseudomonas syringae* pv. *syringae*	pCPP501	Burr *et al.* (1988)
Escherichia coli	*Pseudomonas syringae* pv. *glycinea* and pv. *phaseolicola*	RP1	Lacy & Leary (1975)

antibiotic resistance plasmids into pathogenic bacteria normally has no apparent effect on the pathogenicity of the bacteria in the host, or the ability to induce an HR in tobacco (Panopoulos *et al.*, 1975).

Introduction of novel genes into plant pathogenic bacteria using genetically engineered plasmids

The introduction of recombinant plasmids (containing foreign DNA sequences) into plant pathogenic and saprophytic bacteria provides a useful procedure for altering the phenotype for experimental or biocontrol purposes.

For example, this approach has been used to introduce *lux* genes derived from *Vibrio fischeri* into a range of phytopathogenic and saprophytic bacteria, including pathovars of *Xanthomonas campestris* and *Pseudomonas syringae*, and species of *Agrobacterium*, *Erwinia* and *Rhizobium*. This results in the formation of bioluminescent cells, which can be monitored in terms of their light emission to determine their presence both *in planta* and in the environment (Shaw & Kado, 1986). The use of bioluminescence to detect bacteria has the advantages of high sensitivity and lack of requirement for tissue disruption (maceration) during the infection process. The introduction of *lux* genes by these workers involved cloning of a promoter-less gene set from *Vibrio fischeri* into a broad host range vector molecule (pUCD4) to form a recombinant plasmid (pUCD607), which was then mobilised into recipient bacterial cells by tri-parental mating. Bacteria containing the recombinant plasmid could be selected on the basis of its tetracycline resistance gene, the promoter of which acted as the constitutive promoter of the *lux* operon, resulting in light emission on a continuous basis.

References

Allen C., Stromberg V. K., Smith F. D., Lacy G. H. & Mount M. S. (1986). Completion of an *Erwinia carotovora* subsp. *carotovora* protease mutant with a protease-encoding cosmid. *Mol. Gen. Genet.* **202**, 276–9.

Barras F., Thurn K. K. & Chatterjee A. K. (1987). Resolution of four pectate lyase structural genes of *Erwinia chrysanthemi* (EC16) and characterisation of the enzymes produced in *Escherichia coli*. *Mol. Gen. Genet.* **209**, 319–25.

Bauer D. W. & Beer S. V. (1987). Cloning of a gene from *Erwinia amylovora* involved in induction of hypersensitivity and pathogenicity. In *Plant Pathogenic Bacteria*, ed. E. L. Civerolo, A. Collmer, R. E. Davis and A. G. Gillaspie, pp. 425–9. Dordrecht: Martinus Nijhoff.

Beaty J. S., Powell G. K., Lica L., Regier D. A., MacDonald E. M. S., Hommes N. G. & Morris R. O. (1986). *tzs*, a nopaline Ti plasmid gene from *Agrobacterium tumefaciens* associated with trans-zeatin biosynthesis. *Mol. Gen. Genet.* **203**, 274–80.

Bender C. L., Cooksey D. A. (1986). Indigenous plasmids in *Pseudomonas syringae* pv. *tomato*: Conjugative transfer and role in copper resistance. *J. Bacteriol.* **165**, 534–41.

Bender C. L., Malvick D. K. & Mitchell R. E. (1989). Plasmid-mediated production of the phytotoxin coronatine in *Pseudomonas syringae* pv. *tomato*. *J. Bacteriol.* **171**, 807–12.

Bender C. L., Stone H. E., Sims J. J. & Cooksey D. A. (1987). Reduced pathogen fitness of *Pseudomonas syringae* pv. *tomato* Tn5 mutants defective in coronatine production. *Physiol. Mol. Plant Pathol.* **30**, 273–83.

Boucher C. A., Barberis P. A., Trigalet A. & Demery D. A. (1985). Transposon mutagenesis of *Pseudomonas solanacearum*: isolation of Tn5-induced avirulent mutants. *J. Gen. Microbiol.* **131**, 2449–57.

Boucher C. A., Van Gijsegem F., Barberis P. A., Arlat M. & Zischek C. (1987). *Pseudomonas solanacearum* genes controlling both pathogenicity on tomato and hypersensitivity on tobacco are clustered. *J. Bacteriol.* **169**, 5626–32.

Boyer M. H., Cami B., Kotoujansky A., Chambost J. P., Frixon C. & Cattanco J. (1987). Isolation of the gene encoding the major endoglucanase of *Erwinia chrysanthemi*. *FEMS Microbiol. Lett.* **41**, 351–6.

Broda P. (1979). *The Plasmids*. Oxford: W. H. Freeman.

Brown T. A. (1986). *Gene cloning – an introduction*. Wokingham: Van Nostrand Rheinhold.

Buchmann I., Marner F., Schroder G., Waffenschmidt S. & Schroder J. (1985). Tumour genes in plants: T-DNA encoded cytokinin biosynthesis. *EMBO J.* **4**, 853–9.

Burr T. J., Norelli J. L., Katz B., Wilcox W. F. & Hoying S. A. (1988). Streptomycin resistance of *Pseudomonas syringae* pv. *papulans* in apple orchards and its association with a conjugative plasmid. *Phytopathology* **78**, 410–13.

Chatterjee A., McEvoy J. L., Chambost J. P., Blasco F. & Chatterjee A. K. (1991). Nucleotide sequence and molecular characterisation of *pnl*A, the structural gene for damage-inducible pectin lyase of *Erwinia carotovora* subsp. *carotovora*. *J. Bacteriol.* **173**, 1765.

Cook D. & Sequeira L. (1991). Genetic and biochemical characterisation of a *Pseudomonas solanacearum* gene cluster required for extracellular polysaccharide production and for virulence. *J. Bacteriol.* **173**, 1654–62.

Coplin D. L. (1982). Plasmids in plant pathogenic bacteria. In *Phytopathogenic*

Prokaryotes, ed. M. S. Mount and G. Lacy, vol. 2, New York: Academic Press.

Corotto L. V., Wolber P. K. & Warren G. T. (1986). Ice nucleation activity of *Pseudomonas fluorescens*: Mutagenesis, complementation analysis and identification of a gene product. *EMBO J.* **5**, 231–6.

Daniels M. J., Collinge D. B., Dow J. M., Osbourn A. F. & Roberts I. N. (1987). Molecular cloning of the interaction of *Xanthomonas campestris* with plants. *Plant Physiol. Biochem.* **25**, 353–9.

Daniels M. J., Dow J. M. & Osbourn A. F. (1988). Molecular genetics of pathogenicity in phytopathogenic bacteria. *Annu. Rev. Phytopathol.* **26**, 285–312.

Daniels M. J., Turner P. C., Barber C. E., Cleary W. G. & Reed G. (1984). Towards the genetical analysis of pathogenicity of *Xanthomonas campestris*. In *Molecular Genetics of the Bacteria–plant Interaction*, ed. A. Puhler. Berlin: Springer-Verlag.

Deininger C. A., Mueller G. M. & Wolber P. K. (1988). Immunological characterisation of ice nucleation proteins from *Pseudomonas syringae*, *Pseudomonas fluorescens* and *Erwinia herbicola*. *J. Bacteriol.* **170**, 669–765.

Dolph P. J., Majerczak D. R. & Coplin D. L. (1988). Characterisation of a gene cluster for exopolysaccharide biosynthesis and virulence in *Erwinia stewartii*. *J. Bacteriol.* **170**, 865–71.

Eckhardt T. (1978). A rapid method for the identification of plasmid deoxyribonucleic acid in bacteria. *Plasmid* **1**, 584–8.

Ellingboe A. H. (1985). Prospects for using recombinant DNA technology to study race-specific interactions between host and parasite. In *Genetic Basis of Biochemical Mechanisms of Plant Diseases*, ed. Groth and Bushnell.

Ellis J. G., Kerr A., van Montagu M. and Schell M. (1979). *Agrobacterium*: Genetic studies on agrocin 84 production and its biological control of crown gall. *Physiol. Plant Pathol.* **15**, 311–19.

El-Masry M. H. & Sigee D. C. (1986). *In situ* lysis of *Pseudomonas tabaci*. 1 Electron microscope visualisation of relaxed and supercoiled plasmids. *Microbios* **47**, 193–208.

Fillingham A. J., Wood J., Bevan J. R., Crute I. R., Mansfield J. W., Taylor J. D. & Vivian A. (1992). Avirulence genes from *Pseudomonas syringae* pathovars *phaseolicola* and *pisi* confer specificity towards both host and non-host species. *Physiol. Mol. Plant Pathol.* **40**, 1–15.

Flor H. H. (1955). Host–parasite interactions in flax rust – Its genetics and other implications. *Phytopathology* **45**, 680–5.

Franza T. & Expert D. (1991). The virulence-associated chrysobactin iron uptake system of *Erwinia chrysanthemi* 3937 involves an operon encoding transport and biosynthetic functions. *J. Bacteriol.* **173**, 6874–81.

Gabriel D. W., Burges A. & Lazo G. R. (1986). Gene for gene interactions of five cloned avirulence genes from *Xanthomonas campestris* pv. *malvacearum* with specific resistance genes in cotton. *Proc. Nat. Acad. Sci.* **83**, 6415–19.

Gheysen G., Van Montagu M. & Zambryski P. (1987). Integration of *Agrobacterium tumefaciens* transfer DNA (T-DNA) involves rearrangements of target plant sequences. *Proc. Nat. Acad. Sci.* **84**, 6169–73.

Gillis G. M., Roth D. A., Johnson J. & Rudolph K. (1990). Characterisation by nucleic acids. In *Methods in Phytobacteriology*, ed. Z. Klement, K. Rudolph and D. C. Sands. Budapest: Akademiai Kiado.

Gibbins L. N., Bennett P. M., Saunders J. R., Grinsted J. & Connolly J. C. (1976). Acceptance and transfer of R-factor RP1 by members of the "herbicola" group of the genus *Erwinia*. *J. Bacteriol.* **128**, 309–16.

Glass N. L. & Kosuge T. (1988). Role of indoleacetic acid–lysine synthetase in regulation of indoleacetic acid pool size and virulence of *Pseudomonas savastanoi* subsp. *savastanoi*. *J. Bacteriol.* **170**, 2367–73

Green P. L. & Warren G. J. (1985). Physical and functional repetition of the bacterial ice nucleation gene. *Nature* **317**, 645–8.

Griffith J. D. (1976). Visualisation of prokaryotic DNA in a regularly condensed chromatin-like fiber. *Proc. Nat. Acad. Sci.* **73**, 563–7.

Harding N. E., Cleary J. M., Cabanas D. K., Rosen I. G. & Kang K. S. (1987). Genetic and physical analyses of a cluster of genes essential for xanthan gum biosynthesis in *Xanthomonas campestris*. *J. Bacteriol.* **169**, 2854–61.

Harper S., Zewdie N., Brown I. R. & Mansfield J. W. (1987). Histological, physiological and genetical studies of the responses of leaves and pods of *Phaseolus vulgaris* to three races of *Pseudomonas syringae* pv. *phaseolicola* and to *Pseudomonas syringae* pv. *coronafaciens*. *Physiol. Mol. Plant Pathol.* **31**, 153–72.

Heath M. C. (1991). The role of gene-for-gene interactions in the determination of host species specificity. *Phytopathology* **81**, 127–30.

Hendrick C. A., Haskins W. P. & Vidaver A. K. (1984). Conjugative plasmid in *Corynebacterium flaccumfaciens* subsp. *oortii* that confers resistance to arsenite, arsenate and antimony (III). *Appl. Environ. Microbiol.* **48**, 56–60.

Hitchin F. E., Jenner C. E., Harper S., Mansfield J. W., Barber C. E. & Daniels M. J. (1989). Determinant of cultivar specific avirulence cloned from *Pseudomonas syringae* pv. *phaseolicola* race 3. *Physiol. Mol. Plant Pathol.* **34**, 309–22.

Kahl G. & Schell J. (1982). *Molecular Biology of Plant Tumours*. London: Academic Press.

Kao C. C. & Sequeira L. (1991). A gene cluster required for coordinated biosynthesis of lipopolysaccharide and extracellular polysaccharide also affects virulence of *Pseudomonas solanacearum*. *J. Bacteriol.* **173**, 7841–7.

Keen N. T., Tamaki S., Kobayashi D., Gerhold D., Stayton M., Shen H., Gold S., Lorang J., Thordal-Christensen H., Dahlbeck D. & Staskawicz B. (1990). Bacteria expressing avirulence gene D produce a specific elicitor of the soybean hypersensitive reaction. *Mol. Plant–Microbe Interact.* **3**, 112–21.

Kinscherf T. G., Coleman R. H., Barta T. M. & Willis D. K. (1991). Cloning and expression of the tabtoxin biosynthetic region from *Pseudomonas syringae*. *J. Bacteriol.* **173**, 4132–42.

Knoop V., Staskawicz B. & Bonas U. (1991). Expression of the avirulence gene *avr*Bs$_3$ from *Xanthomonas campestris* pv. *vesicatoria* is not under the control of *hrp* genes and is independent of plant factors. *J. Bacteriol.* **173**, 7142–50.

Kyostio S. R., Cramer C. L. & Lacy G. H. (1991). *Erwinia carotovora* subsp. *carotovora* extracellular protease: Characterisation and nucleotide sequence of the gene. *J. Bacteriol.* **173**, 6537–46.

Lacy G. H. & Leary J. V. (1975). Transfer of antibiotic resistance plasmid RP1 into *Pseudomonas glycinea* and *Pseudomonas phaseolicola in vitro* and *in planta*. *J. Gen. Microbiol.* **88**, 49–57.

Leary J. V. & Trollinger D. (1985). Identification of an indigenous plasmid carrying a gene for trimethoprim resistance in *Pseudomonas syringae* pv. *glycinea*. *Mol. Gen. Genet.* **201**, 485–6.

Letoffe S., Depelaire P. & Wandersman C. (1990). Protease activity by *Erwinia chrysanthemi*: The specific secretion functions are analogous to those of *Escherichia coli* -haemolysin. *EMBO J.* **9**, 1375–82.

Liao C. H. (1991). Cloning of pectate lyase gene *pel* from *Pseudomonas fluorescens* and detection of sequences homologous to *pel* in *Pseudomonas viridiflava* and *Pseudomonas putida*. *J. Bacteriol*. **173**, 4386.

Lindgren P. B., Peet R. C. & Panopoulos N. K. (1986). Gene cluster of *Pseudomonas syringae* pv. *phaseolicola* controls pathogenicity of bean plants and hypersensitivity on non-host plants. *J. Bacteriol*. **168**, 512–22.

Lindgren P. B., Panopoulos N. J., Staskawicz B. J. & Dahlbeck D. (1988). Genes required for pathogenicity and hypersensitivity are conserved and interchangeable among pathovars of *Pseudomonas syringae*. *Mol. Gen. Genet*. **211**, 499–506.

Manulis S., Gafni Y., Clark E., Zutra D., Ophir Y. & Barash I. (1991). Identification of a plasmid DNA probe for detection of strains of *Erwinia herbicola* pathogenic on *Gypsophila paniculata*. *Phytopathology* **81**, 54–7.

Minsavage G. C., Canteros B. I. & Stall R. E. (1990). Plasmid mediated resistance to streptomycin in *Xanthomonas campestris* pv. *vesicatoria*. *Phytopathology* **80**, 719–23.

Mo Y.-Y. & Gross D. C. (1991). Plant signal molecules activate the *syr*B gene, which is required for syringomycin production by *Pseudomonas syringae* pv. *syringae*. *J. Bacteriol*. **173**, 5782–92.

Mogen B. D., Oleson A. E., Sparks R. B., Gudmestad N. C. & Secor G. A. (1988). Distribution and partial characterisation of pCS1, a highly conserved plasmid present in *Clavibacter michiganense* subsp. *sepedonicum*. *Phytopathology* **78**, 1381–6.

Morgan M. K. & Chatterjee A. K. (1988). Genetic organisation and regulation of proteins associated with production of syringotoxin by *Pseudomonas syringae* pv. *syringae*. *J. Bacteriol*. **170**, 5689–97.

Morris R. O. (1986). Genes specifying auxin and cytokinin biosynthesis in phytopathogens. *Annu. Rev. Plant Physiol*. **37**, 509–38.

Mukhopadhyay P., Williams J. & Mills D. (1988). Molecular analysis of a pathogenicity locus in *Pseudomonas syringae* pv. *syringae*. *J. Bacteriol*. **170**, 5479–88.

Napoli C. & Staskawicz B. J. (1987). Molecular characterisation and nucleic acid sequence of an avirulence gene from race 6 of *Pseudomonas syringae* pv. *glycinea*. *J. Bacteriol*. **169**, 572–8.

Orser C., Staskawicz B. J., Panapoulos N. J., Dahlbeck D. & Lindow S. E. (1985). Cloning and expression of bacterial ice nucleating genes in *Escherichia coli*. *J. Bacteriol*. **164**, 359–66.

Palm C. J., Gaffney T. & Kosuge T. (1989). Cotranscription of genes encoding IAA production in *Pseudomonas savastanoi* subsp. *savastanoi*. *J. Bacteriol*. **171**, 1002–9.

Panopoulos N. J. (1978). Genetic nature of streptomycin resistance in *Erwinia amylovora*. In *Proc. IVth Int. Conf. Plant Pathol. Bacteriol*. (Angers), II, pp. 467–9.

Panopoulos N. J., Guimares W. V., Cho J. J. & Schroth M. N. (1975). Conjugative transfer of *Pseudomonas aeruginosa* R factors in plant pathogenic *Pseudomonas* sp. *Phytopathology* **65**, 380–8.

Panopoulos N. J. & Peet R. C. (1985). The molecular genetics of plant pathogenic bacteria and their plasmids. *Annu. Rev. Phytopathol*., **23**, 381–419.

Patil S. S., Hayward A. C. & Emmons R. (1974). An ultraviolet-induced non-toxigenic mutant of *Pseudomonas phaseolicola* of altered pathogenicity. *Phytopathology* **64**, 590–5.

Peet R. C., Lindgren, P. B., Willis D. K., Panopoulos N. J. (1986). Identification and cloning of genes involved in phaseolotoxin production by *Pseudomonas syringae* pv. *phaseolicola*. *J. Bacteriol.* **166**, 1096–105.

Powell G. K. & Morris R. O. (1986). Nucleotide sequence and expression of a *Pseudomonas savastanoi* cytokinin biosynthetic gene: Homology with *Agrobacterium tumefaciens* tmr and tzs loci. *Nucl. Acids Res.* **14**, 2555–63.

Rahme L. G., Mindrinos M. N. & Panopoulos N. J. (1992). Plant and environmental signals control the expression of *hrp* genes in *Pseudomonas syringae* pv. *phaseolicola*. *J. Bacteriol.* **174**, 3499–507.

Randhawa P. S. & Civerolo E. L. (1987). Indigenous plasmids in *Xanthomonas campestris* pv. *pruni*. In *Plant Pathogenic Bacteria*, Proc. Sixth Int. Conf. Plant Pathol. Bacteriol. (Maryland). ed. E. L. Civerolo, A. Collmer, R. E. Davies and A. G. Gillaspie. Dordrecht: Martinus Nijhoff.

Roberts D. P., Denny T. P. & Schell M. A. (1988). Cloning the *egl* gene of *Pseudomonas solanacearum* and analysis of its role in pathogenicity *J. Bacteriol.* **170**, 1445–51.

Ronald P. C. & Staskawicz B. J. (1988). The avirulence gene *avr*Bs1 from *Xanthomonas campestris* pv. *vesicatoria* encodes a 50 kDa protein. *Mol. Plant Microbe Interact.* **1**, 191–8.

Sambrook J., Fritsch E. F. & Maniatis T. (1989). Molecular cloning: A laboratory manual, 2nd edn. Cold Spring Harbor: CSH Laboratory.

Sans N., Schroder G. & Schroder J. (1987). The *Noc* region of Ti plasmid C58 codes for arginase and ornithine cyclodeaminase. *Eur. J. Biochem.* **167**, 81–7.

Schell M. A., Roberts D. P. & Denny T. P. (1988). Analysis of the *Pseudomonas solanacearum* polygalacturonase encoded by *pgl*A and its involvement in phytopathogenicity. *J. Bacteriol.* **170**, 4501–8.

Schroder G., Waffenschmidt S., Weiler E. W. & Schroder J. (1984). The T-region of Ti plasmids codes for an enzyme synthesising indole-3-acetic acid. *Eur. J. Biochem.* **138**, 387–91.

Schroth M. N., Thomson S. V. & Moller W. J. (1979). Streptomycin resistance in *Erwinia amylovora*. *Phytopathology* **69**, 565–8.

Schulte R. & Bonas U. (1992). Expression of the *Xanthomonas campestris* pv. *vesicatoria hrp* gene cluster, which determines pathogenicity and hypersensitivity on pepper and tomato, is plant inducible. *J. Bacteriol* **174**, 815–23.

Shaw J. J. & Kado C. I. (1986). Development of a *Vibrio* bioluminescence gene-set to monitor phytopathogenic bacteria during the ongoing disease process in a non-disruptive manner. *Biotechnology* **4**, 560–4.

Sigee D. C. & El-Masry M. H. (1987). The release and electron microscope visualisation of plasmids in *Erwinia amylovora* using an *in situ* lysis technique. *Acta Hort.* **217**, 183–8.

Somlyai G., Hevesi M., Banfalvi Z., Klement Z. & Kondorosi A. (1986). Isolation and characterisation of non-pathogenic and reduced virulence mutants of *Pseudomonas syringae* pv. *phaseolicola* induced by Tn5 transposon insertions. *Physiol. Mol. Plant Pathol.* **29**, 369–80.

Stall R. E., Loschke D. C. & Jones J. B. (1986). Linkage of copper resistance and avirulence loci on a self-transmissable plasmid in *Xanthomonas campestris* pv. *vesicatoria*. *Phytopathology* **76**, 240–3.

Staskawicz B. J., Dahlbeck D. & Keen N. T. (1984). Cloned avirulence gene of *Pseudomonas syringae* pv. *glycinea* determines race-specific incompatibility on *Glycine max*. *Proc. Nat. Acad. Sci.* **81**, 6024–8.

Surico G. (1986). Indoleacetic acid and cytokinins in the olive knot disease. An overview of their role and their genetic determinants. In *Biology and Molecular Biology of Plant–Pathogen Interactions*, ed. J. A. Bailey. Berlin: Springer-Verlag.

Tamaki S. J., Dahlbeck D., Staskawicz B. & Keen N. T. (1988a). Characterisation and expression of two avirulence genes cloned from *Pseudomonas syringae* pv. *glycinea. J. Bacteriol.* **170**, 4846–54.

Tamaki S. J., Gold S., Robeson M., Manulis S. & Keen N. T. (1988b). Structure and organisation of the *pel* genes from *Erwinia chrysanthemi* EC16 *J. Bacteriol.* **170**, 3468–78.

Tamaki S. T., Kobayashi D. Y. & Keen N. J. (1991). Sequence domains required for the activity of avirulence genes *avr*B and *avr*C from *Pseudomonas syringae* pv. *glycinea. J. Bacteriol.* **173**, 301–7.

Tang J. L., Gough C. L., Barber C. E., Dow J. M. & Daniels M. J. (1987). Molecular cloning of protease gene(s) from *Xanthomonas campestris* pv. campestris: Expression in *Escherichia coli* and role in pathogenicity. *Mol. Gen. Genet.* **210**, 443–8.

Todd G. A., Daniels M. J. & Callow J. A. (1990). *Xanthomonas campestris* pv. *oryzae* has DNA sequences containing genes isofunctional with *Xanthomonas campestris* pv. *campestris* genes required for pathogenicity. *Physiol. Mol. Plant Pathol.* **36**, 73–87.

Turner M. A., Avellano F. & Kozloff L. M. (1991). Components of ice nucleation structures of bacteria. *J. Bacteriol.* **173**, 6515–27.

Turner J. G. & Taha R. R. (1984). Contribution of tabtoxin to the pathogenicity of *Pseudomonas syringae* pv. *tabaci. Physiol. Mol. Plant Pathol.* **25**, 55–69.

Vivian A. (1990). Recognition in resistance to bacteria. In *Recognition* and *Response in Plant–Virus Interactions*, ed. R. S. Fraser. NATO ASI Series. **41**, 17–29.

Vivian A., Atherton G. T., Bevan J. R., Crute I. R., Mur L. A. J. & Taylor J. D. (1989). Isolation and characterisation of cloned DNA conferring specific avirulence in *Pseudomonas syringae* pv. *pisi* to pea (*Pisum sativum*) cultivars, which possess the resistance allele, R2. *Physiol. Mol. Plant Pathol.* **34**, 335–44.

Walden R. (1988). *Genetic Transformation of Plants*. Milton Keynes: Open University Press.

Wandersman C., Delepelaire P., Letoffe S. & Schwartz M. (1987). Characterisation of *Erwinia chrysanthemi* extracellular proteases: Cloning and expression of the protease genes in *Escherichia coli. J. Bacteriol.* **169**, 5046–53.

Warren G., Corotto L. & Wolber P. (1986). Conserved repeats in diverged ice nucleation structural genes from two species of *Pseudomonas. Nucl. Acids Res.* **14**, 8047–60.

Whalen M. C., Stall R. E. & Staskawicz B. J. (1988). Characterisation of a gene from a tomato pathogen determining hypersensitive resistance in a non-host species and genetic analysis of this resistance in bean. *Proc. Nat. Acad. Sci.* **85**, 6743–7.

Xu G-W. & Gross D. C. (1988). Physical and functional analyses of the *syr*A and *syr*B genes involved in syringomycin production by *Pseudomonas syringae* pv. *syringae. J. Bacteriol.* **170**, 5680–8.

Yamada T., Palm C. J., Brooks B. & Kosuge T. (1985). Nucleotide sequences of the *Pseudomonas savastanoi* indoleacetic acid genes show homology with *Agrobacterium tumefaciens* T-DNA. *Proc. Nat. Acad. Sci.* **82**, 6522–6.

9

Disease control

In natural environments, where a particular host species occurs within mixed vegetation, the development and spread of disease is probably limited to some extent by the separation of individual plants within the area. This constraint does not apply in the crop situation, where localised infection and progression of disease within the homogeneous plant population can occur rapidly. In this artificial situation, where the natural balance between pathogen and host does not apply, special control measures often have to be adopted if the large scale occurrence of disease and consequent major crop loss are to be avoided. These measures fall into four main categories: chemical control, biological control, breeding of resistant cultivars and sanitary procedures.

Chemical control

Chemical control agents are of two main types: bactericides (synthetic organic and inorganic compounds) and antibiotics (naturally occurring microbial products). The use of these two types of control agent is considered in the first part of this section, with a final discussion on general aspects of chemical control.

Bactericides

The range of compounds used as bactericides has recently been reviewed by Sekizawa and Wakabayashi (1990) who divide these compounds into four main categories: synthetic bactericides formerly used for crop protection, currently used synthetic bactericides, traditional inorganic compounds and soil nitrification inhibitors. Some of the major properties of currently used

Table 9.1. *Properties and use of some currently employed bactericides*

Bactericide	Target disease and causal organism	Application	*In vitro* anti-bacterial activity
ORGANIC BACTERICIDES			
Bronopol	Blackarm disease of cotton (*Xanthomonas campestris* pv. *malvacearum*)	Seed dressing dust	Bacteriostatic against range of pathogens
Phenazine mono-oxide	Leaf blight of rice (*Xanthomonas campestris* pv. *oryzae*)	Wettable powder	Moderate
Probenazole	Diseases caused by *Xanthomonas campestris.* pv. *oryzae, Pseudomonas glumae, Pseudomonas syringae* pv. *lachrymans* and *oryzae*	Granular	None
Tecloftalam	Leaf blight of rice (*Xanthomonas campestris* pv. *oryzae*)	Wettable powder	None
INORGANIC BACTERICIDES			
Bordeaux mixture (Cu(OH)$_2$). CaSO$_4$	Range of *Xanthomonas campestris* pathovars	Spray suspension	Strong
Basic copper chloride with kasugamycin	Wide range of bacterial diseases	Spray	Strong

Information taken from Sekizawa & Wakabayashi (1990). These authors give further details of bactericide formulations and manufacturers.

bactericides are summarised in Table 9.1 and the chemical structure of these compounds is shown in Fig. 9.1.

Application of bactericides

Bactericides are either applied as a seed dressing (e.g. application of bronopol as a dust to cotton seeds) or as a general environmental treatment (to foliage, soil or irrigation water). An example of foliage application is provided by the use of phenazine mono-oxide as a spray against rice bacterial leaf blight in paddy-fields or nursery beds. The same disease may be controlled via irrigation water by application of probenazole granules to the paddy-field watering system once after the rice transplants have struck root, and once again 4 weeks before earing.

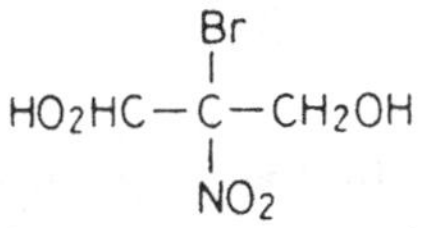

Bronopol
2−Bromo−2−nitropropanol−1,3−diol

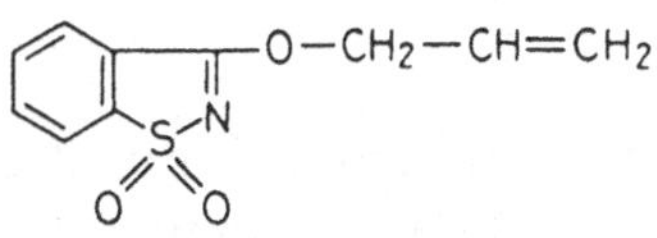

Probenazole
3−allylkoxy−1,2−benzisothiazole1,
1−dioxide

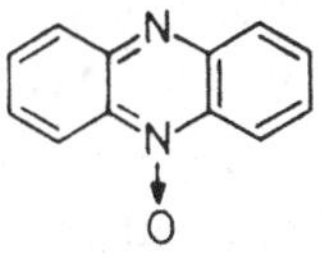

Phenazine mono−oxide

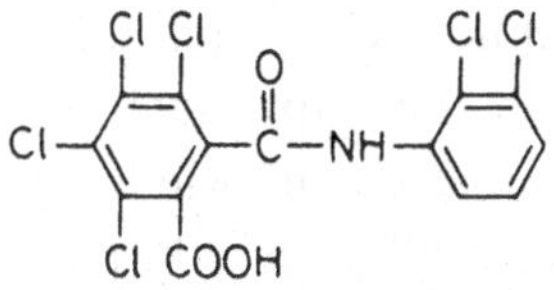

Tecloftalam
2−(2,3−dichlorophenyl)−aminocarbonyl
3,4,5,6−tetrachlorobenzoic acid

Fig. 9.1 Chemical structure of some currently used bactericides.

Screening and activity of bactericides

The use of *in vitro* tests to screen compounds for bacterial acivity has limited use, since there may be no direct relationship between *in vitro* and *in vivo* activity. This is seen in Table 9.1, where some compounds such as the inorganic bactericides and bronopol have strong *in vitro* and *in vivo* activity, but other compounds with limited (phenazine mono-oxide) or no (probenazole, tecloftalam) *in vitro* activity are highly effective in crop application.

Bactericides appear to exert their activity in various specific ways. In the case of the inorganic bactericides, where Cu^{2+} is the toxic agent, there is direct inhibition of bacterial growth leading to cell death. X-ray microanalytical studies by Hodson and Sigee (1991) have shown that treatment of bacterial cells with Cu compounds leads to marked changes in elemental composition and that these changes are consistent with a toxic effect at the cell surface, leading to large scale efflux of K^+ and influx of Ca^{2+} and Cu^{2+}. In other cases, the bactericide does not act directly on the bacterium (no *in vitro* inhibition) but appears to have an indirect effect on disease development, possibly mediated by the plant metabolism. As an example, it has been proposed that probenazole acts on the pathogen–plant recognition process, converting a compatible to an incompatible interaction (Sekizawa & Wakabayashi, 1990).

Antibiotics

The use of antibiotics for control of plant diseases is limited by the relatively small number of compounds that are commercially available, and by practical aspects such as phytotoxicity and rapid degradation. Antibiotics are currently used to control plant disease in two main situations: injection into woody plants to control MLO diseases, and spray application to growing plant parts to control *Erwinia amylovora* and various foliar diseases.

Treatment of woody plants with antibiotic normally involves the use of oxytetracycline-HCl, which is introduced either by gravity infusion or pressure injection. This approach has been effective in the control of some MLO diseases, such as coconut lethal yellows, X-disease of peach and pear decline. Trees are normally treated once a year, between fruit harvest and leaf fall.

Spray applications to foliage, blossoms or fruit most commonly use streptomycin sulphate as the control agent, but other antibiotics such as oxytetracycline and Kasugamycin have also been used. Antibiotic sprays have been particularly effective in the control of *Erwinia amylovora* in the USA, but problems with development of pathogen resistance to streptomycin pose major problems to the continued use of this control method. Maximal retention and absorption of the antibiotic at the plant surface can be achieved by spraying under conditions of no wind, high humidity and the addition of spreader sticker compounds to the formulation. The relatively low persistence of antibiotics on the plant surface under field conditions means that several applications may be necessary to maintain effective concentrations for disease control.

General aspects of chemical control: advantages and disadvantages

In many cases, chemical control agents act mainly at the plant surface, rather than systemically, and control disease by preventing the epiphytic growth and spread of phytopathogenic bacteria. The use of chemical control agents has the advantage that their action is frequently rapid and highly effective, and in some cases may be the only realistic approach to the elimination of plant disease (e.g. control of blister spot disease of apples by streptomycin – see below). The use of chemical control agents does, however, have a number of disadvantages:

Cost-effectiveness The cost of the control agent (particularly antibiotics) and its application clearly have to be economically viable, which may not be the case where the crop has a low cash value or where application of the agent is only partially effective.

The application of bactericides is typically labour-intensive and should ideally coincide with the build-up in epiphytic phytopathogen population. This requires either a few applications (with accurate disease prediction) or a sequence of applications throughout the danger period. The use of bactericides is important, for example, in the control of halo blight and bacterial brown spot of beans in north-east Colorado (USA) by Cu-based bactericides, where inconsistencies in control have been attributed to inappropriate timing of application (Legard & Schwartz, 1987). Even the most efficient spray regime may have limited effect during weather conditions where there is a high disease pressure.

Environmental effects Treatment with chemical control agents is typically non-selective, causing widespread eradication of non-pathogenic as well as pathogenic organisms. There is thus a major effect on the plant surface environment, with the large-scale loss of naturally occurring micro-organisms, some of which may exert their own control influences by antagonistic activity.

Phytotoxicity The use of inorganic treatments such as Cu-containing compounds and antibiotics may cause damage to the crop plant and may lead to surface or internal contamination, rendering the product useless for human or animal consumption.

Development of resistance One major limitation of using either antibiotic or inorganic control agents is that phytopathogenic bacteria frequently appear able to develop resistance to these compounds, thus reducing their effectiveness. The genetic determination of resistance is discussed in Chapter 8, and is normally encoded by plasmids. Such resistance has been observed in field populations of various pathovars of *Pseudomonas syringae* and *Xanthomonas campestris* plus *Curtobacterium flaccumfaciens* and *Erwinia amylovora* (see Table 8.17).

Control of blister spot of apple (caused by *Pseudomonas syringae* pv. *papulans*) in the USA provides a typical example of the economic significance of acquired antibiotic resistance. The use of streptomycin sprays to control blister spot has only been recommended since 1982, yet by 1985 incidents of control failure were reported after proper use of this agent (Burr *et al.*, 1988). Resistance of this bacterium to streptomycin is mediated by a 68 MDa plasmid and presents a serious economic problem since no alternative bactericides are available for control of this disease.

The widespread acquisition of antibiotic resistance by phytopathogenic or non-pathogenic bacteria due to frequent agricultural use of antibiotic sprays also presents a potentially major problem in terms of possible transfer of resistance to animal and human pathogens, and has resulted in a complete ban on their use to control plant disease in various parts of the world (e.g. the European Community). Studies on plasmid-mediated transfer of antibiotic resistance are discussed in Chapter 8 (see Table 8.18).

Inaccessibility of pathogen With some control agents, spray application is only fully effective where bacteria occur on plant surfaces and pathogens occurring within infected tissue will be less accessible to chemical control. This problem will be particularly acute where bacteria are lodged within xylem tissue, and with many bacterial wilt diseases (such as *Xanthomonas campestris* pv. *pelargonii* on geranium) there is no effective chemical control.

Biological control

Biological control of bacterial plant disease normally involves the external application of specific micro-organisms (antagonists) to the plant or its environment to limit the initiation and progression of the disease. Such biological control agents can act at various points in the disease cycle including the survival of the pathogen in the external environment, build-up of pathogen on the host surface, entry of pathogen into the host and transmission of the pathogen between hosts. In some circumstances, the antagonist may also compete against the pathogen within the host tissue, as shown, for example, by the use of avirulent mutants to control *Pseudomonas solanacearum* within the plant vascular system (Trigalet & Trigalet-Demery, 1990).

In practice, most work on the biological control of bacterial plant diseases has aimed to limit the growth and activity of phytopathogenic bacteria at the plant surface (aerial or subterranean), using strains of bacteria that are antagonistic to the pathogen. These bacterial control agents can be either naturally occurring organisms (selected from the environment) or genetically engineered strains.

Unlike chemical control, which results in wholesale environmental eradication of a range of micro- and macro-organisms, biological control has the potential for a much more specific effect on the pathogen and a much more limited effect on the environment as a whole. The introduction of high levels of a biocontrol agent may itself have important environmental effects, however, and it is important to ensure, for example, that a particular control agent is not ice nucleation active (INA^+; causing widespread frost damage) or itself phytopathogenic (inducing an HR in the treated plant or disease on other plants). The use of genetically engineered antagonists has implications for the introduction and spread of new genes in the general microflora, with unknown and possibly unpredictable results. In spite of these limitations, biological control is generally considered a more ecologically acceptable approach compared with chemical control and is more suited to modern concepts of integrated pest management (IPM).

General range of biological control agents

For effective biological control, the bacterial antagonist must be able to survive and grow under natural (field) conditions, where it can successfully compete on a long-term basis with the phytopathogen. For this reason, work on biological control has concentrated particularly on commonly occurring phyllosphere or rhizosphere bacteria, which have an established ability for long-term survival at the plant surface. Successful use of these bacteria in achieving at least partial limitation in population growth and/or diseases of phytopathogenic bacteria is summarised in Table 9.2.

Strategy for biological control

For a particular disease, the development of a successful biological control agent involves initial selection of a suitable antagonist (by laboratory and small-scale field testing), followed by the formulation of an effective strategy of application (including both timing and mode of application) with final large-scale field trials to establish the biological and cost-effectiveness of control under agricultural conditions.

Selection and activity of antagonists

Where naturally occurring antagonists are to be used as biocontrol agents, their selection involves:

1. Isolation of large numbers of strains from different environmental sites.
2. Primary selection of potentially useful strains using a rapid small-scale laboratory

Table 9.2. *Biological control of phytopathogenic bacteria and their diseases*

Disease/pathogen	Control agent	Reference
PHYLLOSPHERE CONTROL		
Fireblight		
(*Erwinia amylovora*)		
Apple and Pear	*Erwinia herbicola*	Beer & Vidaver (1978)
Hawthorn	*Erwinia herbicola*	Wilson *et al.* (1990)
	Pseudomonas syringae	Wilson *et al.* (1992)
Ornamentals	*Erwinia herbicola*	Isenbeck & Schulz (1985)
Bacterial wilt of	*Pseudomonas fluorescens*	Schmidt (1988)
ryegrass	*Erwinia herbicola*	
(*Xanthomonas campestris* pv. *graminis*)		
Bacterial speck of tomato	Fluorescent	Colin & Chafik (1986)
(*Pseudomonas syringae* pv. *papulans*)	pseudomonads	
RHIZOSPHERE CONTROL		
Crown gall	*Agrobacterium radiobacter*	Kerr & Htay (1974)
(*Agrobacterium tumefaciens*)	strain K84	Ellis *et al.* (1979)
	strain K1026	Kerr & Jones (1989)
Soft rot of potato	*Erwinia carotovora*	Axelrod *et al.* (1988)
(*Erwinia carotovora* subsp. *carotovora*)	subsp. *tetravasculorum*	
	Bdellovibrio sp.	Epton *et al.* (1989)
Soft rot and blackleg	Fluorescent pseudomonads	Xu & Gross (1986a)
of potato (*Erwinia carotovora* subsp. *carotovora* & *atroseptica*)	*Pseudomonas fluorescens*	Xu & Gross (1986b)
Bacterial wilt of	Tn5 avirulent mutant	Trigalet & Trigalet-
tomato (*Pseudomonas solanacearum*)		Demery (1990)

assay against the phytopathogen that will effectively screen large numbers of isolates.

3. Small-scale followed by large-scale field testing of selected isolates to determine their activity in the crop situation.

This sequence is illustrated in Fig. 9.2 for the selection of naturally occurring antagonists to the fireblight pathogen *Erwinia amylovora*. In the biological control of this disease, the commonly occurring phylloplane bacteria *Erwinia herbicola* and *Pseudomonas syringae* can be washed off leaf

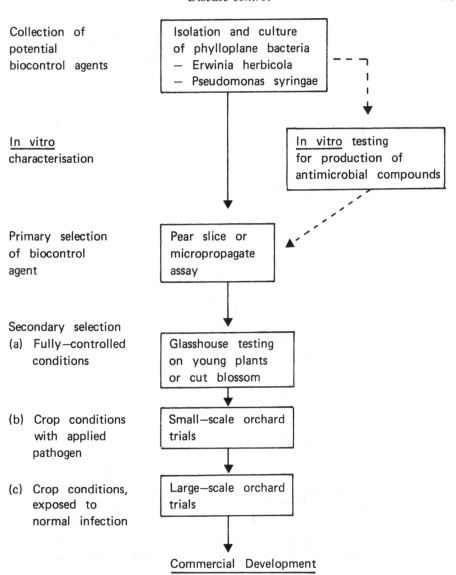

Fig. 9.2 General scheme for the testing and development of biological control agents for fireblight disease.

surfaces, cultured on appropriate media, and isolates from a variety of sites tested for potential control activity. Primary screening of isolates in the laboratory is generally carried out by tissue assay (involving pear slice or micropropagate tests), with agar plate tests providing potentially useful additional information on the *in vitro* characteristics of the antagonists.

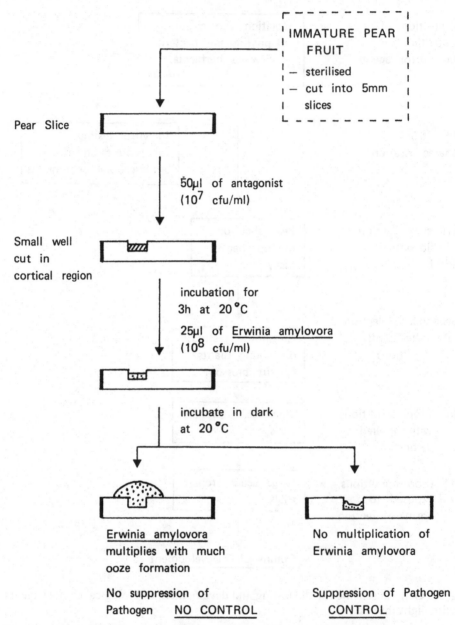

Fig. 9.3 Immature pear fruit assay for laboratory testing of bacterial antagonists to *Erwinia amylovora*.

Selection of antagonists by tissue assay

The pear slice test or immature pear fruit assay (IPFRA) is shown in Fig. 9.3 and involves direct challenge of antagonist and pathogen on the surface of freshly cut immature pear tissue. Although this assay provides a general and useful prediction of antagonist activity on the plant surface (Wilson *et al.*, 1990), the full potential of individual isolates can only be determined by subsequent application to whole plants during secondary selection procedures (Fig. 9.2).

As an alternative to the IPFRA, potential antagonists can be tested in combination with the pathogen using cut leaves of pear micropropagates (see Fig. 6.4), giving comparable results to the IPFRA (Nicholson *et al.*, 1990).

In vitro *testing of antagonists using an agar plate assay*

The *in vitro* antagonistic potential of bacterial isolates can be tested either by direct challenge of antagonist and pathogen cells or by the indirect effect of antagonist exudates on a pathogen culture. Direct interaction, with clear antagonistic activity, is shown in Fig. 9.6, where a colony of *Pseudomonas syringae* has a well-defined inhibition zone on a lawn of *Erwinia amylovora*. The use of *in vitro* testing to characterise the type of chemical interaction between pathogen and antagonist is described below (see Fig. 9.4).

Mode of action of biological control agents

Biological control agents are able to limit the growth and activity of bacterial phytopathogens in two main ways: production of anti-microbial substances and competition for space and nutrients at specific sites on the plant surface (site competition). A third type of interaction, antagonism by predation, is also of potential importance but has so far received little attention.

Production of anti-microbial substances

The importance of anti-microbial compound production in bacterial competition (antibiosis) can be demonstrated during both *in vitro* (agar plate) and *in vivo* (plant surface) antagonism.

In vitro *antibiosis* The role of anti-microbial compounds during *in vitro* antagonism is indicated by the clear (cell-free) inhibition zone around the central antagonist colony (Fig. 9.4) which represents the expanding circle of diffusing inhibitory metabolites from the central source. The production of anti-bacterial metabolites without direct pathogen–antagonist cell interaction can be shown experimentally using a cellophane overlay procedure, as a modification of the technique shown in Fig. 9.4. This two-step process involves:

1. Initial deposition and growth of the antagonist on a cellophane disk

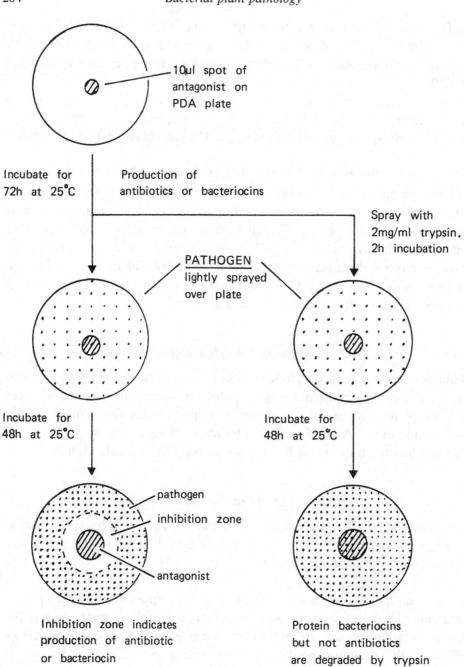

Fig. 9.4 Experimental procedures for *in vitro* testing of antagonists for antibiotic and bacteriocin production. Bacteria are grown on potato dextrose agar (PDA), which does not promote siderophore production.

overlying the agar medium. The cellophane allows diffusion of nutrients from the agar to the test organism, and diffusion of low to medium molecular weight anti-microbial products in the opposite direction.

2. Removal of the cellophane disk (plus test organism) followed by spray inoculation of the phytopathogen onto the exposed agar surface. Growth of the pathogen leads to uniform coverage of the whole agar surface with the exception of a central clear inhibition zone where localised accumulation of anti-microbial compounds occurred. Active diffusion of these compounds through the agar typically leads to a lateral spread of the inhibition zone well beyond the original boundaries of the central colony.

In vivo *antibiosis* The role of anti-microbial substances at the plant surface is discussed later in relation to the major types of compound concerned, but some general indication of their activity can be obtained from measurements of antagonist and pathogen populations following separate and mixed inoculations. This is shown in Fig. 9.7, where competition solely by antibiosis would theoretically result in a minimal change to the antagonist population following co-inoculation but have a major effect on the pathogen.

Although the activity of anti-microbial compounds has been clearly demonstrated for interactions between organisms in the rhizosphere, little is known of their importance for microbial interactions on aerial plant surfaces. However, a number of commonly occurring phylloplane bacteria have been shown actively to produce these compounds *in vitro*.

Anti-microbial compounds are of three main types: antibiotics, bacteriocins and siderophores. These are distinguished in terms of their chemical nature, anti-microbial activity and means of detection during *in vitro* culture (Table 9.3).

Antibiotics

Antibiotic production has been looked at in particular detail in the commonly occurring phylloplane bacterium *Erwinia herbicola*, which is able to compete with the indigenous microflora in a wide variety of natural habitats and is important as a potential biocontrol agent.

In vitro studies by a number of workers (Table 9.4) have shown that this organism is remarkable in terms of the variety of anti-microbial compounds that it is able to generate, either by simple breakdown of external molecules (sugars, arbutin), or by direct synthesis of specific antibiotics. The production of different antibiotics is not only characterised by variation in the chemical nature of the compound, but also differences in the strain of producing

Table 9.3. *General features of antibiotics, bacteriocins and siderophores*

Compound	Chemical nature	Activity	Detection
Antibiotic	Low MW	Typically broad spectrum	Protease-resistant
Bacteriocin	Typically protein	Limited to closely related organisms	Protease-sensitive Induced by UV light or mitomycin C
Siderophores	Wide range	Reduce level of free Fe^{3+} ions by chelation	Only produced and effective at low levels of Fe

Table 9.4. *Antibiotic production by* Erwinia herbicola

Anti-microbial compound	Culture requirement	Organisms inhibited	Reference
Organic acids	High sugar medium	*Erwinia amylovora*	Riggle & Klos (1972)
β-glucosidase (converts host-cell arbutin to hydroquinones)	Arbutin broth	*Erwinia amylovora*	Chatterjee *et al.* (1969)
Herbicolins A + B (acylated peptide antibiotics)	Carbohydrate-rich media	Yeasts and filamentous fungi	Winkelman *et al.* (1980)
β-Lactam (very unstable)	Defined medium	Wide range of bacteria	Parker *et al.* (1982)
Herbicolin 0	Defined medium	Wide range of bacteria	Ishimaru *et al.* (1988)
Herbicolin I	Defined medium	Narrow range	
Herbicolin 112Y (non-protein)		Narrow range	Stein & Beer (1980)

organism, dependence on a specific growth medium for production, and the identity and range of organisms that are inhibited.

The production of peptide antibiotics by *Erwinia herbicola* (referred to as herbicolins) was initially reported by Winkelman *et al.* (1979). These were considered to be analogous to the peptide antibiotics produced by organisms such as *Streptomyces* and *Bacillus* and could be isolated from culture filtrate by column or thin layer chromatography, with further characterisation using techniques such as ultraviolet and NMR spectroscopy. Peptide antibiotics are typically cyclic compounds, and are resistant to peptidases or proteases of animal and plant origin.

The synthesis and activity of herbicolins 0 and 1 by *Erwinia herbicola* strain C9-1 (Ishimaru *et al.*, 1988) has been investigated in particular detail. This isolate produces very large inhibition zones on lawns of test organisms such as *Erwinia amylovora* (*in vitro* antibiosis) when growth is on a defined solid medium (containing D-gluconate and L-asparagine as sources of energy) buffered at neutral pH. The *in vitro* activity of these antibiotics differs markedly in terms of the range of bacteria that are inhibited, rates of diffusion in agar, inhibitory effect of L-histidine and senstivity to pH. Specific differences in the effects of these antibiotics are further emphasised by the fact that some mutants of *Erwinia amylovora* are inhibited by herbicolin 0 but not herbicolin 1, and vice versa.

Role of antibiotics at the plant surface The production of antibiotics by potential antagonists during liquid medium or solid agar *in vitro* culture does not imply that they will necessarily produce these compounds on the aerial or subterranean plant surface. Indications that anti-biosis may be of some importance *in vivo* have been obtained from studies on tissue slices and with whole plants. Ishimaru *et al.* (1988) showed that treatment of pear slices with *Erwinia herbicola* strain C9-1 or with partially purified herbicolins led to a significant reduction in the pathogenic effect of *Erwinia amylovora* suggesting that herbicolins may be an important factor in control of the pathogen on the plant surface under natural conditions. Vanneste *et al.* (1992) found that antibiotic-deficient mutants of *E. herbicola* had a reduced but not negligible effect on *E. amylovora in vivo*, suggesting that the biocontrol activity of this organism is at least partly attributable to antibiotic production.

Bacteriocins

Bacteriocins may be defined as non-replicating anti-bacterial substances that have a specific inhibitory effect on closely related organisms.

Bacteriocin diversity Originally, the term bacteriocin was applied to antibiotic-like compounds produced by *Escherichia coli* (the 'colicins'), which were specific in killing closely related strains and were protein in nature. The specificity and chemical composition of these compounds distinguished them from classical antibiotics such as streptomycin, a low molecular weight, broad spectrum glycoside.

Today, bacteriocins are defined primarily in terms of their specificity of inhibition (see above). In identifying bacteriocins it is important to make sure

Table 9.5. *Major types of phytopathogen bacteriocins*

Bacterial group	Chemical nature	Specificity	Taxonomic use
Coryneform bacteria	Low MW proteins	Inhibit wide range of bacteria	Differentiation of species and subspecies
Pseudomonas Erwinia	High MW proteins	Inhibit narrow range of bacteria	Differentiation of strains and pathovars
Agrobacterium	Low MW nucleotides	Specific to *A. tumefaciens* strains with a nopaline Ti plasmid	Differentiation of *A. tumefaciens* strains

that the antagonism really is specific, and not due to non-specific compounds such as hydrogen peroxide or lactic acid. This can be achieved by testing the compounds against the producer organism, as with genuine bacteriocins the producer is rarely sensitive to its own bacteriocin. Within the pathogenic bacteria and related organisms, bacteriocins are produced by widely differing taxonomic groups, and show corresponding diversity in their chemical nature and specificity (Table 9.5).

Bacteriocin activity The mode of action of bacteriocins has been discussed by Vidaver (1983), and appears to involve an initial attachment to the bacterial cell. Several bacteriocins require specific receptor sites on the cell surface, and particulate bacteriocins can be seen under the electron microscope as attached, virus-like particles. Agrocin 84 attaches instead to a specific binding protein within the periplasmic space of *Agrobacterium tumefaciens*.

After initial binding and possible entry into the bacterium, bacteriocins are then generally able to kill their target cells in a number of ways: by perturbing protein synthesis, DNA stability, energy flux or membrane integrity. In the particular case of bacteriocins that are active against plant pathogenic bacteria, Agrocin 84 specifically inhibits DNA synthesis, while carotovoricin has phospholipase activity and acts on the cytoplasmic membrane (Vidaver, 1983).

Experimental uses of bacteriocins Bacteriocins are experimentally important in plant pathology in a number of ways, including epidemiology, taxonomy and biological control (see later).

As far as taxonomy is concerned, the ability to use these compounds as precise tools for identification has great potential, but depends to a large extent on their specificity (Table 9.5). Although bacteriocin typing is now an important technique for identifying coryneform bacteria and *Pseudomonas syringae* pathovars (Gross & Vidaver, 1990), the usefulness of this approach is clearly limited for pathogens such as *Xanthomonas campestris* pv. *phaseoli*, where only 2% of the strains produce bacteriocins, and where only 22% are sensitive to bacteriocins.

Bacteriocin production in vitro *and at the plant surface* Plant pathogenic bacteria that produce bacteriocins normally do so in culture (on ordinary complex media) under conditions that favour optimal growth. The production of particulate bacteriocins by *Pseudomonas syringae* pathovars and *Erwinia carotovora* is considerably enhanced by exposure of the producer cells to ultraviolet light or mitomycin C, both of which cause DNA damage. An experimental scheme for the demonstration of *in vitro* production of bacteriocins is shown in Fig. 9.4, where the presence of these molecules is monitored by their antagonistic effect on a target bacterium. In this diagram, the production of proteinaceous bacteriocin by the antagonist is distinguished from antibiotic production by its sensitivity to trypsin digestion.

The production and effect of these molecules under natural conditions is difficult to establish, but *in vitro* production has been demonstrated for residents of both the phyllosphere (e.g. *Pseudomonas syringae*, *Erwinia herbicola*), and the rhizosphere (e.g. *Pseudomonas solanacearum*, *Erwinia carotovora* subsp. *carotovora*, *Agrobacterium radiobacter*), where they may have a role to play in biocontrol.

Correlation between *in vitro* bacteriocin production and *in vivo* ability of the bacterium to control disease provides a useful criterion for assessment of the role of these molecules at the plant surface. On this basis, herbicolin production by *Erwinia herbicola* appears to be of limited importance in competition with *Erwinia amylovora* (Beer, 1981), while bacteriocin production by *Erwinia carotovora* and *Agrobacterium radiobacter* is important in competition with related pathogens (Table 9.6).

Bacteriocins and biological control The restriction of bacteriocin production to closely related organisms imposes a limitation in terms of choice of potential antagonists for disease control, since only non-pathogenic (or weakly pathogenic) close relatives to the pathogen concerned will be suitable as biocontrol agents. Other disadvantages include the limited susceptibility of some pathogens (e.g. *Xanthomonas campestris* pv. *phaseoli* – see above) to bacteriocins, the limited range of bacteriocins (with limited range of target

Table 9.6. *Examples of biocontrol involving bacteriocin-producing organisms*

Bacteriocin producer	Bacteriocin	Target bacterium	Reference
Pseudomonas syringae pv. *syringae*	Syringacin	*Pseudomonas syringae* pv. *syringae* pathogenic strain	Smidt & Vidaver (1982)
Erwinia herbicola	Herbicolacins	*Erwinia amylovora*	Beer (1981) Stein & Beer (1980)
Erwinia carotovora	Carotovoricins	Pathogenic strains	Axelrod *et al.* (1988)
Agrobacterium radiobacter	Agrocin 84	*Agrobacterium tumefaciens*	Kerr & Jones (1989)

organisms) that a single antagonist will produce, and the possibility of transfer of bacteriocin-producing ability (with correlated resistance) from the antagonist to the pathogen. This last possibility has been of particular importance in the use of *Agrobacterium radiobacter* for the control of *Agrobacterium tumefaciens* (see below).

The major advantage of using bacteriocin-producing bacteria as biocontrol agents is the similarity in ecological niche that the close relationship with the pathogen frequently confers. Some examples of the use of bacteriocin-producing bacteria in biocontrol are shown in Table 9.6, and details of the use of *Erwinia carotovora* and *Agrobacterium radiobacter* as biocontrol agents are given below.

Erwinia carotovora. Many strains of *Erwinia carotovora* subsp. *carotovora* (pathogen of potato) produce bacteriocins (carotovoricins) *in vitro*, the majority of which are high MW compounds resembling defective phage particles. The closely related organism *Erwinia carotovora* subsp. *tetravasculorum* (pathogen of sugar beet) also produces a purported bacteriocin which is inhibitory to *E.c.* subsp. *carotovora* and *Erwinia amylovora* (Axelrod *et al.*, 1988). Production of bacteriocins and antibiotics by *E.c.* subsp. *carotovora* appears to be significant in the competition of this phytopathogen with other micro-organisms in the soil environment since there is a good correlation between production of these anti-microbial compounds *in vitro* and inhibition of competing *Erwinia* strains in potato tuber wound infections (Axelrod *et al.*, 1988).

Agrobacterium radiobacter. In 1973, Kerr and Htay showed that rhizosphere control of the crown gall pathogen *Agrobacterium tumefaciens* by the closely related non-pathogenic soil organism *Agrobacterium radiobacter* strain 84 was mediated by the production of a bacteriocin, subsequently

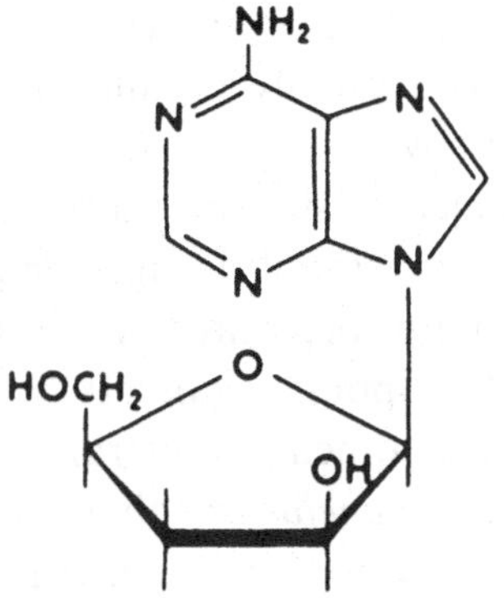

Fig. 9.5 Nucleoside component of agrocin 84, a bacteriocin produced by *Agrobacterium radiobacter*.

referred to as agrocin 84. This bacteriocin is an adenine nucleotide, the nucleoside component of which is shown in Fig. 9.5.

Sensitivity of *Agrobacterium tumefaciens* to the bacteriocin is closely related to the virulence of the pathogen (Watson *et al.*, 1975), with both characteristics being determined by the Ti plasmid (see Chapter 8). The production of agrocin 84 by *Agrobacterium radiobacter* is mediated by a small plasmid (see Chapter 8), the transfer of which to a non-producing organism results in bacteriocin synthesis and the formation of a potential biocontrol agent (Ellis *et al.*, 1979).

Strain 84 of *Agrobacterium radiobacter* has been used since 1973 by commercial stone fruit and rose growers in Australia to control the incidence of crown gall. Plant material is dipped in a suspension of the bacterial control agent prior to planting, achieving nearly complete protection of the crop against crown gall. The development of this biological control system represented the first use of a specific micro-organism to control a plant pathogen in soil, and the first commercial use of a bacterium to control any plant disease.

With any control agent involving the production of anti-microbial compounds there is the potential danger of breakdown of control due to the development of genetic resistance in the pathogen. In the case of the *Agrobacterium radiobacter–tumefaciens* interaction, this might arise either due to a defect in genetic expression of the Ti plasmid (e.g. plasmid loss or deletion mutation) or by acquisition by the pathogen of the agrocin plasmid, since this confers agrocin resistance as well as agrocin synthesis.

Ti plasmid loss or defect. The close correlation between agrocin 84 sensitivity and pathogenicity in *Agrobacterium tumefaciens* is probably the major reason for the success of this control system. Resistant colonies of *A.*

tumefaciens do arise due to Ti plasmid loss or defect, but the resulting breakdown in biocontrol is normally paralleled by a loss in virulence, so disease does not develop anyway.

Transfer of agrocin plasmid. Using the naturally occurring strain 84 of *Agrobacterium radiobacter*, transfer of the agrocin plasmid into the pathogen presents a major problem for biocontrol in certain situations. This was initially demonstrated by Panopoulos *et al.* (1978), who showed during field experiments in Greece that treatment with strain 84 gave rise to galls caused by strains of *Agrobacterium tumefaciens* that were themselves agrocin 84 producers and were insensitive to the bacteriocin. Transfer of the agrocin plasmid (see Chapter 8) occurs during the process of conjugation, which depends on the presence of nopaline. In a situation where the nopaline level is high (due to existing gall tissue, or where there is incomplete control due to pathogen excess), conjugation will be promoted and resistant pathogenic strains of *Agrobacterium tumefaciens* will arise. This may explain geographic differences in the effectiveness of strain 84, since the high level of control in Australia may relate to the relatively low level of disease there, while in Greece the high endemic level of virulence leads to nopaline production and rapid loss of control. The loss of biological control arising from plasmid transfer appears to be quite complex, depending to some degree on the host plant and related rhizosphere aspects of microbial interactions, soil salts, moisture, nutrient uptake and exudation by the root system.

Transfer of plasmid pAgK84, the plasmid encoding synthesis of agrocin 84, is controlled by a defined segment of DNA on the plasmid, the *tra* region. Recently, Kerr & Jones (1989) have used recombinant DNA techniques to construct a new bacterial strain (K1026) that is identical to K84 apart from a 5.9 kb deletion overlapping the *tra* region – which is therefore inactivated. This strain is unable to transfer its modified agrocin-producing plasmid (pAgK1026), thus eliminating the risk of generating resistant pathogens and the resulting breakdown of control. Permission has now been given in Australia for the use of this new strain as a biocontrol agent, providing the first example of disease control in the field using a genetically engineered organism.

Siderophores

The role of these iron-chelating compounds in the activities of plant pathogenic bacteria has been discussed previously in relation to their occurrence as pigments (Chapter 3), effects on the microenvironment (Chapter 4) and their genetic determination (Chapter 8). The molecular structure of pyoverdine, a siderophore produced by fluorescent pseudomonads, is shown in Fig. 3.3. The

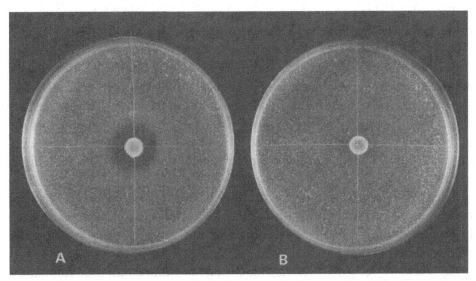

Fig. 9.6 *In vitro* inhibition of *Erwinia amylovora* by an isolate of *Pseudomonas syringae* under iron-limiting conditions. Ten µl. of *Pseudomonas syringae* suspension was spotted onto the centre of two agar plates, which were subsequently sprayed with a suspension of pathogen (*Erwinia amylovora*) as shown in Fig. 9.4. (*a*) Bacteria growing under iron-limiting conditions (King's medium B). The zone of inhibition around the central colony is due to siderophore production. (*b*) Bacteria growing on King's medium B, supplemented with ferric ions. Conditions are not iron-limiting, and there is no siderophore production by the central colony. (Photograph in collaboration with H. Epton.)

potential importance of siderophores in bacterial competition on both aerial and subterranean plant surfaces has clear implications for biological control and the use of siderophore-producing organisms as antagonists.

Demonstration of siderophore production and antagonism in vitro The *in vitro* production of siderophores on low iron media can be determined in several ways, including the visible production of fluorescent pigment (see Fig. 3.2) and the demonstration of antagonistic activity towards other bacteria. The latter can be achieved using similar *in vitro* procedures to those shown in Figs. 9.4 and illustrated in Fig. 9.6, with the development of a clear inhibition zone under conditions of low iron availability (King's medium B). Confirmation that the central inhibition zone is attributable to siderophore production and not other causes should be established by showing that the zone extends well outside the limits of the test organism colony (which would not be the case if inhibition were caused by direct nutrient depletion) and that inhibition only occurs on iron-limiting media (i.e. not on King's medium B supplemented with Fe^{3+} ions).

The role of siderophores in microbial interactions and biological control has been established mainly in the rhizosphere environment.

Siderophore production and biological control in the rhizosphere Siderophores are produced by a wide range of rhizosphere bacteria under appropriate *in vitro* and natural conditions, including general soil pseudomonads (Kloepper *et al.*, 1980), saprophytic bacteria and pathogenic bacteria. Examples of the latter include *Erwinia rhapontici*, which produces ferrorasamine A (Feistner *et al.*, 1983) and *Agrobacterium tumefaciens*, which produces agrobactin (Ong *et al.*, 1978).

Siderophore production is important in biocontrol within the rhizosphere both in terms of the suppression of general deleterious soil bacteria and the inhibition of specific soil pathogens, and has recently been reviewed by Bakker *et al.* (1991).

Suppression of general deleterious soil bacteria. Fluorescent pseudomonads, such as *Pseudomonas fluorescens and Pseudomonas putida*, are particularly important as biological control agents in the rhizosphere, since they are widely occurring and highly active in the production of siderophores. These water soluble, rapidly diffusing compounds are referred to either as pyoverdines (see Fig. 3.2) or pseudobactins.

Studies by Kloepper *et al.* (1980) have shown that application of naturally occurring fluorescent pseudomonad strains to the roots of crops such as potato and sugar beet can result in improved plant growth by the suppression of deleterious soil micro-organisms, with up to 90% loss of existing Gram-negative bacteria (including pseudomonads producing hydrogen cyanide (HCN)) and 65% loss of rhizosphere fungi. The beneficial (antagonistic) bacteria were referred to as plant growth promoting rhizobacteria (PGPR), and their suppressive effect was attributed to the production of fluorescent siderophores, resulting in the inhibition of non- (or less efficient) siderophore-producing organisms.

Critical confirmation that siderophore production is directly involved in plant growth stimulation comes from several lines of evidence (Bakker *et al.*, 1991):

1. *In vitro* characterisation of PGPR, and the effects of mutagenesis. Growth promotion by pseudomonads is directly related to their *in vitro* siderophore-producing activity. Although siderophore-producing wild-type pseudomonads increase plant growth, UV, transposon or chemically induced mutants that no longer have antagonistic activity on Fe^{3+}-deficient media have lost their growth stimulation effect.

2. Chemical treatment of rhizosphere. Addition of purified siderophores

mimics the effect caused by PGPR, while addition of ferric ions nullifies the growth-promoting effects of PGPR.

Populations of PGPR are highly variable under field conditions, but may be detected at naturally occurring levels of up to 10^6 CFU/cm on roots of potato and sugar beet, where their presence has a marked effect on the occurrence of other bacteria (Kloepper, 1983).

Several strains of PGPR have been isolated that are highly effective in promoting plant growth, including *Pseudomonas putida* WCS358 (Bakker *et al.*, 1991), which gives significantly increased potato root growth and tuber yield in intensively cropped soil. This strain is believed to be particularly effective in iron uptake (and thus competition) since its siderophore–iron complex can only be retrieved by this strain, but siderophore complexes derived from other bacteria can also be taken up by the organism.

Suppression of major soil pathogens. The use of siderophore-producing antagonists to control major soil diseases has been demonstrated for bacterial pathogens *Erwinia carotovora* and *Pseudomonas solanacearum*, and pathogenic fungi (Bakker *et al.*, 1991).

Under field conditions, it seems likely that the existing microflora will be of major importance in the suppression of *Erwinia carotovora*, and that fluorescent pseudomonads have a particular role in this respect. Kloepper (1983) reported that introduction of plant growth-promoting rhizobacteria to potato soils not only resulted in the suppression of generally deleterious bacteria, but also had a suppressive effect on *Erwinia carotovora*. Application of some fluorescent pseudomonad strains to potato seed pieces resulted in a reduction of the population of this pathogen on roots and daughter tubers by 30–100%, with a subsequent reduction in blackleg disease.

In a more extensive study, Xu and Gross (1986a) screened approximately 300 fluorescent pseudomonad strains for *in vitro* production of siderophores and for growth inhibition of *Erwinia carotovora* subsp. *atroseptica*. Approximately 90% of these showed some inhibitory activity, in the majority of cases by siderophore production (i.e. specific *in vitro* inhibition on King's medium B) but also by production of antibiotics (*in vitro* antagonism on potato dextrose agar). Further selection was carried out by small-scale greenhouse testing of potential antagonists (in combination with high levels of the pathogen) on the growth of potatoes. Effective control required levels of antagonist which were high (10^6 CFU per seed piece) but which were comparable to naturally occurring fluorescent pseudomonad levels in the field.

Role of site competition in biological control

Although work on the mechanism of biological control has concentrated largely on interactions involving anti-microbial substances, direct competition for space and nutrients (site competition) may also be an important factor. For successful competition to occur, the antagonist must occupy the same micro-habitats and have similar growth characteristics to the pathogen. Such biocontrol agents will either be ecologically related to the pathogen (i.e. naturally occurring) or specifically genetically engineered. One way to achieve the latter is to use avirulent mutants as the antagonist, differing from the pathogen only in the absence of key pathogenicity genes. Trigalet and Trigalet-Demery (1990) have suggested that mutagenesis within any *hrp* gene cluster should produce mutants which are potential biocontrol agents, able to give some protection against the corresponding virulent bacteria without themselves causing disease (see 'site competition *in planta*' below).

Site competition may either occur on the plant surface or within infected tissue.

Site competition at the plant surface

Site competition may occur on regions of the plant surface that are important for bacterial multiplication, long-term survival or entry into the host, thereby limiting the epiphytic phase of the pathogen and the onset of disease.

The ability of potential antagonists to outcompete pathogens at specific sites on the plant surface is difficult to investigate by *in vitro* laboratory (agar or tissue) tests, and evidence has largely been obtained from observations of bacterial distribution and measurement of population changes occurring on the plant.

The relative importance of site competition and chemical inhibition in biological control may be investigated by measurement of bacterial population changes following co-inoculation of pathogen and control agent. Interactions involving site competition would lead to quite different population changes compared with those involving chemical inhibition, as summarised in Fig. 9.7. In the first case, there will be mutual exclusion, limiting both populations, while in the second only the pathogen will be affected. In practice, both types of competition may operate together, which explains why antibiotic-deficient mutants frequently retain a degree of control activity. It also provides a further reason why the *in vitro* selection of antagonists on the sole basis of antibiotic production may not lead to the development of the most effective biocontrol agent.

The effectiveness of a naturally occurring biocontrol agent in site compe-

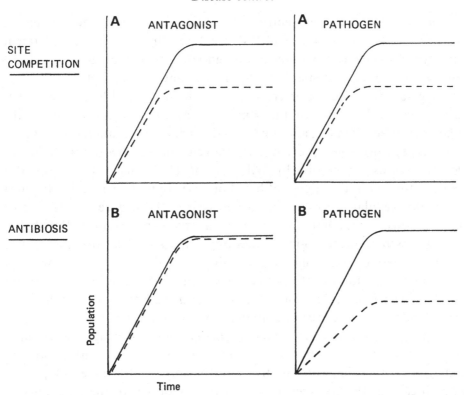

Fig. 9.7 Comparison of the effects of site competition and antibiosis on pathogen and antagonist populations. Graphs show theoretical growth curves for bacteria on the plant surface resulting from inoculation with pure (continuous line) or mixed (broken line) suspension. A, site competition. Where antagonist and pathogen are equally competitive, with no antibiosis, there is equal depression of bacterial growth following co-inoculation. B, antibiosis. Where anti-microbial substances are produced solely by the antagonist, in the complete absence of site (including nutrient) competition, there is depression of pathogen but not of antagonist population following co-inoculation. (In collaboration with H. A. S. Epton and M. Wilson.)

tition with the pathogen will relate to their habitats and microenvironments under natural conditions, and will depend on such factors as nutrient requirements, growth rate parameters, chemotaxis and general motility, ability to withstand desiccation and adhesion to plant surfaces. Although site competition has so far been implicated in only a few cases involving plant pathogenic bacteria, including interaction between *Erwinia herbicola/Erwinia amylovora* in blossoms and *Pseudomonas syringae* strains on leaf surfaces, it is probably of wide significance in biological control.

Site competition between Erwinia herbicola *and* E. amylovora Infection of blossom by *Erwinia amylovora* involves bacterial multiplication on stigmatic and nectarial surfaces, with entry of bacteria into the receptacle tissue via nectarthodes (see Figs. 4.2b and 5.4). Evidence suggests that pretreatment with *Erwinia herbicola* provides a substantial level of biological control of blossom blight and that site competition is a contributory factor in this.

Using scanning electron microscopy, Hattingh *et al.* (1986) showed that *Erwinia herbicola* colonised the same sites as *Erwinia amylovora* on the stigmatic surface of apple, and suggested that prior colonisation of the stigma by the antagonist would prevent the pathogen entering these sites. Light and electron microscope studies by Wilson *et al.* (1990) have demonstrated a similar situation in the stigma of hawthorn, where pathogen and antagonist invade intercellular spaces and dead host cells in the surface papillary zone, both occupying a similar niche within this microenvironment. Population studies on separate and co-inoculated populations revealed an intermediate position between the extremes noted in Fig. 9.7, indicating the occurrence of some site competition (both populations depressed by co-inoculation) but with major chemical inhibition of *E. amylovora* (pathogen population much more depressed than control agent after co-inoculation).

The occupation of key surface sites by *Erwinia herbicola* may either occur in direct competition with the pathogen (competitive antagonism) or prior to pathogen colonisation (pre-emptive antagonism). Competitive antagonism would occur in the co-inoculation experiment or where the antagonist is sprayed onto pathogen-infected surfaces, while pre-emptive antagonism would occur where antagonist is applied before arrival of the pathogen.

Site competition between strains of Pseudomonas syringae Competition between isogenic strains of *Pseudomonas syringae* (differing only in ice nucleation activity) has been investigated both on blossom (Lindemann & Suslow, 1987) and phylloplane (Lindow, 1988; Kinkel & Lindow, 1989) surfaces. The largely unidirectional depression of one bacterial population seen previously in co-inoculation experiments was not observed in these situations, indicating that chemical inhibition is not an important factor in these circumstances. The isogenic strains had no effect on each other until the carrying capacity of the environment was approached, suggesting that site competition was important at high population levels.

Site competition in planta

Avirulent mutants of *Pseudomonas solanacearum* have been used as biocontrol agents for the virulent pathogen (Trigalet & Trigalet-Demery, 1990), and are only effective where the mutant is able to infect host plants from the soil and spread within the host (without causing disease). In this situation the antagonist simply appears to occupy the same internal niche (vascular tissue) as the pathogen, thereby limiting the spread of the latter. Spontaneous avirulent mutants are not able to survive and multiply *in planta*, but Tn5-induced mutants (with insertions in the *hrp* cluster) are invasive and achieve significant protection against disease.

Biological control involving predatory antagonists

A range of microbes are predators of plant pathogenic bacteria, involving both parasitic interactions (bacteriophages and bacteria in the genus *Bdellovibrio*) and direct ingestion (e.g. of rhizosphere bacteria by protozoa). Some of these are potentially useful as biocontrol agents, and two in particular – bacteriophages and *Bdellovibrio* – have been used to control disease.

Bacteriophages

Although bacteriophages have clear advantages in terms of long-term survival in the soil environment and the potential for selective destruction of particular bacterial species and pathovars, they have received relatively little attention in recent years as biocontrol agents. This may relate in part to a number of limitations in their use, including their ability to mutate from virulent to temperate phages, ability of phages to carry bacterial genes (including virulence genes) from one bacterial cell to another and the susceptibility of phage host range and infection ability to changes in environmental conditions.

Vidaver (1976) compared the activity of bacteriophages and bacteriocins as biocontrol agents, and noted that bacteriophages had been successfully used against various pathogenic bacteria, including *Agrobacterium tumefaciens*, *Xanthomonas campestris* pv. *pruni* and *X.c.* pv. *oryzae*. In these studies, infection of tissue was typically reduced either when bacteriophage was mixed with a suspension of the pathogen (compared with control with pathogen only) or when the phage was applied to the plant surface (by spraying or dipping) before inoculation with pathogen.

Civerolo and Keil (1969) used bacteriophage to control leaf infection by *Xanthomonas campestris* pv. *pruni*, and considered the most effective method of control to be foliar application of phage-infected pathogen cells. This appeared to result in the release of phage particles within the intercellular spaces at internal sites of pathogen multiplication and infection.

More recent studies by Tanaka and Maeda (1988) have confirmed these earlier ideas and extended the carrier cell approach, using phage in combination with avirulent pathogen cells to control plant disease. The use of avirulent carrier cells clearly reduces the risk of bacterial infection, and the avirulent cells themselves may also act as biocontrol agents. In this particular study, involving control of bacterial wilt of tobacco (caused by *Pseudomonas solanacearum*), primary inoculation of roots of tobacco with the phage-infected avirulent mutant limited the development of disease that would have occurred after subsequent inoculation with pathogen. Protection was less effective after pretreatment with avirulent mutant only, and no protection was obtained after inoculation with bacteriophage culture only, suggesting that the use of bacteriophage in this instance may require simultaneous employment of host cells to be effective. The most effective bacteriophages that were used lysed both the avirulent and virulent strains of pathogen, and appeared to exert their activity inside the vascular system of the host.

Bdellovibrio

Bacteria in the genus *Bdellovibrio* were initially isolated (and shown to be active predators of other bacteria) by Stolp and Petzold in 1962, and were appropriately placed in the species *Bdellovibrio bacteriovorus*. These bacteria occur widely in soil and aquatic (marine and freshwater) environments, and can be readily isolated by plating out environmental samples onto lawns of 'host' bacteria, with the formation of localised transparent plaques (where *Bdellovibrio* is causing lysis of the host bacterial cells). The plaques caused by *Bdellovibrio* are similar in general appearance to those caused by bacteriophage, but develop after a longer time period.

Parasitic activity Since the early studies characterising the predatory nature of this organism, it has become clear that not all *Bdellovibrio* strains are active parasites. There is wide variation between strains, ranging from saprophytic (host-independent) to facultative and obligate parasitic (host-dependent) organisms.

Parasitic *Bdellovibrio* strains that have so far been characterised are only active against Gram-negative bacteria, including members of the genera *Azotobacter*, *Erwinia*, *Escherichia*, *Pseudomonas*, *Shigella*, *Rhizobium* and *Xanthomonas*. Host specificity varies considerably, with some strains of *Bdellovibrio* only being active against one host species, while others are able to lyse bacteria belonging to a range of genera.

Life cycle The life cycle of an obligate parasitic strain of *Bdellovibrio* is shown in Fig. 9.8, and involves alternating phases of growth (and multiplication) in the host and predation (with little multiplication) in the general environment. Interaction with the target organism involves a well-defined sequence of

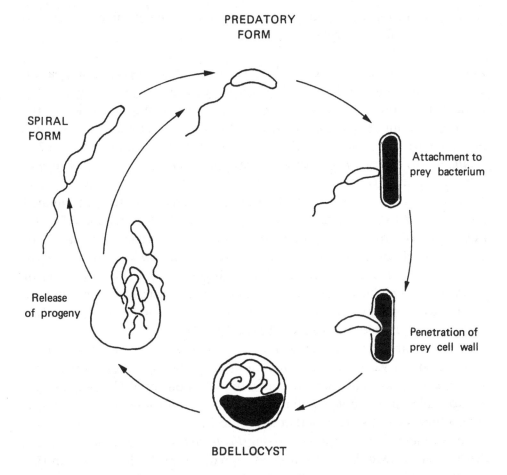

Fig. 9.8 Life cycle of *Bdellovibrio*. The protoplast of the host bacterium is shown in black and is pushed to one side of the cell in the bdellocyst stage. The spiral form of the organism can be observed in established cultures of some strains and is able to form the predatory stage by division.

events, including initial encounter, attachment to the host cell, penetration into the periplasmic space, intraperiplasmic growth of *Bdellovibrio* and release of the *Bdellovibrio* progeny.

The process of host cell penetration is enzymatic, with the production by *Bdellovibrio* of glycanases and peptidases which rapidly solubilise the host cell wall peptidoglycan (Thomashow & Rittenberg, 1978). Penetration results in the host cell becoming a spherical body (the bdelloplast). The parasite develops within the periplasm to form a spiral-shaped cell, which subsequently breaks up into separate vibrioid units – each of which develops its own flagellum. The *Bdellovibrio* cells are subsequently released by solubilisation of the bdelloplast peptidoglycan, and are immediately able to attach to and parasitise new host cells.

Biological control Isolation of *Bdellovibrio* strains that parasitise plant patho-genic bacteria could result in the development of useful biocontrol agents. Up to the present time, the use of laboratory *in vitro* techniques has led to the isolation of strains that are individually active against a variety of phytopath-ogenic bacteria, including *Pseudomonas syringae* pv. *glycinea* (Scherff, 1973), *Xanthomonas campestris* pv. *oryzae* (Uematsu, 1980) and *Erwinia carotovora* subsp. *carotovora* (Epton *et al.*, 1990).

Some of these organisms have been used for disease control with limited success in small-scale laboratory experiments. Scherff (1973) used *Bdellovibrio* isolates to control the development of local and systemic symptoms of bacterial blight when inoculated with *Pseudomonas syringae* pv. *glycinea*. Effective control was only achieved at high cell ratios (9 : 1 and 99 : 1) of *Bdellovibrio* to pathogen, and the ability of particular strains to limit disease was related to the average 'burst size' – which is the number of *Bdellovibrio* progeny that are released from the host bdelloplast.

More recently, Epton *et al.* (1990) have used various strains of *Bdellovibrio* for the laboratory control of *Erwinia carotovora* subsp. *atroseptica* on potato, with effective control of the pathogen in the development of blackleg (using a potato leaflet screen) but only limited success in control of soft rot (using a potato tuber screen similar to that shown in Fig. 7.2).

One of the major drawbacks in using *Bdellovibrio* as a biocontrol agent is that interaction with host cells follows a predator–prey relationship. In this situation, addition of *Bdellovibrio* to phytopathogen cells results in cyclic oscillations in host–parasite populations (Epton *et al.*, 1990) in which there is never complete eradication of the host cells, even at the local level. This limitation is made worse by the fact that cell contact (leading to penetration) occurs by random collision rather than chemotactic attraction, with the result that *Bdellovibrio* is only efficient as a predator when the target bacterium is at high population density. Varon and Zeigler (1978) have calculated that the minimum population density required to support an active host–parasite equilibrium is about 10^5 to 10^6 host cells/ml, which is much higher than the minimal inoculum level of pathogen required to initiate some diseases (e.g. soft rot of potato). In considering the use of *Bdellovibrio* as a biocontrol agent, it should also be borne in mind that this organism is essentially adapted to a subterranean (soil water) environment. It would only be expected, therefore, to be effective in the long term against soil (not aerial) pathogens and requires the presence of free water for effective activity.

Genetic control of plant disease

The use of genetic resistance to limit bacterial infection has been of major importance for the control of plant disease, in many cases offering the main long-term control strategy.

In some situations genetic control may be the only effective measure available. Most wilt diseases, for example, cannot be controlled chemically, and breeding for resistance is particularly important in these cases. Breeding tomato cultivars resistant to *Pseudomonas solanacearum* has enabled the production of this crop in India on fields that were heavily infested with the pathogen (Rao *et al.*, 1975).

Another example where breeding for resistance has particular importance is provided by control of bacterial leaf spot of bell pepper (*Capsicum annuum*), caused by *Xanthomonas campestris* pv. *vesicatoria*. Control of this pathogen is difficult in warm, humid environments in which the pathogen is endemic. With chemical control, even the most intensive spray regime has proved inadequate under conditions of high disease pressure, and sanitation measures have been of limited value. In the absence of effective chemical or biological control agents, use of genetic resistance represents the only feasible approach to control of this disease at the present time.

Genetic control involves the use of resistance genes that are already present within the host population or in populations of closely related plants. The first stage in developing genetic control of disease is to identify varieties of plant that are resistant to infection (primary sources of resistance). These may then either be cultivated directly, or their resistance genes combined with other genotypes in a breeding programme which may involve sexual hybridisation, somatic hybridisation or molecular genetics.

The search for resistant germplasm

With many diseases caused by plant pathogenic bacteria, the identification and selection of new resistance genes for crop breeding programmes has involved screening a wide range of cultivated varieties, naturally occurring (wild) progenitors of the crop plant and closely related species from different geographic locations. In addition to testing new sources of plants for resistance, the development of improved tissue culture techniques has also opened up the possibility of identifying and generating novel resistant genotypes by somaclonal variation.

Plant collection and screening for resistance

Plant collection and screening of primary sources for resistance provide the genetic basis for the origin of present-day and future resistant cultivars. The ability to find genes conferring immunity or high levels of resistance against a particular pathogen varies considerably. No genes have been found, for example, that confer complete resistance in potato against the soft rot pathogen *Erwinia carotovora*, and breeding programmes have to differentiate between degrees of susceptibility in this particular crop.

In other crops, genes conferring high levels of resistance to particular pathogens have been found. In some of these cases, the acquisition of new sources of resistance has involved world-wide collection of plants for primary screening (e.g. fireblight disease of pear), while in others primary screening has so far been carried out on more limited populations of host plants (e.g. corky root of lettuce).

Fireblight disease

During the early 1900s, a number of expeditions were made to the Orient, and Reimer (1925) reported a high degree of resistance to fireblight in various species of pear collected from Japan, China, Korea and Manchuria, including *Pyrus ussuriensis*, *P. calleryana*, *P. betulaefolia* and *P. pyrifolia*. These acquisitions subsequently became the parent species of many oriental cultivars. More recently, plant exploration trips have been made in search of resistant germplasm throughout eastern Europe (Van der Zwet & Bell, 1990), resulting in the introduction of new resistant genes into North American breeding programmes. This part of the world is close to the centre of diversity where the domesticated pear is thought to have originated, and provides a valuable source of resistant germplasm which is absent from North American and European pear collections.

Corky root of lettuce

Even within modern collections of cultivars, where genetic variation is not so diverse, resistance to new or more virulent forms of disease may be usefully tested by screening. Recent outbreaks of corky root disease of lettuce, newly identified as bacterial in origin (see Fig. 1.1), led Brown and Michelmore (1988) to screen 550 lettuce acquisitions, of which 19 contained high levels of resistance to the pathogen. Genetic analysis demonstrated that in most of the resistant accessions evaluated, resistance was conferred by a recessive allele at a single locus (cor).

Somaclonal variation

Plants that have been regenerated from tissue cultures exhibit variation in a range of features (somaclonal variation), some of which may have considerable agricultural significance. These include crop yield, protein content, salt tolerance, herbicide and disease resistance. It is not entirely clear whether somaclonal variation results simply by selection from pre-existing variation within the cultured cells, or whether there is some specific induction by the culture process itself. Somaclonal variation has been used to generate novel plants of pear showing resistance to fireblight.

Breeding for resistance

Sexual hybridisation

Sexual hybridisation and subsequent gene segregation have been widely successful in the introduction of resistant genes into economically important crops, leading to commercially available cultivars. In some cases, individual cultivars show specific resistance to particular races of pathogen due to the presence of plant resistance genes that are complementary to the bacterial avirulence genes (Chapter 8). This is shown in relation to the interaction between *Pseudomonas syringae* pv. *phaseolicola* with bean (Table 3.3) and *P.s.* pv. *glycinea* with soybean (Table 8.2). A general summary showing the differentiation of *Pseudomonas syringae* and *Xanthomonas campestris* pathovar races on host lines is shown in Table 3.4.

The range of bacterial diseases for which there is now some degree of genetic control have recently been reviewed by Lozano *et al.* (1990) and, in addition to those mentioned above, include leaf blights (caused by *Pseudomonas syringae* pv. *tabacum*, *Xanthomonas campestris* pv. *malvacearum*), canker and die-back diseases (*Erwinia amylovora*, *P.s.* pv. *mors-prunorum*), wilt diseases (*P. solanacearum*, *E. stewartii*, *E. tracheiphila*), soft rots (*E. carotovora*) and tumour diseases (*Agrobacterium tumefaciens*). Further details on the introduction of resistance to specific pathogens in some economically important crops (tobacco, cotton and pepper) are given below.

Tobacco

Wildfire disease of tobacco, caused by *Pseudomonas syringae* pv. *tabacum*, has been of major importance in North America and Zimbabwe during the first half of this century, causing serious epidemics and economic loss.

Clayton transferred genetic high resistance from *Nicotiana longiflora* to *Nicotiana tabacum* breeding lines in 1947, but the first commercially available

Table 9.7. *Genetic control of bacterial leaf spot in pepper, caused by* Xanthomonas campestris *pv.* vesicatoria

(*a*) Resistance genes and bacterial races

Pathogen	Resistant genes		
	Bs1	Bs2	Bs3
Race 1	S	R	R
Race 2	R	R	S

(*b*) Genotypes of resistant pepper cultivars

Cultivar	Genotype	Resistance
271-4	Bs1 bs2 Bs3	Race 1 and 2
Florida XVR	Bs1 Bs2 bs3	Race 1 and 2
Delray Bell	Bs1 bs2 bs3	Race 1
Early Calwonder 10R		

Data taken from Hibberd *et al.* 1987

wildfire-resistant tobacco variety was not released until 1955. Although the use of resistant varieties contributes greatly to the control of wildfire disease, it is not a permanent solution since some bacterial strains pathogenic to resistant varieties have been reported. Additional resistant varieties must be developed continuously and used in conjunction with other control measures.

Cotton

Considerable variety exists in cotton cultivars in terms of resistance to bacterial blight caused by *Xanthomonas campestris* pv. *malvacearum*. Some cultivars, such as AC44, are completely susceptible, with no major genes for resistance to the disease, while other cultivars, such as Im266, are highly resistant. Resistance in the latter case is probably due to three major resistance genes B2, B3 and b7 plus a complex of minor genes (Brinkerhoff *et al.* 1984).

Pepper

Resistance in pepper to *Xanthomonas campestris* pv. *vesicatoria* (causing black spot disease) has been investigated in terms of both single and multiple gene activity (Hibberd *et al.*, 1987, 1988).

Single gene action Resistance that is determined by single resistant genes leads to a clear-cut hypersensitive reaction, as described in Chapter 6. Three resistant genes – Bs1, Bs2 and Bs3 – have been reported to control hyper-sensitivity in pepper, against two races of the pathogen (Table 9.7*a*). Genes B1 and B3 are relatively weak genes, while B2 confers resistance to all known races in the field (though bacterial mutants have been isolated in culture that are virulent on B2 plants). Races 1 and 2 of the pathogen occur equally frequently in Florida (USA), where bacterial leaf spot is a major problem, but race 1 predominates in other pepper production areas of the world. These genes segregate independently during crossing and occur in resistant cultivars to different degrees (Table 9.7*b*).

Multiple gene action In some cases, resistance to a particular phytopathogen appears more like a slow form of hypersensitivity, and can only be detected and evaluated by quantitative means (see 'Screening of resistant cultivars' below). This type of resistance was observed by Hibberd *et al.* (1988) in certain strains of pepper and was attributed to additive or polygenic genetic activity, conferring resistance to all three races of *Xanthomonas campestris* pv. *vesicatoria*.

Somatic hybridisation

Somatic hybridisation involves induced fusion between plant protoplasts, leading to hybrid cells (heterokaryons, then synkaryons) which can be grown into mature plants and which can combine novel genotypic characteristics originally separated in the two parent cells. Somatic hybridisation has the advantage over sexual hybridisation in not being limited by the usual species compatibility barriers, though a certain degree of compatibility is required for complete development of the hybrid cells to intact plants.

Interspecific somatic hybridisation has been used to generate partially resistant strains of potato (*Solanum tuberosum*) to soft rot diseases caused by species of *Erwinia* (Austin *et al.*, 1988). Although wild species of potato, such as *Solanum brevidens*, are a potential source of desirable agronomic character-istics for the development of potato tubers, full use of this potential has been limited by sexual incompatibility between gamete cells.

Austin *et al.* (1988) produced somatic hybrids between *Solanum brevidens*, a diploid non-tuber-bearing wild species and *Solanum tuberosum* (tetraploid cultivated potato). Unlike the *Solanum tuberosum* fusion parent, somatic hybrids had tubers with a substantial degree of resistance to *Erwinia caroto-vora* subsp. *carotovora*, *subsp. atroseptica*, and *Erwinia chrysanthemi*. The resistance incorporated from *Solanum brevidens* by somatic fusion was sexually transmissible, since some of the sexual progeny from crosses with other cultivars had the same level of resistance as the somatic hybrids. In this work, interspecific somatic hybridisation has thus facilitated access to completely new sources of disease resistance.

Screening and evaluation of resistant cultivars

Efficient and accurate methods for assessing the response of plants to phytopathogens are an important aspect of both primary screening for resistance and evaluation of cultivars derived from sexual or somatic hybridisation.

Evaluation of resistant cultivars may involve either laboratory trials (with artificial inoculation and maintenance of the plant under controlled conditions) or field trials (where cultivars are assessed under crop conditions). The use of laboratory trials to assess plant cultivars for resistance is making increasing use of *in vitro* experimental systems (see Chapter 6), and involves artificial inoculation with pathogen. Assessment of resistance under field conditions can either involve monitoring the plant response following artificial inoculation of pathogen or under conditions of natural occurrence.

Artificial inoculation and disease assessment

Where screening for resistance is based on artificial inoculation of plants, a mixture of pathogen strains should be used that is representative of the complete range of differential virulence within the geographic region where the plants will be cultivated. In testing for resistance to bacterial spot disease of stone fruits, for example, a mixture of strains of *Xanthomonas campestris* pv. *pruni* should be used since these show considerable differential virulence to cultivars of peach, plum and apricot.

Other aspects that are also important in screening methodology include appropriate environmental conditions, use of plant material at a proper developmental stage, use of inoculation methods that yield dependable symptom expression and use of symptom assessment methods that clearly distinguish susceptible from resistant reactions.

The level of inoculum used and the visual assessment of the resistant response depends to a large extent on the genetic nature of the resistance (Hibberd *et al.*, 1988), as summarised in Table 9.8. With single gene action, high levels of inoculum should be used, leading to a clear hypersensitive reaction. With additive gene action, where resistance is determined by a multiplicity of genes, low levels of inoculum can be used, leading to a quantitative response.

Table 9.8. *Assessment of genetic resistance*

	Single gene action	Additive gene action
Genes	Single incompatibility genes	Polygenes
Assessment	Qualitative: HR	Quantitative: e.g. number and size of lesions
Detection	Inoculate high concentrations of pathogen	Can use low concentrations of pathogen
Examples	Many examples	Bacterial leaf spot of pepper: *Xanthomonas campestris* pv. *vesicatoria*

Assessment of cultivar resistance in field trials

Assessment of cultivars in field trials is typically a long-term project, leading ultimately to assessment of plants in the crop situation in different geographic and environmental conditions. Where assessment relates to the development of disease by natural infection, it is clearly important to carry out the tests at a site where the disease is endemic and where symptoms routinely occur on susceptible varieties.

The field procedures involved in screening for resistance vary with the particular host and pathogen. The protocol adopted by the International Rice Research Institute (Mew *et al.*, 1990) for screening rice for resistance to bacterial blight (caused by *Xanthomonas campestris* pv. *oryzae*) involves primary screening of rice germplasm twice a year (rainy and non-rainy season) followed by further nursery testing of plants showing resistance the following season. The more thorough assessment that takes place in the Bacterial Blight General Screening Nursery involves testing under defined conditions of soil fertility and plot design, with computer evaluation of results.

Sanitary procedures in the control of plant disease

Sanitary procedures involve all aspects of plant culture and treatment that are important in limiting the spread of plant disease and include primary use of disease-free plants and seed, new types of propagation procedure, application of optimal growth and culture conditions in the field and use of correct procedures in equipment and product handling. General reviews of methods used in agronomy to reduce bacterial diseases are given by Dinesen *et al.* (1990) and Okon (1990).

Use of disease-free plants and seed

The use of uncontaminated plants and seed during the initial planting operation is an important aspect of disease control. In practice, this involves careful inspection and testing of the plants and seed for the presence of phytopathogenic bacteria, with planting only of certified stock. In the USA and elsewhere, specific certification programmes are in force to control bacterial diseases of particular importance.

Plant certification: control of **Pseudomonas syringae** *pv.* **tomato**

A good example of the importance of plant certification is provided by the control of bacterial diseases of tomato in the USA. The production of tomato transplants is important in Florida and South Georgia, where a strict certification programme is in force to ensure that these are disease-free prior to export. In these States two fluorescent pseudomonads are of particular importance, *Pseudomonas syringae* pv. *tomato* and *P.s.* pv. *syringae*. Of these *P.s.* pv. *syringae* is the most commonly isolated of the two phytopathogens, but because its survival and disease effects decline as the tomato crop matures, it is not of major importance on this host, and plants with this bacterium retain certification (Denny, 1988). *P.s.* pv. *tomato*, on the other hand, is able to survive epiphytically on tomatoes and non-host plants over extended time periods, with the potential to reach epidemic levels and cause serious disease at the time of fruit production, so the presence of this bacterium results in a loss of certification. The major problem encountered in the certification programme is that symptoms produced by these bacteria are often indistinguishable, and although these pathovars can be distinguished by several phenotypic characteristics, the necessary testing may take a week or more to complete. The need for a rapid test to distinguish the pathogens has lead to the use of serological and bacteriophage typing, and the use of DNA probes (see page 67) has proved particularly rapid and sensitive (Denny, 1988).

Certification of seed

Testing seeds for the presence of plant pathogenic bacteria is an important aspect of controlling seedborne bacterial pathogens. Until recently, relatively little seed testing occurred due to the lack of established assay techniques and also to the belief that chemical seed treatments could effectively control seedborne diseases. Although this is correct for a number of fungal diseases, chemical treatment of seed has been of relatively limited use with bacterial pathogens (Schaad, 1982). The development of new diagnostic techniques that are rapid, reliable and highly sensitive has meant that large-scale testing of seeds is now feasible, and assays have been developed for a number of seedborne bacteria.

Contaminated seed is an important source of primary inoculum in the case of *Pseudomonas syringae* pv. *syringae* and *P.s.* pv. *phaseolicola*, which cause brown spot and halo blight respectively in bean plants. In the USA, most of the bean seed has traditionally been grown in the arid region of southern Idaho, where bacterial blights and production of contaminated seed were not normally a problem. However, in the 1960s substantial losses occurred from halo blight, and stringent rules for laboratory testing of seeds for both pathovars were introduced. Several methods have been used for the detection of halo blight and brown spot pathogens in seed, including:

1. Host inoculation test: inoculating bean plants with liquid from soaked seeds.
2. Plating seed soak liquid onto semi-selective media.
3. Serology: including agar gel diffusion, immunofluorescence and ELISA.

Host inoculation tests are very time consuming and require a considerable amount of greenhouse space. At the present time, procedures 2 and 3 provide the best tests for this pathogen (Mohan & Schaad, 1987). Immunofluorescence provides a particularly sensitive test for the detection of seedborne bacteria, but has the disadvantages that the cells detected may be non-viable, unable to infect or maybe misidentified. Problems of viability are not encountered in seed assays (e.g. procedures 1 and 2 above) that involve recovery and growth of the pathogen.

One very important aspect of seedborne pathogens is the quantitative role of inoculum in the development of field epidemics, and the determination of levels of seed contamination above which the risk of epidemic becomes probable. This will determine what level of seed infection (if any) can be accepted in a seed certification programme (tolerance level) and also which diagnostic tests are sufficiently reliable and sensitive. In the case of halo blight, for example, it has been shown in commercial batches of seed that a level of 5 naturally infected seeds per 10 000 (0.05%) tends to result in disease epidemics, while one per 20 000 (0.005%) does not (Schaad, 1982). With such a virulent pathogen, sampling and testing of seed samples must be very carefully carried out, and the USA seed health certification programme has established zero tolerance.

A similar situation has been demonstrated in the transmission of *Xantho-monas campestris* pv. *campestris* by seed of crucifers (Schaad, 1982), where contamination levels of 0.03% or above lead to a high field incidence of black rot, while 0.01% contamination does not. With this pathogen, laboratory (agar plating) assays are used to detect seed contamination down to the 0.01% level of infection.

New types of propagation procedure

An alternative to selecting whole disease-free plants as a source of propaga-tion and planting is to propagate from parts of plants that are pathogen-free. Culture from meristems has the advantage that these parts of the plant are probably less heavily infected than other regions and meristems also provide a convenient means of raising multiple propagates from a mother plant.

Meristem culture has been particularly useful in the production of potato tubers free of infection by *Clavibacter michiganensis* subsp. *sepedonicum*. Work by Sletten (reported in Dinesen *et al.*, 1990) has involved using meristem tips to produce plants which are then subcultured by stem cuttings to produce a clone of plantlets. Single plantlets within the clone are tested for the presence of pathogenic bacteria and infected clones (derived from an infected meristem tip) are discarded. Non-infected clones are raised to normal plants in the greenhouse and the seed potatoes harvested and used for further propagation.

Crop growth and culture conditions

Any general aspects of plant culture which promote the infection and spread of bacterial plant pathogens should be avoided. These aspects include irrigation with water that has not been tested for bacterial contamination, accumulation of infected material, crop damage and overcrowding. Crop rotation is important for the control of pathogens such as *Pseudomonas solanacearum* which are able to survive in the soil over a long period.

Specific aspects of culture, such as watering, may also be important. This is illustrated by the studies of Dickey and Zumoff (1987) on bacterial blight of *Syngonium*, caused by *Xanthomonas campestris* pv. *syngonii*, where infection is mediated largely by watersplash, and overhead watering constitutes the primary means of dispersal of the pathogen within the crop. These authors recommend removal of infected leaves at the junction of the petiole and stem as soon as symptoms occur, and avoidance of overcrowding and overhead watering of stock plants.

Wounds caused by cultural practices may also be a major source of bacterial infection. Cut leaf tips or broken roots during transplantation of rice, for example, permit entry of *Xanthomonas campestris* pv. *oryzae*, leading to bacterial blight.

With some bacterial diseases, particularly of perennial crops, removal of infected and surrounding plants may be necessary. As an example of this, current USA regulations on the eradication of citrus canker by *Xanthomonas campestris* pv. *citri*, require the destruction of all citrus plants within 40 m of symptomatic plants (Timmer *et al.*, 1987).

Equipment and product handling

The use of uncontaminated equipment is important in disease control for both crop and post-harvest situations, and is particularly relevant to the handling of seeds and potatoes, where bacterial spread can easily occur from primary sources of infection.

Bacterial ring rot of potatoes, caused by *Corynebacterium michiganense* subsp. *sepedonicum*, is especially prone to dissemination by this means, with bacteria spreading via contaminated equipment from diseased to healthy tubers during the cutting and planting process (Secor *et al.*, 1988). Sterilisation of equipment to control this pathogen can be achieved using a wide range of disinfectants, including hypochlorite, iodine, phenol, ethanol and Cu groups. Factors such as the presence of additional organic matter and the degree of hydration of the equipment surface had variable effects on the activity of the disinfectant, and the most important limiting factor affecting sterilisation was treatment time.

References

Austin S., Lojkowska E., Ehlenfeldt M. K., Kelman A. & Helgeson J. P. (1988). Fertile interspecific somatic hybrids of *Solanum*: A novel source of resistance to *Erwinia* soft rot. *Phytopathology* **78**, 1216–20.

Axelrod P. E., Manuelle R. & Schroth M. N. (1988). Role of antibiosis in competition of *Erwinia* strains in potato infection courts. *Appl. Environ. Microbiol.* **54**, 1222–9.

Bakker P., Van Peer R. & Schippers B. (1991). Suppression of soil-borne plant pathogens by fluorescent pseudomonads: Mechanisms and prospects. In *Biotic Interactions and Soil-Borne Diseases*, ed. A. Beemster, G. J. Bollen, M. Gerlagh, M. A. Ruissen, B. Schippers and A. Tempel. Amsterdam: Elsevier.

Beer S. V. (1981). Biological control of fireblight. *Acta Hort.* **117**, 123.

Beer S. V. & Vidaver A. K. (1978). Bacteriocins produced by *Erwinia herbicola* inhibit *Erwinia amylovora*. *3rd Int. Congr. Plant Pathol.*, Munich, FRG.

Brinkerhoff L. A., Verhalen L. A., Johnson W. M., Essenberg M. & Richardson

P. E. (1984). Development of immunity to bacterial blight of cotton and its implications for other diseases. *Plant Dis.* **68**, 168–74.

Brown P. R. & Michelmore R. W. (1988). The genetics of corky root resistance in lettuce. *Phytopathology* **78**, 1145–50.

Burr T. J., Norelli J. L., Katz B., Wilcox W. F. & Hoying S. A. (1988). Streptomycin resistance of *Pseudomonas syringae* pv. *papulans* in apple orchards and its association with a conjugative plasmid. *Phytopathology* **78**, 410–13.

Chatterjee A. K., Gibbins L. N. & Carpenter J. A. (1969). Some observations on the physiology of *Erwinia herbicola* and its possible implications as a factor antagonistic to *Erwinia amylovora* in the fireblight syndrome. *Can J. Microbiol.* **15**, 640–2.

Civerolo E. L. & Keil H. L. (1969). Inhibition of bacterial spot of peach foliage by *Xanthomonas pruni* bacteriophage. *Phytopathology* **59**, 1966–7.

Colin J. E. & Chafik Z. (1986). Comparison of biological and chemical treatments for control of bacterial speck of tomato under field conditions in Morocco. *Plant Dis.* **70**, 1048–50.

Denny T. P. (1988). Differentiation of *Pseudomonas syringae* pv. *tomato* from *Pseudomonas syringae* pv. *syringae* with a DNA hybridisation probe. *Phytopathology* **78**, 1186–93.

Dickey R. S. & Zumoff C. H. (1987). Bacterial leaf blight of *Syngonium* caused by a pathovar of *Xanthomonas campestris. Phytopathology* **77**, 1257–62.

Dinesen I. G., Zeller W., Hirano S. S., Upper C. D., Caristi J. & Sands D. C. (1990). Epidemiology. In *Methods in Phytopathology*, ed. Z. Klement, K. Rudolph and D. C. Sands, pp. 281–96. Budapest: Akademiai Kiado.

Ellis J. G., Kerr A., van Montagu M. & Schell J. (1979). *Agrobacterium*: Genetic studies on agrocin 84 production and the biological control of crown gall. *Physiol. Plant Pathol.* **15**, 311–19.

Epton H. A. S., Walker N. M. & Sigee D. C. (1990). *Bdellovibrio*: A potential control agent for soft rot and blackleg of potato. *Proc. 7th. Int. Conf. Plant Pathogenic Bacteria.* Budapest: Akademiai Kiado.

Feistner G., Korth H., Ko H., Pulverer G. & Budzikiewicz H. (1983). Ferrorosamine A from *Erwinia rhapontici. Curr. Microbiol.* **8**, 239–43.

Gross D. C. & Vidaver A. K. (1990). Bacteriocins. In *Methods in Phytopathology*, ed. Z. Klement, K. Rudolph and D. C. Sands, pp. 245–9. Budapest: Akademiai Kiado.

Hattingh M. J., Beer S. V. & Lawson E. W. (1986). Scanning electron microscopy of apple blossom colonised by *Erwinia amylovora* and *Erwinia herbicola. Phytopathology* **76**, 900–4.

Hibberd A. M., Bassett M. J. & Stall R. E. (1987). Allelism tests of three dominant genes for hypersensitive resistance to black spot of pepper. *Phytopathology* **77**, 1304–7.

Hibberd A. M., Stall R. E. & Bassett M. J. (1988). Quantitatively assessed resistance to bacterial leaf spot in pepper that is simply inherited. *Phytopathology* **78**, 607–12.

Hodson N. & Sigee D. C. (1991). Copper toxicity in *Erwinia amylovora*: An X-ray microanalytical study. *Scan. Microsc.* **5**, 427–38.

Isenbeck M. & Schulz F. A. (1985). Biological control of fireblight (*Erwinia amylovora* [Burr] Winslow *et al.*) on ornamentals. I. Control of the pathogen by antagonistic bacteria. *Phytopathol. Z.* **113**, 324–33.

Ishimaru C. A., Klos E. J. & Brubaker R. R. (1988). Multiple antibiotic production by *Erwinia herbicola*. *Phytopathology* **78**, 746–50.

Kerr A. & Htay K. (1973). Biological control of crown gall through bacteriocin production. *Physiol. Plant Pathol.* **4**, 37–44.

Kerr A. & Jones D. A. (1989). Construction, testing and registration of a genetically engineered strain of *Agrobacterium* for the control of crown gall. *7th Int. Conf. Plant Pathogenic Bacteria*. Budapest: Akademiai Kiado.

Kinkel L. L. & Lindow S. E. (1989). Dynamics of *Pseudomonas syringae* strains co-existing on leaf surfaces. *Phytopathology* **79**, 1162.

Kloepper J. W. (1983). Effect of seed piece inoculation with plant growth-promoting rhizobacteria on the population of *Erwinia carotovora* on potato roots and daughter tubers. *Phytopathology* **73**, 217–19.

Kloepper J. W., Teintze J. & Schroth M. N. (1980). *Pseudomonas* siderophores: A mechanism explaining disease suppressive soils. *Curr. Microbiol.* **4**, 317–20.

Legard D. E. & Schwartz H. F. (1987). Sources and management of *Pseudomonas syringae* pv. *phaseolicola* and *Pseudomonas syringae* pv. *syringae* epiphytes on dry beans in Colorado. *Phytopathology* **77**, 1503–9.

Lindemann J. & Suslow T. V. (1987). Competition between ice nucleation active wild-type and ice nucleation deficient mutant strains of *Pseudomonas syringae* and *Pseudomonas fluorescens* biovar I and biological control of frost on strawberry blossoms. *Phytopathology* **77**, 882–6.

Lindow S. E. (1988). Lack of correlation of *in vitro* antibiosis with antagonism of ice nucleation active bacteria on leaf surfaces by non-ice nucleation active bacteria. *Phytopathology* **78**, 444–50.

Lozano J. C., Zeigler R., Mew T. W., Reyes R. C., Vera-Cruz C. M., Aldwinckle H. S., Mavridis A., Rudolph K., Vidaver A., Dorke F., Burr T. & Kennedy B. W. (1990). Breeding for resistance. In *Methods in Phytobacteriology*, ed. Z. Klement, K. Rudolph and D. C. Sands, pp. 333–68. Budapest: Akademiai Kiado.

Mew T. W., Reyes R. C. & Vera-Cruz C. M. (1990). Screening rice for resistance to bacterial blight. In *Methods in Phytobacteriology*, ed. Z. Klement, K. Rudolph and D. C. Sands. Budapest: Akademiai Kiado.

Mohan S. K. & Schaad N. W. (1987). An improved agar plating assay for detecting *Pseudomonas syringae* pv. *syringae* and *Pseudomonas syringae* pv. *phaseolicola* in contaminated seed. *Phytopathology* **77**, 1390–5.

Nicholson S., Sigee D. C. & Epton H. A. S. (1990). Biological control of fireblight of perry pear: Comparative evaluation of antagonists on immature fruit slices, micropropagated shoots and orchard blossom. *Acta Hort.* **273**, 397–403.

Okon J. (1990). Methods in agronomy to reduce bacterial diseases. In *Methods in Phytobacteriology*, ed. Z. Klement, K. Rudolph and D. C. Sands. Budapest: Akademiai Kiado.

Ong S. A., Peterson T. & Neilands J. B. (1978). Agrobactin, a siderophore from *Agrobacterium tumefaciens*. *J. Biol. Chem.* **254**, 1860–5.

Panopoulos C. G., Psallidas P. G. & Alivizatos A. S. (1978). Evidence of a breakdown in the effectiveness of biological control of crown gall. *8th Int. Congr. Plant Pathol.*, Munich, FRG.

Parker W. L., Rathnum M. L., Wells J. S., Trejo W. H., Principe P. A. & Sykes R. B. (1982). SQ 27,860, a simple carbapenem produced by species of *Serratia* and *Erwinia*. *J. Antibiot.* **35**, 653–60.

Rao M. V., Sohi H. S. & Tikoo S. K. (1975). Reaction of wilt-resistant tomato varieties and lines to *Pseudomonas solanacearum* in India. *Plant Dis. Reptr.* **59**, 734–6.

Reimer F. C. (1925). Blight resistance in pears and characteristics of pear species and stocks. *Oregon Agr. Expt. Sta. Bull.* 214.

Riggle J. R. & Klos E. J. (1972). Relationship of *Erwinia herbicola* to *Erwinia amylovora. Can. J. Bot.* **50**, 1077–83.

Schaad N. W. (1982). Detection of seedborne bacterial plant pathogens. *Plant. Dis.* **66**, 885–90.

Scherff R. H. (1973). Biological control of bacterial blight of soybean by *Bdellovibrio bacteriovorus. Phytopathology* **63**, 400–2.

Schmidt D. (1988). *Pseudomonas fluorescens* and *Erwinia herbicola* reduce wilt of grasses caused by *Xanthomonas campestris* pv. *graminis. J. Phytopathol.* **122**, 245–52.

Secor G. A., DeBuhr L. & Gudmestad N. (1988). Susceptibility of *Corynebacterium sepedonicum* to disinfectants *in vitro. Plant Dis.* **72**, 585–8.

Sekizawa Y. & Wakabayashi K. (1990). Bactericides. In *Methods in Phytobacteriology*, ed. Z. Klement, K. Rudolph and D. C. Sands. Budapest: Akademiai Kiado.

Smidt M. & Vidaver A. K. (1982). Bacteriocin production by *Pseudomonas syringae* PSW-1 in plant tissue. *Can. J. Microbiol.* **28**, 600–4.

Stein J. I. & Beer S. V. (1980). Partial purification of a bacteriocin from *Erwinia herbicola. Phytopathology* **70**, 469.

Stolp H. & Petzold H. (1962). Untersuchungen über einen obligat parasitischen Mikro-organismus mit lytischer aktivat für *Pseudomonas* backterien. *Phytopathol. Z.* **45**, 364–90.

Tanaka H. & Maeda H. (1988). Biological control of bacterial wilt of tobacco by an avirulent mutant strain of *Pseudomonas solanacearum* and its bacteriophages. *Proc. 5th Int. Congr. Plant Pathol.*, Kyoto, Japan, p. 347.

Thomashow M. F. & Rittenberg S. C. (1978). Intraperiplasmic growth of *Bdellovibrio bacteriovorus* 109J: Solubilisation of *Escherichia coli* peptidoglycan. *J. Bacteriol.* **135**, 998–1007.

Timmer L. W., Marois J. J. & Achor D. (1987). Growth and survival of xanthomonads under conditions non-conducive to disease development. *Phytopathology* **77**, 1341–5.

Trigalet A. & Trigalet-Demery D. (1990). Use of avirulent mutants of *Pseudomonas solanacearum* for the biological control of bacterial wilt of tomato plants. *Physiol. Mol. Plant Pathol.* **36**, 27–38.

Uematsu T. (1980). Ecology of *Bdellovibrio* parasitic to rice bacterial leaf blight pathogen, *Xanthomonas oryzae. Rev. Plant Prot. Res.* **13**, 12–26.

Van der Zwet T. & Bell R. L. (1990). Fireblight susceptibility in *Pyrus* germplasm from eastern Europe. *Hort. Sci.* **25**, 566–8.

Vanneste J. L., Yu J. & Beer S. V. (1992). Role of antibiotic production by *Erwinia herbicola* Eh252 in biological control of *Erwinia amylovora. J. Bacteriol.* **174**, 2785–96.

Varon M. & Zeigler B. P. (1978). Bacterial predator–prey interaction at low prey density. *Appl. Env. Microbiol.* **36**, 11–17.

Vidaver A. K. (1976). Prospects for control of phytopathogenic bacteria by bacteriophages and bacteriocins. *Ann. Rev. Phytopathol.* **14**, 451–65.

Vidaver A. K. (1983). Bacteriocins: The lure and the reality. *Plant Dis.* **67**, 471–5.

Vidaver A. K., Mathys M. L., Thomas M. E. & Schuster M. L. (1972).

Bacteriocins of the phytopathogens *Pseudomonas syringae, Pseudomonas glycinea* and *Pseudomonas phaseolicola. Can. J. Microbiol.* **18**, 705–13.

Watson B., Currier T. C., Gordon M. P., Chitton M. D. & Neuter E. W. (1975). Plasmid required for virulence of *Agrobacterium tumefaciens. J. Bacteriol.* **123**, 255–64.

Wilson M., Epton H. A. S. & Sigee D. C. (1989). *Erwinia amylovora* infection of hawthorn blossom. II. The stigma. *J. Phytopathol.* **127**, 15–28.

Wilson M., Epton H. A. S. & Sigee D. C. (1990). Biological control of fireblight of hawthorn (*Cratoegus monogyna*) with *Erwinia herbicola* under protected conditions. *Plant Pathol.* **39**, 301–8.

Wilson M., Epton H. A. S. & Sigee D. C. (1993). Biological control of fireblight of hawthorn (*Crataegus monogyna*) with fluorescent *Pseudomonas* spp. under protected conditions. (In press.)

Winkelmann G., Gluck Ch., Lupp R. & Jung G. (1979). Herbicolins – new peptide antibiotics from enterobacteria. In *Structure and Activity of Natural Peptides: Selected Topics.* Proc. Fall Mtg. Gesellschaft für Biologische Chemie, Tubingen, FRG.

Winkelmann G., Lupp R. & Jung G. (1980). Herbicolins – new peptide antibiotics from *Erwinia herbicola. J. Antibiot.* **33**, 353–8.

Xu G. W. & Gross D. C. (1986a). Selection of fluorescent pseudomonads antagonistic to *Erwinia carotovora* and suppressive of seed piece decay. *Phytopathology* **76**, 414–22.

Xu G. W. & Gross D. C. (1986b). Field evaluation of the interaction among fluorescent pseudomonads, *Erwinia carotovora* and potato yields. *Phytopathology* **74**, 423–30.

Index